Enzymatic Transformation

Soundar Divakar

Enzymatic Transformation

Springer

Soundar Divakar
Central Food Technological Research Institute
Mysore, Karnataka, India

ISBN 978-81-322-0872-3 ISBN 978-81-322-0873-0 (eBook)
DOI 10.1007/978-81-322-0873-0
Springer New Delhi Heidelberg New York Dordrecht London

Library of Congress Control Number: 2012951119

Printed on acid-free paper

Springer is part of Springer Science+Business Media (www.springer.com)

Viky and Sasi

Preface

This book is a summary of the research work carried out by me with my students and collaborators at the Central Food Technological Research Institute, Mysore, India. The work was carried out in a period spanning 18 years or so from 1995 onwards. The essence of the work from 70 or so publications is presented here in brief under 11 chapters.

The objective of this book is to explain the various techniques and strategies involved in enzymatic transformation reactions. Esterification of acids using lipases and glycosylation of alcohols and phenols using glycosidases are the two enzymatic reactions highlighted in this book. Both esterification and glycosylation reactions involve multifunctional substrates. Challenges using underivatised polar and nonpolar substrates in terms of product specificity/selectivity, yields and feasibility in such enzymatic reactions carried out in nonpolar media are explained with several examples of divergent multifunctional substrates. Hence, the information provided would be a source of useful information for not only established researchers but also for beginners who eye this as a potential and fruitful research area.

The author is sure that this book has brought out the trials and tribulations of carrying out enzymatic transformations clearly.

About the Author

Dr. Divakar is working as a Chief Scientist in the Central Food Technological Research Institute, Mysore, where he joined as a Scientist. Before this, he served as CSIR Pool Officer at Molecular Biophysics Unit, Indian Institute of Science, Bangalore. After completing his post-graduation in 1976 from Pachaiyappa's College, Chennai, he had a brief stint in a polymer industry, M/S Reichhold Chemicals, Ind. Ltd., Chennai. He completed his Ph.D. degree in Chemistry from the Australian National University, Canberra, Australia in 1982.

The author is actively involved in research for the past 36 years. While working as a Research and Development chemist at Ms. Reichhold Chemicals (Ind.) Ltd., he was actively involved in research in the fields of alkyd resins, unsaturated polyesters, phenol-formaldehyde resins (resoles and novolaks), urea-formaldehyde resins and epoxy resins.

After completing his Ph.D., he was involved in research work on three important areas, namely,

1. NMR investigations of some biological systems
2. Host-guest complexation chemistry with special reference to cyclodextrins
3. Enzyme catalyzed transformation reaction employing lipases and glycosidases

He has so far guided various Ph.D. and M.Sc. students and has published about 149 research articles including reviews. His work on Enzymatic Transformations has resulted in about 70 publications, which prompted him to write this book on Enzymatic Transformation.

Acknowledgements

The author gratefully acknowledges Central Food Technological Research Institute, Mysore, for providing the facilities and sanctioning the projects. Acknowledgement is also due to the Department of Biotechnology, India, and Department of Science and Technology, India, for providing the financial assistance.

Author expresses his gratitude and appreciation to the following students and the other contributors: Dr. B. Manohar, Dr. P. Ravi, Dr. K.R. Kiran, Dr. C.V. Suresh Babu, Dr. H.H. Pattekhan, M.S. Pramila Rao, Dr. K. Lohith, Dr. G.R. Vjayakumar, Dr. B.R. Somashekar, Dr. R. Sivakumar, Mr. Mallikarjuna, Mr. Swaminathan and Mr. G. Vadivelan.

The author also acknowledges Springer for their prompt response and support in publishing this work.

Contents

List of Abbreviations

List of Abbreviations and Symbols

A	Absorbance
α	Alpha
ANOVA	Analysis of variance
ACE	Angiotensin-converting enzyme
Å	Angstrom
β	Beta
BSA	Bovine serum albumin
CRL	Candida rugosa lipase
^{13}C	Carbon-13
CCRD	Central composite rotatable design
J	Coupling constant
CMC	Critical micellar concentration
°C	Degree centigrade
d	Delta
DMSO-d_6	Deuterated dimethyl sulphoxide
CH_2Cl_2	Dichloromethane
DMSO	Dimethyl sulphoxide
eV	Electron volt
EC	Enzyme commission
γ	Gamma
g	Gram
Hz	Hertz
HMQCT	Heteronuclear multiple quantum coherence transfer
HSQCT	Heteronuclear single quantum coherence transfer
HPLC	High-performance liquid chromatography
h	Hour
IR	Infrared
K_{cat}	Catalytic rate constant
Ki	Inhibitor constant
v	Initial velocity
KDa	Kilodalton
MS	Mass spectroscopy
V_{max}	Maximum velocity
MHz	Mega hertz
K_M	Michaelis–Menten constant

K_{mA}	Michaelis–Menten constant for the lipase–L-alanine complex
$K_{m\ L\text{-}alanine}$	Michaelis–Menten constant for the lipase–L-alanine complex
K_{mB}	Michaelis–Menten constant for the lipase–D-glucose complex
$K_{m\ D\text{-}glucose}$	Michaelis–Menten constant for the lipase–D-glucose complex
K_i	Dissociation constant for the lipase–inhibitor (D-glucose) complex
$K_{i\ D\text{-}glucose}$	Dissociation constant for the lipase–inhibitor (D-glucose) complex
mM	Millimolar
mol	Mole
[M]+	Molecular ion
nm	Nanometre
N	Normality
NMR	Nuclear magnetic resonance
[a]	Optical rotation
ppm	Parts per million
%	Percentage
p	Pi
PAGE	Polyacrylamide gel electrophoresis
PPL	Porcine pancreas lipase
KBr	Potassium bromide
^{1}H	Proton
RSM	Response surface methodology
R_f	Retention factor
RT	Retention time
RML	Rhizomucor miehei lipase
σ	Sigma
SDS	Sodium dodecyl sulphate
TMS	Tetra-methyl silane
TLC	Thin layer chromatography
2D	Two-dimensional
UV	Ultraviolet
v/v	Volume by volume
a_w	Water activity
cm^{-1}	Wave per centimetre
w/w	Weight by weight

1 Introduction

Abstract

General concept on enzymes and enzymatic transformation along with the importance of carrying out enzymatic reactions in organic solvents is mentioned in this introductory chapter. A concluding Scope of the Book indicates that the book deals mainly with the transformations mediated by two important hydrolytic enzymes – glycosidases and lipases.

1.1 Introduction

The diverse biochemical reactions taking place in living systems are all mediated by series of enzymes (Michal 1999). Almost all the biochemical reactions are catalysed by enzymes, from both within and outside the cells. Hence, the subject of enzymatic transformation is not new. However, the term *enzymatic transformation* largely refers to those reactions conducted in vitro, using enzymes either in the form of microbial cells or isolated from intact cells or. Only about 2,800 enzymes have been characterised of the estimated 25,000 enzymes present in nature. Of which, 400 enzymes, mainly hydrolases, transferases and oxido-reductases have been identified as commercially potential ones. However, only 50 different kinds of enzymes find application on an industrial scale (Winterhalter and Schreier 1993; Schreier and Winterhalter 1993; Berger 1995). The enzymes of commercial importance in food industry are amylases, proteases, pectinases, cellulases, hemicellulases, lipases and lactases.

Enzymes are protein specific in their catalytic action, simple or conjugated capable of catalysing biochemical reactions. An enzyme in a reaction lowers the amount of activation energy required by the reaction. Enzymes are thermolabile and like proteins are inactivated at certain temperatures in a manner characteristic for protein denaturation. Substances on which enzymes act specifically are substrates, and their specificity plays a role in enzyme classification.

In the past two decades, the application of enzymes in organic synthesis has increased enormously as an extremely important synthetic strategy, with a wide variety of enzymes as tool in enzymatic and chemo-enzymatic synthesis (Ward and Singh 2000; Davis and Boyer 2001). As biocatalysts save additional reaction steps compared to chemical synthesis, most of the organic chemists use biocatalysts. Highly chemo-, regio- and stereospecific reactions under safer

S. Divakar, *Enzymatic Transformation*,
DOI 10.1007/978-81-322-0873-0_1,

and eco-friendly conditions are possible with enzymes and can be used for the preparation of wide range of organic compounds, especially in food and pharmaceutical preparations. Oxido-reductases, hydralases (lipases, esterases, glycosidases, transglycosidases, peptidases, acylases, amidases, epoxide hydralases, nitrilases and hydantoinases), lyases and isomerases are used in organic synthesis (Drauz and Waldmann 2002; Faber 2004; Buchholz et al. 2005).

Synthetic chemistry involving chemical routes requires drastic conditions, like use of acids or alkali, high temperature, hazardous chemicals, toxic metals and catalysts leading to high energy consumption, colouring of products, low selectivity and environmental pollution. It is also tedious to synthesise regio- and stereoselective compounds, as they require continuous protection and deprotection steps. Thus, chemical syntheses include multistep processes and result in a number of by-products and hence economically inefficient. Some chemically synthesised products also require exhaustive and cost-intensive purification steps.

Enzymes give better solutions to all these problems. Enzymes can function under milder reaction conditions, without requiring high temperatures and use of hazardous chemicals, and they are totally 'eco friendly' in nature. Enzymatic activity in nonaqueous solvents offers a new methodology for the production of many useful compounds which are not feasible in aqueous media. Klibanov (1986) initiated this novel approach and an outstanding synthetic strategy which now find use in a large number of applications in organic synthesis. Enzymes are employed in organic solvents for the synthesis of esters (Santaniello et al. 1993), chiral compounds (Orrenius et al. 1995), surfactants (Plou et al. 1999; Sarney and Vulfson 1995; Sarney et al. 1996), pharmaceutical intermediates (Duan et al. 1997), biotransformations of oils and fats (Bosley and Clayton 1994) and sugar-based polymers (Patil et al. 1991). Thus, the use of enzymes as biocatalysts in biotechnology has found potential applications in pharmaceutical, food, cosmetic, flavour and fragrance and beverage industries.

1.2 Scope of the Book

In the last two decades, literature on biotransformations has been ever growing enriching our knowledge on the subject. A lot of information on the behaviour of enzymes under diverse reaction conditions is available. However, only very few general rules have emerged from these studies, indicating that the field is wide open for creative minds to explore.

This book intends to throw some more light in this ever-growing area with a large number of examples on structurally diverse substrates and reaction conditions through investigations involving different strategies dealing with random parametric studies, optimization, response surface methodology and kinetics. Attempts have been made in this documentation to bring out the broad specificity of such enzyme-catalysed transformations involving two commercially important class of enzymes – lipases and glycosidases.

Both glycosidase and lipases, among all the known enzymes, have come a long way in establishing themselves as important synthetic tools to bio-organic reactions. The main scope of the present book is to discuss the use of glycosidases and lipases in reactions carried out in our laboratory for more than a decade or so, to prepare commercially important compounds and the underlying problems associated with the same. Thus, all efforts have been made to create interest in researchers about certain intricate details of the approaches in order to reap fruitful results as the problems are many and complex.

References

Berger RG (1995) Aroma biotechnology. Springer, New York/Berlin/Heidelberg

Bosley JA, Clayton JC (1994) Blueprint for a lipase support: use of hydrophobic controlled-pore glasses as model system. Biotechnol Bioeng 43:934–938

Buchholz K, Kasche V, Bornscheuer UT (2005) Biocatalysts and enzyme technology. Willey-VCH, Weinheim

Davis BG, Boyer V (2001) Biocatalysis and enzymes in organic synthesis. Nat Prod Rep 18:618–640

Drauz K, Waldmann H (eds) (2002) Enzyme catalysis in organic synthesis, 2nd edn, vols 1–3. Willey-VCH, Weinheim

Duan G, Ching CB, Lim E, Ang CH (1997) Kinetic study of enantioselective esterification of ketoprofen with *n*-propanol catalysed by an lipase in an organic medium. Biotechnol Lett 19:1051–1055

Faber K (2004) Enzymes in organic solvents. In: Faber K (ed) Biotransform. Organic chemistry, 5th edn. Springer, Berlin

Klibanov AM (1986) Enzymes that work in organic solvents. Chem Technol 16:354–359

Michal G (1999) Biochemical pathways. Wiley, New York

Orrenius C, Norin T, Hult K, Carrea G (1995) The *Candida antartica* lipase B catalysed kinetic resolution of seudenol in non-aqueous media of controlled water activity. Tetrahedron Asym 12:3023–3030

Patil DR, Rethwisch DG, Dordick JS (1991) Enzymatic synthesis of sucrose containing liner polyester in nearly anhydrous organic media. Biotechnol Bioengg 37:639–646

Plou FJ, Cruces MA, Pastor E, Ferrer M, Bernabe M, Ballesterose A (1999) Acylation of sucrose with vinyl esters using immobilized hydrolysis demonstration that chemical catalysis may interfere with enzymatic catalysis. Biotechnol Lett 21:635–639

Santaniello E, Ferraboschi P, Grisenti P (1993) Lipase-catalyzed transesterification in organic solvents. Applications to the preparation of enantiomerically pure compounds. Enzyme Microb Technol 15:367–382

Sarney DB, Vulfson EN (1995) Application of enzymes to the synthesis of surfactants. Trends Biotechnol 13:164–172

Sarney DB, Barnard MJ, MacManus DA, Vulfson EN (1996) Application of lipases to the regioselective synthesis of sucrose fatty acid monoesters. J Am Oil Chem Soc 73:1481–1487

Schreier P, Winterhalter P (1993) Progress in flavor precursor studies. Allured, Carol Stream

Ward OP, Singh A (2000) Enzymatic asymmetric synthesis by decarboxylases. Curr Opin Biotechnol 11:520–526

Winterhalter P, Schreier P (1993) Biotechnology – challenge for the flavor industry. In: Acree TE, Teranishi R (eds) Flavor science sensible principle and techniques. American Chemical Society, Washington, DC, pp 225–258

Glycosidases

2

Abstract

Glycosidases catalyse transformations leading to the attachment of carbohydrate molecules to aglycons. Hence, a detailed description of glycosidases is made in this chapter which includes their classification, nature, source, structural features, mechanism of glycosylation and advantages of such reactions. Also mentioned are examples of glycosylation reactions involving a wide variety of aglycons with different carbohydrate molecules in the form of a table.

2.1 Introduction

In order to carry out an enzymatic transformation reaction, one requires a profound knowledge on enzymes themselves. Although several enzymes have been employed in such reactions, this book will deal mainly with two most well-known hydrolytic enzymes, glycosidases and lipases. A detailed description on glycosidases is outlined in this chapter.

Among the enzymes dealing with carbohydrates, glycosidases and transglycosidases play an important role in the synthesis of glycosides. They belong to the group of carbohydrate-processing enzymes, widely employed in the regio- and stereoselective glycosylation reactions. Glycosidases are carbohydrases – enzymes that catalyse the hydrolysis of glycosidic bonds to liberate monosaccharides and oligosaccharides of lower molecular weight than the native simple as well complex carbohydrate substrates. These enzymes are very widely distributed in nature and found in all organisms. These large and important groups of enzymes, now known as amylase, were investigated long back by Payen and Persoz (1833), who were probably the first to recognise this enzyme in 1833 as 'diastase' (1833). Subsequently, a detailed study on glycosidases was carried out by many eminent chemists and biochemists including Fischer (1894).

2.2 Amylolytic Enzymes

Starch-degrading enzymes have been broadly classified into two groups – endo-acting enzymes or endohydralases and exo-acting enzymes or exohydralases (Berfoldo and Anthranikian 2001). α-Amylase (α-1,4-glucan-4-glucanohydralse; EC 3.2.1.1) is an endo-acting enzyme which hydrolyses linkages in the starch polymer chain randomly, leading to the generation of linear and

S. Divakar, *Enzymatic Transformation*,
DOI 10.1007/978-81-322-0873-0_2,

branched oligosaccharides. Most starch-hydrolyzing enzymes belong to the α-amylase family containing a characteristic catalytic $(\beta/\alpha)_8$ barrel domain. Exo-acting starch hydrolases such as β-amylase, glucoamylase, α-glucosidase and isoamylase attack the substrate from the nonreducing end, producing oligosaccharides. β-Amylase (EC 3.2.1.2), also referred to as α-1,4-D-glucan maltohydrolase or saccharogen amylase, hydrolyses α-1,4-glucosidic linkages of the starch chain to liberate successive maltose units from the nonreducing end, thereby producing β-maltose units by an inversion of configuration. α-Glucosidase (EC 3.2.1.20) attacks α-1,4 linkages of oligosaccharides and liberates glucose by retaining α-anomeric configuration.

2.3 Glucoamylase

Glucoamylase (E.C 3.2.1.3) is a fungal enzyme which goes under the names amyloglucosidase, 1,4-α-D-glucan hydrolase and γ-amylase. Enzyme code assigned for this enzyme by the Enzyme Commission (IUBMB 1992) is EC 3.2.1.3 where number 3 denotes hydrolases, referring to catalytic hydrolytic cleavage of large molecules with the addition of water; number 2 indicates glycosidic bond-cleaving glucosidases, and number 1 refers to hydrolysis of O-glycosyl compounds. There are several enzymes under the group 3.2.1, of which glucoamylase is number 3 which forms the fourth number in the nomenclature. Glucoamylase refers to hydrolysis of terminal α-1,4-linked-D-glucose residues successively from nonreducing ends of the carbohydrate chains from starch and malto-oligosaccharides, releasing D-glucose with inversion of configuration to β-D-glucose (Fogarty 1983).

When the next bond sequence is α-1,4, most forms of the enzyme can hydrolyse α-1,6-D-glucosidic bonds also. However, in vitro, this enzyme hydrolyses α-1,6- and α-1,3-D-glucosidic bonds also in other polysaccharides with high molecular weights. Since this enzyme is capable of completely hydrolyzing starch under long incubation periods, it is also called the saccharifying enzyme. Glucoamylases have the capacity to degrade large oligosaccharides up to about 90% α-1,6 linkages depending on the size of the substrate and the position of the α-1,6 linkages. Reverse reactions involving synthesis of saccharides and glycosides from D-glucose occur with a very high glucoamylase concentration for prolonged incubation periods and high concentrations of substrates.

2.4 Sources of Glucoamylases

The main source of glucoamylases is fungi although they are derived from a wide variety of plants, animals and microorganisms. Commercial enzymes originate from strains of either *Aspergillus niger* or *Rhizopus* sp. where they are used for the conversion of malto-oligosaccharides into glucose (Fogarty 1983). Since the discovery of two forms of glucoamylase from black koji mould in the 1950s, many reports have appeared on the multiplicity of glucoamylases, envisaged to be the result of several mechanisms, namely, mRNA modifications, limited proteolysis, variation in carbohydrate content or presence of several structural genes (Pretorius et al. 1991).

Fungal glucoamylases are usually one to five forms of glycoproteins. *Aspergillus niger* is being used widely in the commercial production of an extracellular glucoamylase. Two forms of glucoamylase – AG-I (glucoamylase I,99 kDa) and AG-II (glucoamylase II (112 kDa) – isolated from *A. niger* differed in their carbohydrate content, pH, temperature stabilities and activity (Williamson et al. 1992; Stoffer et al. 1993).

Gucoamylase from *Aspergillus terreus* strains was examined for the production of D-glucose and corn syrups (Ghosh et al. 1990; Ali and Hossain 1991). A glucoamylase from *Rhizopus* sp. released glucose from starch with 100% efficiency (Yu and Hang 1991). Takahashi et al. (1985) isolated three forms of glucoamylase from *Rhizopus* sp., GA-I (74 kDa), GA-II (58.6 kDa) and GA-III (61.4 kDa). Glucoamylases from other mould strains are *Humicola lanuginosa* (Taylor et al. 1978), *Thermomyces lanuginosa* (Haasum et al. 1991), *Myrothecium* sp. M1 (Malek and Hossain 1994) and a phytopathogenic fungus *Colletotrichum gloeosporiodes* (Krause et al. 1991).

There are several reports on the production of yeast glucoamylases (Saha and Zeikus 1989; Pretorius et al. 1991). Glucoamylase has been identified in *Saccharomyces cerevisiae* (Pugh et al. 1989), *Saccharomyces cerevisiae* var. *diastaticus* (Kleinman et al. 1988; Pretorius et al. 1991), *Saccharomycopsis fibuligera* (Itoh et al. 1989), *Schwanniomyces castellii* (Sills et al. 1984), *Schwanniomyces occidentalis* (Gellissen et al. 1991), *Pichia burtonii* and *Talaromyces* sp.

Bacterial glucoamylases have also been identified from aerobic strains such as *B. stearothermophilus* (Srivastava 1984), *Flavobacterium* sp. (Bender 1981), *Halobacterium sodamense* (Chaga et al. 1993) and *Arthrobacter globiformis* I42 (Okada and Unno 1989). Anaerobic strains include *Clostridium thermohydrosulfuricum* (Hyun and Zeikus 1985), *Clostridium* sp. G0005 (Ohinishi et al. 1991), *Clostridium acetobutylicum* (Chojecki and Blaschek 1986; Soni et al. 1992), *Clostridium thermosaccharolyticum* (Specka et al. 1991) and the microaerophile, *Lactobacillus amylovorus* (James and Lee 1995).

2.5 Sources of Other Glycosidases

Among the thermostable glycosidases used in the synthesis of glycosides, the most remarkable one is the β-glucosidase from the hyperthermophilic archeon *Pyrococcus furiosus* (Kengen et al. 1993) which is relatively easy to grow, and the enzyme is stable for 85 h at 100 °C. The enzyme has been cloned and over-expressed in *Escherichia coli* (Voorhorst et al. 1995). β-Galactosidase from *Aspergillus oryzae* was efficient towards alkylation (Stevenson et al. 1993). β-Galactosidase from *Streptococcus thermophilus* (Stevenson and Furneaux 1996) was employed for the synthesis of ethyl glycoside. β-Galactosidase from *Bacillus circulans* was also exploited by number of workers for synthetic purposes (Kojima et al. 1996). Enzymatic synthesis of butylglycoside via a transglycosylation reaction of lactose was carried out using β-galactosidase from *A. oryzae* (Ismail et al. 1999a). With primary as well as secondary alcohols, β-xylosidase from *A. niger* is an efficient glycosyl transfer catalyst that gave high (> 80%) yields of alkyl xylosides (Shinoyama et al. 1988) from methanol up to butanol. Almond glucosidase has been widely employed for the synthesis of alkyl and phenolic glycosides (Ljunger et al. 1994; Vic and Crout 1995; Vic et al. 1995; Ducret et al. 2002).

2.6 Structural Features of Glucoamylase

The structure of different glucoamylases showed a common subsite arrangement with seven in total and the catalytic site was located between subsite 1 and 2 (Hiromi et al. 1973; Ohinishi 1990; Fagerstrom 1991; Ermer et al. 1993). Subsite 2 has the highest affinity for oligomeric substrates and glucose, followed by decreasing affinity towards subsites 3–7 (Fagerstrom 1991). The glucoamylase G1 of *A. niger* consists of three parts: (1) Ala-1-Thr-440, containing the catalytic site; (2) Ser-441-Thr-551, a highly O-glycosylated linker segment; and (3) Pro-512-Arg-616, a C-terminal domain responsible for substrate binding (Stoffer et al. 1995; Svensson et al. 1983). Functionally important carboxyl groups in glucoamylase G2 from *A. niger* were identified to be Asp176, Glu179 and Glu180 in the catalytic site (Svensson et al. 1990). Tryptophan residues have been proposed to be essential for enzymatic activity (Rao et al. 1981) in *A. niger* glucoamylase and essentially tryptophan120 is reported to be responsible for binding of substrate and maintaining the structural integrity necessary for catalysis (Clarks and Svensson 1984).

Aspergillus awamori GA-I has also three catalytic domains (Svensson et al. 1983) like *A. niger*, a catalytic domain (residues 1–440), an O-glycosylated domain (residues 441–512) and a starch-binding domain (residues 513–616). Aleshin et al. (1994) produced a structural model for the catalytic domain of glucoamylase from *A. awamori* from a 2.2-Å resolution crystal structure of a proteolized form of GA-I *A. awamori* var *X100,* which contained the complete catalytic domain plus GA-II domain the N-terminal half of the O-glycosylated domain (residue 1–471). Amino acid sequence of three glucoamylases from

Rhizopus, Aspergillus and *Saccharomyces* were compared (Tanaka et al. 1986), of which the glucoamylases from *Rhizopus* and *Aspergillus* were highly homologous in both the nucleotide sequence and the amino acid sequence suggesting that these two glucoamylases were the most closely related among the three. The catalytic site in glucoamylase is believed to consist of two carboxyl groups (Hiromi et al. 1966a, b), where one acts as a general acid, protonating the glucosidic oxygen, while the other in the ionised carboxylate form stabilises the substrate intermediary oxonium ion (Braun et al. 1977; Matsumura et al. 1984; Post and Karplus 1986; Rantwijk et al. 1999).

Itoh et al. (1989) reported that in *S. fibuligera* glucoamylase, Ala-81, Asp-89, Trp-94, Arg-96, Arg-97 and Trp-166 were required for wild-type levels of activity, and Ala-81 and Asp-89 were not essential for catalytic activity which however played a role in thermal stability.

Complexes of glucoamylase from *A. awamori* with acarbose and D-gluco-dihydroacarbose indicate hydrogen bonds between sugar OH groups and Arg-54, Asp-55, Leu-177, Try-178, Glu-180 and Arg-305 of subsites 1 and 2 (Aleshin et al. 1994; Stoffer et al. 1995). Glu-179 (Sierks et al. 1990) and Glu-400 are positioned geometrically for general acid and base catalysis, ideal for the glucoside bond cleavage and assistance in the nucleophilic attack of water at the anomeric centre of the carbohydrate (Harris et al. 1993; Frandsen et al. 1994). Both the active sites of *A. niger* and *Rhizopus oryzae* glucoamylases are very much identical (Stoffer et al. 1995). In the active site of *R. oryzae,* the amino acid residues Arg-191, Asp-192, Leu-312, Trp-313, Glu-314, Glu-315 and Arg-443 are responsible for substrate binding through hydrogen bonds, whereas Glu-314 and Glu-544 are for glucosidic bond cleavage (Ashikari et al. 1986; Sierks et al. 1990).

2.7 Structural Features of β-Glucosidase

Sweet almond β-glucosidase has been known to hydrolyse glycosides resulting in the net retention of anomeric configuration (Eveleigh and Perlin 1969). It has followed the standard mechanism of such retaining glycosidases (McCarter and Withers 1994; Sinnot 1990). Assignment of sweet almond β-glucosidase as a family 1 glycosidase and identification of its active site nucleophilic residues sequence Ile-Thr-Glu-Asn-Gly were done by He and Withers (1997).

The primary structures of maize and sorghum β-glucosidases possess highly conserved peptide motifs TENEP and ITENG, which contain the two glutamic acids (Glu-191 and Glu-406) involved in the general acid/base catalysis and the respective family 1 β-glucosidases nucleophiles (San-Aparicio et al. 1998). A part slot-like active site (Davies and Henrissat 1995) was formed by these residues necessary for the substrate hydrolysis (Withers et al. 1990).

In the glycosylation step, the nucleophile Glu-406 attacks the anomeric carbon (C-1) of the substrate and forms a covalent glycosyl–enzyme intermediate with concomitant release of the aglycon after protonation of the glucosidic oxygen by the acid catalyst Glu-191 (Withers et al. 1990). In the next deglycosylation step, Glu-191 acts as a base, and a water molecule functions as the nucleophile and attacks the covalent glycosyl–enzyme, releasing the glucose and regenerating the nucleophilic Glu-406. In maize β-glucosidase isozyme Glu-1, these two catalytic glutamic acids are positioned within the active site at expected distances of ~5.5 Å for this mechanism (Czjzek et al. 2001). Verdoucq et al. (2003) from co-crystals of enzyme substrate and enzyme aglycon complexes of maize β-glucosidase isozyme Glu1 (ZmGlu1) have shown that five amino acid residues – Phe-198, Phe-205, Try-378, Phe-466 and Ala-467 – are located in the aglycon-binding site of ZmGlu1 which form the basis of aglycon recognition and binding and hence the substrate specificity. Kaper et al. (2000) have studied the substrate specificity of a family 1 glycosyl hydrolase – the β-glucosidase (CelB) from the hyperthermophilic archean *Pyrococcus furiosus,* at a molecular level exhibiting a homotetramer configuration, with subunits having a typical $(\beta\alpha)_8$-barrel fold. Comparison of the 3D model of the *Pyrococcus furiosus* β-glucosidase and the 6-phospho-β-glycosidase (LacG) from the mesophillic bacterium *Lactococcus lactis* (Kaper et al. 2000) showed that the positions of the active site residues in LacG and CelB are very well conserved, and the conserved residues involved in substrate binding are Asn-17, Arg-77, His-150, Asn-206, Tyr-307 and Trp-410. The average distance between the oxygen atoms of these glutamate carboxylic acids is 4.3 Å (±1 Å) in CelB, which is very much in the range of the general observed distance in retaining glycosyl hydrolases (McCarter and Withers 1994).

Investigation by Hays et al. (1998) of the catalytic mechanism, substrate specificity and transglycosylation acceptor specificity of guinea pig liver cytosolic β-glucosidase (CBG) indicated that CBG employed a two-step catalytic mechanism with the formation of a covalent enzyme–sugar intermediate and that CBG transferred sugar residues to primary hydroxyls and equatorial but not axial C-4 hydroxyls of aldopyranosyl sugars (Hays et al. 1998). Also the specificity of CBG for transglycosylation reactions was different from its specificity for hydrolytic reactions (Hays et al. 1998) and that CBG possessed a single active site nucleophile, specifically the glutamate residue in the sequence TITENG.

2.8 Glycosylation

Hydrolysis is the natural reaction for glucosidases and glucoamylases, whereas glycosylation is a forced, reversed reaction. Glycosides are asymmetric mixed acetals formed by the reaction of the anomeric carbon atom of the intermolecular hemiacetal or pyranose/furanoses form of the aldohexoses or aldoketoses with a hydroxyl group furnished by an alcohol (Lehinger 1975; Ernst et al. 2000). The bond formed is called glycosidic bond, and the reaction is called glycosylation. Because of multiple hydroxyl groups of similar reactivity, controlled glycosylation remains a challenge to organic chemists. Classical chemical approaches inevitably require quite a number of protection, activation, coupling and deprotection steps (Igarashi 1977; Konstantinovic et al. 2001). In contrast, enzymes (glycosidases and transglycosidases) offer one-step synthesis under mild conditions in a regio- and stereoselective manner (Vic and Thomas 1992). Enzyme-catalysed glycoside and oligosaccharide synthesis involves two types of reaction – a reverse hydrolytic glycosidase and a glycosyl-transferase-catalysed glycoside bond formation. A sugar donor and acceptor are incubated with the appropriate glycosidase or glycosyl-transferase that catalyses the efficient and selective transfer of the glycosyl residue to the acceptor. Glycosyl-transferases are often difficult to obtain (Auge et al. 1990), while, in contrast, the glycosidase approach uses simpler glycosyl donors, the free monosaccharide itself. This method has the advantage of using relatively simple glycosyl donors and readily available commercial enzymes at the expense of the absence of region selectivity in some instances (Trincone et al. 2003).

There are three types of reactions catalysed by glycosidases such as hydrolysis, reverse hydrolysis and transglycosylation (Scheme 2.1). In aqueous media, when there is large excess of water, glycoside or oligosaccharide or polysaccharide, hydrolysis is the dominant reaction (Scheme 2.1A). Other two reactions, namely, reverse hydrolysis and transglycosylation, lead to synthesis of glycosides, and the difference depends on the nature of the glycosyl donor.

The reverse hydrolytic approach is an equilibrium-controlled synthesis where the equilibrium is shifted towards synthesis (Panintrarux et al. 1995; Vic et al. 1997; Rantwijk et al. 1999) of a glycoside from a carbohydrate and an alcohol (Scheme 2.1B). This can be achieved by reducing the water activity, increasing the substrate concentrations and removing, if possible, the products of reaction (Vic and Crout 1995). This is a widely employed method for the enzymatic synthesis of alkyl glycosides and phenolic glycosides in an organic co-solvent (Vic and Crout 1995; Vic et al. 1997; Ducret et al. 2002).

The transglycosylation method is a kinetically controlled synthesis where the enzyme catalyses the transfer of a glycosyl residue from a glycosyl donor to the glycosyl acceptor (Scheme 2.1C). The reaction yield depends on the relative rate of product synthesis to that of hydrolysis. An efficient acceptor used in a high concentration

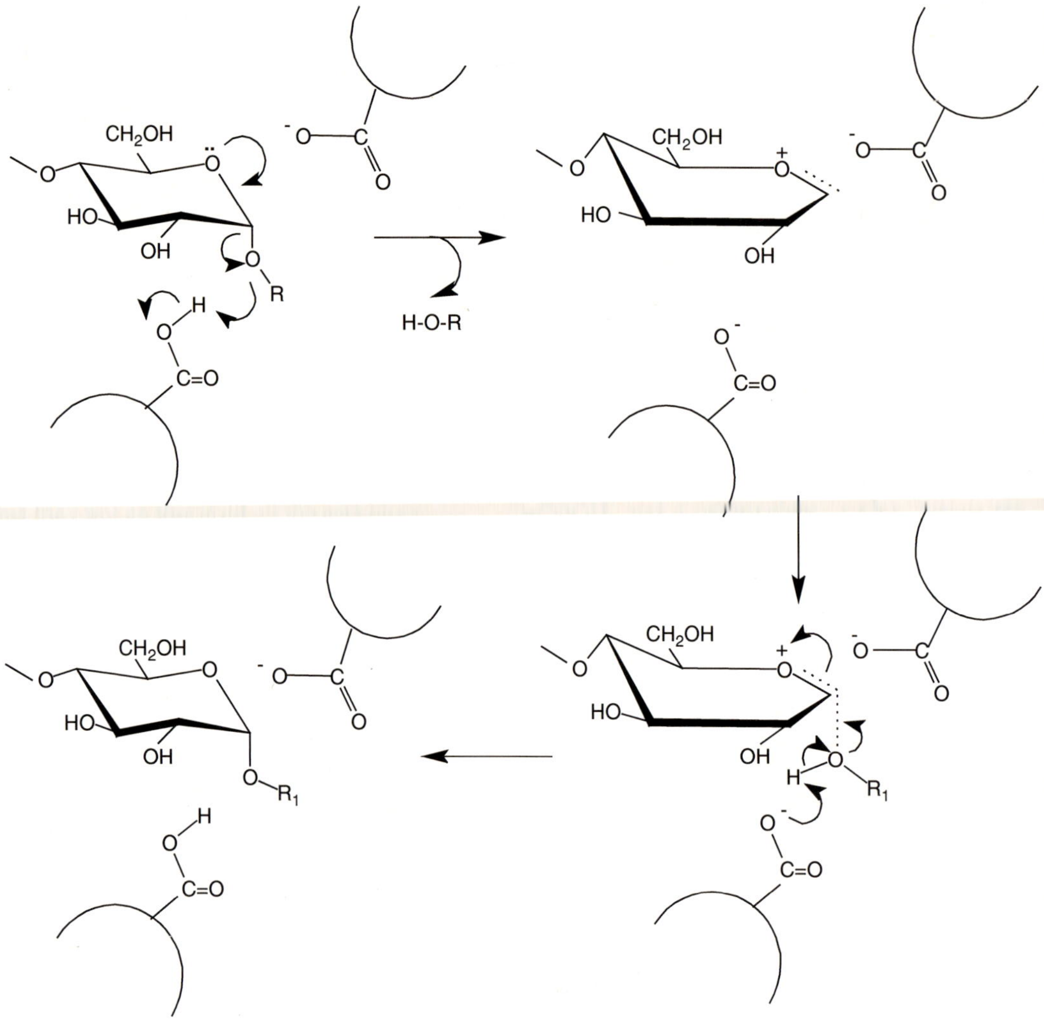

Scheme 2.3 Oxo-carbenium ion intermediate mechanism (Chiba 1997)

EC food regulations, and therefore, chemical preparation of glycosides is not applicable in the food industry.

Many glycosides are used in broad range of applications as surfactants (Busch et al. 1994), as food colourants and flavouring agents (Sakata et al. 1998), sweeteners (Shibata et al. 1991), antioxidants, anti-inflammatory (Gomes et al. 2002), antitumor (Kaljuzhin and Shkalev 2000), antibiotics (Ikeda and Umezawa 1999), antifungal (Tapavicza et al. 2000), antimicrobial (Zhou 2000) and cardiac-related drugs (Ooi et al. 1985). Glycosylation renders lipophilic compounds more water-soluble and thereby increases bioavailability of biologically active compounds besides imparting stability to the aglycon (Kren and Martinkova 2001). Alkyl glycosides are mainly used as nonionic surfactants in food, pharmaceuticals, chemical, cosmetic and detergent industries. These types of nonionic surfactants exhibit several interesting properties in detergency, foaming, wetting, emulsification and antimicrobial effect (Matsumura et al. 1990; Balzar 1991). Alkyl glycosides are non-toxic, non-skin-irritating and biodegradable (Matsumura et al. 1990; Busch et al. 1994; Madsen et al.

1996). Further alkyl glycosides are used as raw materials for sugar fatty acid ester synthesis (Mutua and Akoh 1993).

2.11 Advantages of Enzymatic Glycosylation over Chemical Methods

There are many advantages of using glycosidases (Vijayakumar 2007; Sivakumar 2009):

1. Exploitation of regio- and stereospecificity and selectivity
2. Milder reaction conditions
3. Non-generation of by-products associated with the use of several chemical procedures
4. Improved product yield and better product quality
5. Use of nonpolar solvents which impart stability to glycosidases, renders insolubility of the enzyme, solubility of alcohols and products in organic solvents and easy product workout procedures
6. No protection activation and deprotection required
7. Less environmental pollution

The use of organic solvent in enzyme catalysis has attracted much attention due to several desirable factors such as solubilities of the organic compounds, shifting equilibrium towards the synthesis, increasing the enzyme stability and recovery of the enzyme (Rubio et al. 1991; Mohri et al. 2003). Poor solubility of the carbohydrate substrate in the organic phase is a limiting factor especially when hydrophobic alcohol (glycosyl acceptor) itself is used as a substrate and in some cases as a solvent media (Laroute and Willemot 1992; Vic and Crout 1995; Crout and Vic 1998). There are reports where glycosylations were carried out either in biphasic systems of a water-immiscible alcohol and water (that maintains sugar substrate and enzyme) or water and water-miscible monophasic system (Monsan et al. 1996). The process of glycosylation can be effected under nonaqueous, solvent-free, high-substrate, high-temperature and moderate to high water activity conditions to achieve good yield of glycosides (Nilsson 1987; Roitsch and Lehle 1989; Gygax et al. 1991; Laroute and Willemot 1992; Vic and Thomas 1992; Shin et al. 2000).

Table 2.1 lists some of the important surfactants, phenolic, flavonoid, terpinyl, sweetener and medicinal glycosides, which have been prepared by the use of glycosidases, glucoamylases and glycosyl-transferases.

Table 2.1 Glycosides from enzymatic glycosylations

Name of the compound	Source of enzyme	Applications	References
A. Surfactant glycosides			
(1) β-D-Glycopyranosides of n-heptanol, n-octanol, 2-phenyl hexanol, 3-phenyl propanol, 4-phenyl butanol, 5-phenyl petanol, 6-phenyl hexanol, 2-pyridine methanol, isobutanol, isopentanol, p-methoxy cinamyl alcohol, isopropanol, cyclohexanol, 1-phenyl ethanol, 1,5-pentanediol, 1,6-hexanediol, 1,7-heptanediol, 1,8-octanediol, 1,9-nonane-diol, salicyl alcohol and 4-nitrophenol	β-Glucosidase from almonds	As nonionic surfactants, in detergents and cosmetics	Katusumi et al. (2004)
(2) β-D-Glucopyranosides of propanol, hexanol and octanol	Raw almond meal	In detergents and cosmetics	Chahid et al. (1992, 1994)
(3) α/β-Glucopyranosides of ethanol, 1-propanol, 2-propanol, 2-methyl 2-propanol, 1-butanol, 2-butanol, 1-pentanol, 1-hexanol, 1,3-butanediol, 1,4-butanediol, 2,3-butane-diol, 1,2-pentanediol, 1,5-pentanediol	Glucoamylase and β-glucosidase	In detergents and cosmetics	Laroute and Willemot (1992)
(4) Allyl and benzyl β-D-glucopyranoside, allyl-β-D-galactopyranoside	Almond β-D-glucosidase	Used in the synthesis of glycopolymers, as temporary anomeric-protected derivatives in carbohydrate chemistry	Vic and Crout (1995)
(5) n-Octyl glucoside, n-octyl galactoside	β-Galactosidase from *A. oryzae*, almond meal	In detergents and cosmetics	Chahid et al. (1994)
(6) n-Octyl -β-D-glucoside, 2-hydroxy benzyl glucopyranoside.	Almond β-glucosidase	In detergents and cosmetics	Vic et al. (1997)
(7) n-Octyl-β-D-glucoside, n-octyl-β-D-xylobioside, n-octyl-β-D-xyloside		As biological detergents and as emulsifying agents in cosmetics	Nakamura et al. (2000)
B. Phenolic glycosides			
(1) Eugenol-α-glucoside	α-Glucosyl transfer enzyme of *Xanthomonas campestris* WU-9701	As a prodrug of a hair restorer, as a derivative of spices	Sato et al. (2003)
2. Eugenol-β-glucoside	Biotransformation by cultured cells of *Eucalyptus perriniana*	As a prodrug of a hair restorer	Orihara et al. (1992)
(3) Vanillin-β-D-monoglucopyranoside	By suspension-cultured cells of *Coffea arabica*	As a food additive flavour	Kometani et al. (1993a)

(4) Capsaicin-β-D-glucopyranoside	By suspension-cultured cells of *Coffea arabica*	Food ingredient and pharmacological applications	Kometani et al. (1993b)
	By cultured cells of *Phytolacca americana*		Hamada et al. (2003)
(5) α-Salicin, α-isosalcin, β-salicin	*Bacillus macerans* cyclodextrin glucanyl transferase and *Leuconostoc mesenteroides* B-742CB dextransucrase	Anti-inflammatory, analgesic antipyretic prodrug	Yoon et al. (2004)
(6) Curcumin glycosides	By cell suspension cultures of *Catharanthus roseus*	Food colourant, as antioxidant	Kaminaga et al. (2003)
C. Flavonoid glycosides			
(1) Quercetin-3-O-β-D-xylopyranosyl (1→2)-β-D-galactopyranoside	Isolated from *Trifolium repens L*	UV-B radiation protection	Hofmann et al. (2000)
(2) Kaempferol-3-O-β-D-xylopyranosyl (1→2)-β-D-galactopyranoside			
D. Sweetener glycosides			
(1) Stevioside, steviobioside, rebaudioside A, rebaudioside B	Isolated from the leaves of *Stevia rebaudiana*	As a natural sweetener, utilised in beverages	Kohda et al. (1976)
(2) Steviol-13-O-glucopyranoside, steviobioside, stevioside and rebaudioside	Enzyme fractions prepared from the soluble extracts of stevia	As a natural food sweeteners	Shibata et al. (1991)
E. Terpinyl glycosides			
(1) Gereniol β-glucoside, nerol β-glucoside, citroniol β-glucoside	β-Glucosidase from *A. niger, Trichoderma reesei, Candida molischiana* and almond	Good bioavailability, antifungal and antimicrobial activity	Gunata et al. (1994)
(2) Gereniol β-galactoside, nerol β-galctoside, citroniol β-galactoside	β-Galactosidase from *A. oryzae*	Good bioavailability, antifungal and antimicrobial activity	Donho et al. (1996)
F. Glycosides in medicine			
(1) Enediyne antibiotics – calicheamicin	Isolated from the cultivation broth of *Micromonospora echinospora*	Antitumor agents	Lee et al. (1987), Golik et al. (1987)
(2) Vitamin glycosides			
5′-O-(β-D-galactopyranosyl)-thiamin	A. oryzae β-galactosidase	Excellent nutritional efficiencies, more stable against UV and light.	Suzuki and Uchida (1994)

(continued)

Table 2.1 (continued)

Name of the compound	Source of enzyme	Applications	References
5′-O-(β-D-glucopyranosyl)-thiamin	Cyclomaltodextrin glucanotransferase from *Bacillus stearothermophilus*	Pleasant taste and odour, good bioavailability. More stable towards oxidative stress and UV irradiation	Uchida and Suzuki (1998)
4-α-D-glucopyranosyl rutin 2-O-α-glucopyranosyl-L-ascorbic acid	Cyclomaltodextrin glucanotransferase from *Bacillus stearothermophilus*		Suzuki and Suzuki (1991), Aga et al. (1990)
Alkaloid glycosides – elymoclavine-O-β-D-fructofuranoside	Isolated from a saprophytic culture of *Claviceps* sp.	In the treatment orthostatic circulatory disturbances, hypertension, hyperprolactinaemia, antibacterial and cytostatic effects and hypolipaemic activity	Kren and Cvak (1999)
Steroidal glycosides – glycosides of diosgenin, solasodine, solasonine	Isolated from *Solanum* sp.	Anticarcinogenic activity	Nakamura et al. (1996)

References

Aga H, Yoneyama M, Sakai S, Yamamoto I (1990) Synthesis of 2-O-α-D-glucopyranosyl L-ascorbic acid by cyclomaltodextrin glucotransferase from *Bacillus stearothermophilus*. Agric Biol Chem 55:1751–1756

Akao T, Yoshino T, Kobashi K, Hatlori M (2002) Evaluation of salicin as an antipyretic prodrug that does not cause gastric injury. Planta Med 68:714–718

Aleshin AE, Firsov LM, Honzatko RB (1994) Refined structure for the complex of acarbose with glucoamylases from *Aspergillus awamori* var. X100 to 2.4 Å resolution. J Biol Chem 269:15631–15639

Ali S, Hossain Z (1991) Characteristics of glucoamylase from *Aspergillus terreus*. J Appl Bacteriol 71:144–146

Ashikari T, Nakamura N, Tanaka Y, Kiuchi N, Shibano Y, Tanaka T, Amachi T, Yoshizumi H (1986) *Rhizopus* raw-starch-degrading glucoamylase Its cloning and expression in yeast. Agric Biol Chem 50:957–964

Auge C, Fernandez RF, Gautheron CM (1990) The use of immobilized glycosyltransferases in the synthesis of sialyl oligosaccharides. Carbohydr Res 200:257–268

Balzar D (1991) Alkylglucosides, their physico-chemical properties and their uses. Tenside Surf Det 28:419–427

Bender H (1981) A bacterial glucoamylase degrading cyclodextrins. Eur J Biochem 115:287–291

Berfoldo C, Anthranikian G (2001) Amylolytic enzymes from hyperthermophiles. Methods Enzymol 330:269–289

Braun H, Cogoli A, Semenza G (1977) Carboxyl groups at the two active centers of sucrose-isomaltoase from rabbit small intestine. Eur J Biochem 73:437–442

Busch P, Hensen H, Khare J, Tesmann H (1994) Alkylpolyglycosides-a new cosmetic concept for milderness. Agro-Food-Ind Hi-Tech 5:20–28

Chaga G, Porath J, Illeni T (1993) Isolation and purification of amyloglucosidase from *Halobacterium sodomenase*. Biomed Chromatogr 7:256–261

Chahid Z, Montet D, Pina M, Graille J (1992) Effect of water activity on enzymatic synthesis of alkylglycosides. Biotechnol Lett 14(4):281–284

Chahid Z, Montet D, Pina M, Bonnot F, Graille J (1994) Biocatalyzed octylglycoside synthesis from a disaccharide. Biotechnol Lett 16:795–800

Chiba S (1995) In: The Amylase Research Society of Japan (ed) Enzyme chemistry and molecular biology of amylase and related enzymes. CRC Press, Boca Raton/Ann arbor/London/Tokyo, pp 68–82

Chiba S (1997) Molecular mechanism in α-glucosidase and glucoamylase. Biosci Biotech Biochem 61:1233–1239

Chojecki A, Blaschek HP (1986) Effect of carbohydrate source on alpha-amylase and glucoamylase formation by *Clostridium acetobutylicum* SA-1. Ind Microbiol 1:63–67

Clarks AJ, Svensson B (1984) Identification of an essential tryptophanyl residue in the primary structure of glucoamylase G2 from *Aspergillus niger*. Carlesberg Res Commun 49:559–566

Crout DHG, Vic G (1998) Glycosidases and glycosyl transferases in glycoside and oligosaccharides synthesis. Biocatal Biotransform 2:98–111

Czjzek M, Cicek M, Zamboni V, Bevan DR, Henrissat B, Esen A (2001) The mechanism of substrate (aglcone) specificity in β-glucosidase –DIMBOA, – DIMBOA Glc and –dhurrin complexes. Proc Natl Acad Sci USA 97:13555–13560

Davies G, Henrissat B (1995) Structures and mechanisms of glycosyl hydrolases. Structure 3:853–859

Donho M, Kimura T, Hara H (1996) Methods of producing geranyl β-D-galactopyranoside as flavoring material by enzymatic galactosylation of citronellol. Jpn Kokai Tokkyo Konho JP 8188589–8188591 (CA 125 222344)

Ducret A, Carriere JF, Trani M, Lortie R (2002) Enzymatic synthesis of octyl glucoside catalysed by almond β-glucosidase in organic media. Can J Chem 80:653–656

Ermer J, Rose K, Huber G, Schhellenenberger A (1993) Subsite affinities of *Aspergillus niger* glucoamylase II determined with *p*-nitophenylmaltooligosaccharides. Biol Chem Hoppe Seyler 374:123–128

Ernst B, Hart GW, Sinay P (2000) Carbohydrates in chemistry and biology, vol 1. Willey-VCH, Weinheim, pp 177–193

Eveleigh DE, Perlin AS (1969) A proton magnetic resonance study of the anomeric species produced by D-glucosidases. Carbohydr Res 10:87–95

Fagerstrom R (1991) Subsite mapping of *Hormoconis resinae* glucoamylase and their inhibition by gluconolactone. J Gen Microbiol 137:1001–1008

Fischer E (1894) Einfluss der konfiguration auf die wirkung der enzyme. Ber Chem Ges 27:2985–2993

Fogarty WM (ed) (1983) Microbial amylases. Microbial enzymes and biotechnology. Appl Science Publishers, London, pp 1–92

Frandsen TP, Dupont C, Lehmbeck J, Stoffer B, Sierks MR, Honzatko RB, Svensson B (1994) Site-directed mutagenesis of the catalytic base Glutamic acid 400 in glucoamylase from *Aspergillus niger* and of Tyrosine 48 and Glutamine 401, both hydrogen bonded to the gamma-carboxylate group of Glutamic acid 400. Biochemistry 33:13808–13816

Gellissen G, Janowicz ZA, Merckelbach A, Piontek M, Keup P, Weydemann U, Hollenberg CP, Srasser AWM (1991) Heterologus gene expression in *Hansenula polymorpha*: efficient secretion of glucoamylase. Biotechnology 9:291–295

Ghosh A, Chatterjee BS, Das A (1990) Characterization of glucoamylase from *Aspergillus terreus* 4. FEMS Microbial Lett 66:345–349

Golik J, Clardy J, Dubay G, Groenewold G, Kawaguchi H, Konishi M, Krishnan B, Ohkuma H, Saitoh K, Dobye TW (1987) Esperamicins, a novel class of potent antitumor antibiotics. 3. Structures of esperamicins A1, A2 and A1b. J Am Chem Soc 109:3461–3464

Gomes DCF, Alegrio LV, Leon LL, de Lima MEF (2002) Total synthesis and anti-leishmanial activity of some curcumin analogues. Arzneim-Forsch 52:695–698

Gunata Z, Vallier MJ, Sapis JC, Baumes R, Bayonove C (1994) Enzymic synthesis of monoterpeny β-D-glucosides by various β-glucosidases. Enzyme Microb Technol 16:1055–1058

Gygax D, Spies P, Winkler T, Pfaar U (1991) Enzymatic synthesis of β-D-glucuronides with *in situ* regeneration of uridine 5'-diphosphoglucuronic acid. Tetrahedron 47:5119–5122

Haasum I, Ericksen SH, Jensen B, Olsen J (1991) Growth and glucoamylase production by the thermophilic fungus *Thermophilus lanuginose* in a synthetic medium. Appl Microbiol Biotechnol 34:656–660

Hamada H, Nishida K, Furuya T, Ishihara K, Nakajima N (2003) Preparation of a new pepper: chemoenzymatic synthesis of capsaicin oligosaccharide and 8-nordihydrocapsaicin. J Mol Catal B: Enzym 16:115–119

Harris EMS, Aleshin AE, Firsov LM, Honzatko RB (1993) Refined structure of the complex of 1-deoxynojirimycin with glucoamylase from *Aspergillus awamori* var X100. Biochemistry 32:1618–1626

Hays WS, Vander Jagt DJ, Bose B, Serianni AS, Glew RH (1998) Catalytic mechanism and specificity for hydrolysis and transglycosylation reactions of cytosolic β-glucosidase from guinea pig liver. J Biol Chem 273:34941–34948

He S, Withers SG (1997) Assignment of sweet almond β-glucosidase as a family 1 glycosidase and identification of its active site nucleophile. J Biol Chem 272:24864–24867

Hiromi K, Kawai M, Ono S (1966a) Kinetic studies on glucoamylase IV. Hydrolysis of isomaltose. J Biochem 59:476–480

Hiromi K, Takahashi K, Hamazu Z, Ono S (1966b) Kinetic studies on glucoamylase III. The influence of pH on the rates of hydrolysis of maltose and panose. J Biochem 59:469–475

Hiromi K, Nitta Y, Numata C, Ono S (1973) Subsite affinities if glucoamylase examination of the validity of the subsite theory. Biochem Biophys Acta 302:362–375

Hofmann RW, Swinny EE, Bloor SJ, Markham KR, Ryan KG, Campbell BD, Jordan BR, Fountain DW (2000) Responses of nine *Trifolium repens L.* populations to ultraviolet-B radiation. Differential flavonol glycoside accumulation and biomass production. Ann Bot 86:527–537

Hyun HH, Zeikus JG (1985) General biochemical characterization of thermostable pullulanase and glucoamylase from *Clostridium thermohydrosulfuricum*. Appl Environ Microbiol 49:1168–1173

Igarashi K (1977) The Koenigs-Knorr reaction. Adv Carbohydr Chem Biochem 34:243–283

Ikeda D, Umezawa S (1999) Aminoglycoside antibiotics. In: Ikan R (ed) Naturally occurring glycosides. Wiley, England, pp 1–42

Ismail A, Linder M, Ghoul M (1999a) Optimization of butylgalactoside synthesis by β-galactosidase from *Aspergillus oryzae*. Enzyme Microb Technol 25:208–213

Ismail A, Soultani S, Ghoul M (1999b) Enzymatic-catalyzed synthesis of alkylglycosides in monophasic and biphasic systems. I. The transglycosylation reaction. J Biotechnol 69:135–143

Itoh T, Sakata Y, Akada R, Nimi O, Yamshita I (1989) Construction and characterization of mutant glucoamylases from the yeast *Saccharomycopsis fibuligera*. Agric Biol Chem 53:3159–3168

IUBMB (1992) Enzyme nomenclature. Academic Press, San Diego, California, ISBN 0-12-227164-5

Jacobson RH, Zhang X-J, DuBose RF, Matthews BW (1994) Three dimensional structure of β-galactosidase from *E. Coli*. Nature 369:761–766

Jacobson RH, Kuroki R, Weaver LH, Zhang X-J, Matthews BW (1995) In: Saddler JN, Penner MH (eds) Enzymatic degradation of insoluble carbohydrates, vol 618. ACS Symposium Series, Washington, DC, pp 38–50

James JA, Lee BH (1995) Cultural conditions for production of glucoamylase from *Lactobacillus amylovorus* ATCC 33621. J Appl Bacteriol 79:499–505

Kaljuzhin OV, Shkalev MV (2000) Immunomodulator and pharmaceutical compositions with antitumor properties, and a food additive. Patent EP1038532 (CA 129 335732)

Kaminaga Y, Nagatsu A, Akiyama T, Sugimoto N, Yamazaki T, Maitani T, Mizukami H (2003) Production of unnatural glucosides of curcumin with drastically enhanced water solubility by cell suspension cultures of *Catharanthus roseus*. FEBS Lett 555:311–316

Kaper T, Lebbink JHG, Pouwels J, Kopp J, Schulz GE, Oost JV, Vos WM (2000) Comparative structural analysis and substrate specificity engineering of the hyperthermostable β-glucosidase CelB from *Pyrococcus furious*. Biochemistry 39:4963–4970

Katusumi K, Mikio F, Yoshiteru I, Hiroyuki A (2004) Simple synthesis of β-D-glycopyranosides using β-glycosidase from almonds. Chem Pharm Bull 52:270–275

Kengen SWM, Luesink EJ, Stams AJM, Zehnder AJB (1993) Purification and characterization of an extremely thermostable β-glucosidase from the hyperthermophilic archaeon *Pyrococcucus furious*. Eur J Biochem 213:305–312

Kleinman MJ, Wilkinson AE, Wright IP, Evans IH, Bevan EA (1988) Purification and properties of an extracellular glucoamylase from a diastatic strain of *Saccharomyces cerevisiae*. Biochem J 249:163–170

Kohda H, Kasai R, Yamasaki K, Tanaka O (1976) New sweet diterpene glucosides from *Stevia rebaudiana*. Phytochemistry 15:981–983

Kojima M, Maruo S, Ohgi T, Ezure Y (1996) Enzymatic synthesis of 4-O-β-D-glactopyranosylmoranoline and 3-O-β-D-galactopyranosylmoranoline by using β-galactosidase from *Bacillus circullans*. Biosci Biotech Biochem 60:694–696

Kometani T, Tanimoto H, Nishimura T, Kanbara I, Okada S (1993a) Glucosylation of capsaicin by cell suspension cultures of *Coffea arabica*. Biosci Biotech Biochem 57:2192–2193

Kometani T, Tanimoto H, Nishimura T, Okada S (1993b) Glucosylation of vanillin by cultured plant cells. Biosci Biotech Biochem 57:1290–1293

Konstantinovic S, Predojevic J, Gojkovic S, Ratkovic Z, Mojsilovic B, Pavlovic V (2001) Synthesis of C7-C16 alkyl 2,3 dideoxy glucosides from glucose and fatty acids. Ind J Chem 40B:1242–1244

Krause DR, Wood CJ, MacLean DJ (1991) Glucoamylase (exo-1,4-α-D-gluanohydrolase, E.C. 3.2.1.3) is the major starch-degrading enzyme secreted by the phytopathogenic fungus *Colletotrichm gloeosporiodes*. J Gen Microbiol 137:2463–2468

Kren V (2001) Chemical biology and biomedicine of glycosylated natural compounds. In: Fraser-Reid B, Tatsuta K, Thiem J (eds) Glycoscience chemistry and chemical biology, vol 3. Springer, Berlin, pp 2471–2529

Kren V, Cvak L (1999) Ergot genus *Claviceps*, medicinal and aromatic plants-industrial profiles. Harwood Publ. Ltd., Amsterdam/London

Kren V, Martinkova L (2001) Glycosides in medicine: the role of glycosidic residue in biological activity. Curr Med Chem 8:1313–1338

Laroute V, Willemot RM (1992) Glucoside synthesis by glucoamylase or β-glucosidase in organic solvents. Biotechnol Lett 14:169–174

Lee MD, Dunne TS, Chang CC, Ellestad GA, Siegel MM, Morton GO, McGahren WJ, Borders DB (1987) Calichemicines, a novel family of antitumor antibiotics 2. Chemistry and structure of calichemicin, γI. J Am Chem Soc 109:3466–3468

Lehinger AL (1975) Sugars, storage polysaccharides and cell walls. In: Biochemistry. Worth Publishers Inc., New York, pp 249–276

Ljunger G, Adlercreutz P, Mattiasson B (1994) Enzymatic synthesis of octyl-β-glucoside in octanol at controlled water activity. Enzyme Microb Technol 16:751–755

Madsen T, Petersen G, Seiero C, Torslov J (1996) Biodegradability and aquatic toxicity of glycoside surfactants and a nonionic alcohol etherate. J Am Oil Chem Soc 73:929–933

Malek SAS, Hossain Z (1994) Purification and characterization of a thermostable glucoamylase from *Myrothecium* isolate. J Appl Bacteriol 76:210–215

Matsumura Y, Kasunoki M, Harada W, Kakudo M (1984) Structure and possible catalytic residues of taka amylase A. J Biochem 95:697–702

Matsumura S, Imai K, Yoshikawa S, Kawada K, Uchibori T (1990) Surface activities, biodegradability and antimicrobial properties of n-alkyl glucosides, manosides and galactosides. J Am Oil Chem Soc 67:996–1001

McCarter J, Withers SG (1994) Mechanisms of enzymatic glycoside hydrolysis. Curr Opin Struct Biol 4:885–892

Mohri K, Watanabe Y, Yoshida Y, Satoh M, Isobe K, Sugimoto N, Tsuda Y (2003) Synthesis of glycosylcurcuminoids. Chem Pharm Bull 51:1268–1272

Monsan PF, Paul F, Pelenc P, Bouler E (1996) Enzymatic production of α-butyl glucoside and its fatty acid esters. Ann NY Acad Sci 799:633–641

Mutua LN, Akoh CC (1993) Synthesis of alkyl glucoside fatty acid esters in non aqueous media by *Candida* sp. lipase. J Am Oil Chem Soc 70:43–46

Nakamura T, Komori C, Lee Y-Y, Hashimoto F, Yohara S, Nohara T, Ejima A (1996) Cytotoxic activities of solanum steroidal glycosides. Biol Pharm Bull 19:546–566

Nakamura T, Toshima K, Matsumura S (2000) One-step synthesis of n-octyl β-D-xylotrioside, xylobioside and xyloside from xylan and n-octanol using acetone powder of *Aureobasidium pullulans* in supercritical fluids. Biotechnol Lett 22:1183–1189

Nilsson KGI (1987) A simple strategy for changing the regio selectivity of glycosidase catalyzed formation of disaccharides. Carbohydr Res 167:95–103

Ohinishi M (1990) Subsite structure of *Rhizopus niveus* glucoamylase, estimated with the binding parameters for maltooligosaccharides. Starch/Starke 42:311–313

Ohinishi H, Sakai H, Ohta T (1991) Purification and some properties of a glucoamylasse from *Clostridium sp.* G0005. Agric Biol Chem 55:1901–1902

Okada G, Unno T (1989) A glucodextranase accompanied by glucoamylase activity from *Arthrobacter globiformis* I 42. Agric Biol Chem 53:223–228

Ooi Y, Hashimoto T, Mitsuo N, Satoh T (1985) Enzymatic formation of β-galactosidase from *Aspergillus oryzae* and its application to the synthesis of chemically unstable cardiac glycosides. Chem Pharm Bull 33:1808–1814

Orihara Y, Furuya T, Hashimoto N, Deguchi Y, Tokoro K, Kanisawa T (1992) Biotransformation of isoeugenol and eugenol by cultured cells of *Eucalyptus perriniana*. Phytochemistry 31:827–831

Panintrarux C, Adachi S, Araki Y, Kimura Y, Matsuno R (1995) Equilibrium yield of n-alkyl-β-D-glucoside through condensation of glucose and n-octanol by β-galactosidase in a biphasic system. Enzyme Microb Technol 17:32–40

Payen A, Persoz JF (1833) Mémoire sur la diastase, les principaux produits de ses reactions et leur applications aux arts industriels. Annales de chimie et de physique 53:73–92

Post CB, Karplus M (1986) Does lusozyme follow the lusozyme pathway? An alternative based on dynamic structural and stereoelectronic considerations. J Am Chem Soc 108:1317–1319

Pretorius IS, Lambrechts MG, Marmur J (1991) The glucoamylase multigene family in *Saccharomyces cerevisiae var. diastaticus* an overview. CRC Crit Rev Biochem Mol Biol 26:53–76

Pugh TA, Shah JC, Magee PT, Clancy MJ (1989) Characterization and localization of the sporulation glucoamylase from *Saccharomyces cerevisiae*. Biochem Biophys Acta 994:200–209

Rantwijk FV, Oosterom MW, Sheldon RA (1999) Glycosidase-catalyzed synthesis of alkyl glycosides. J Mol Catal B: Enzym 6:511–532

Rao VB, Sastri NVS, Rao PVS (1981) Purification and characterization of a thermostable glucoamylase from the thermophilic fungus *Thermomyces lanuginose*. Biochem J 193:379–385

Robyt JF (1998) Essentials of carbohydrate chemistry. Springer, New York, pp 64–68

Roitsch T, Lehle L (1989) Structural requirements for protein N-glycosylation. Influence of acceptor peptides on cotranslational glycosylation of yeast influence and site-directed mutagenesis around a sequon sequence. Eur J Biochem 181:525–529

Rubio E, Fernandez MA, Klibanov AM (1991) Effect of the solvent on enzyme regio selectivity. J Am Chem Soc 113:695–696

Saha BC, Zeikus JG (1989) Microbial glucoamylases biochemical and biotechnological features. Starch/Starke 41:57–64

Sakata I, Maruyama I, Kobayashi A, Yamamoto I (1998) Production of phenethyl alcohol glycoside. Jpn Kokai Tokkyo Konho, Japan Patent JP 10052297 (CA 128 229438)

San-Aparicio J, Hermoso JA, Martinz-Ripoll M, Laquerica JL, Polaina J (1998) Crystal structure of β-glucosidase A from *Bacillus polymyxa* insights into the catalytic activity in family1 glycosyl hydrolases. J Mol Biol 275:491–502

Sato T, Takeuchi H, Takahashi K, Kurosu J, Yoshida K, Tsugane T, Shimura S, Kino K, Kirimura K (2003) Selective α-glucosylation of eugenol by α-glucosyl transfer enzyme of *Xanthomonas campestris* WU-9701. J Biosci Bioeng 96:199–202

Schmid B, Kotter I, Heide L (2001) Pharmacokinetics of salicin after oral administration of a standard willow bark extract. Eur J Clin Pharmacol 57:387–391

Shibata H, Sonoke S, Ochiai H, Nishihashi H, Yamada M (1991) Glucosylation of steviol and steviol glucosides in extracts from *Stevia rebaudiana Bertoni*. Plant Physiol 95:152–156

Shin HK, Kong JY, Lee JD, Lee TH (2000) Synthesis of hydroxy benzyl-α-glucosides by amyloglucosidase-catalysed transglycosylation. Biotechnol Lett 22:321–325

Shinoyama H, Kamiyama Y, Yasui T (1988) Enzymatic synthesis of alkyl β-xylosides from xylobiose by application of the transxylosyl reaction of *Aspergillus niger* β-xylosidase. Agric Biol Chem 52:2197–2202

Sierks MR, Ford C, Reilly PJ, Svensson B (1990) Catalytic mechanism of fungal glucoamylases as defined by mutagenesis of Asp 176, Glu179, and Glu180 in the enzyme from *Aspergillus awamori*. Protein Eng 3:193–198

Sills AM, Saunder ME, Stewart GG (1984) Isolation and characterization of the amylolytic system of *Schwanniomyces castellii*. J Inst Brew 90:311–316

Sinnot ML (1990) Catalytic mechanism of glycosyl transfer. Chem Rev 90:1171–1202

Sivakumar R (2009) Enzymatic synthesis of selected phenolic and vitamin glycosides. PhD thesis, University of Mysore

Soni BK, Kapp C, Goma G, Soucaille P (1992) Solvent production from starch effect of pH on α-amylase and glucoamylase localization and synthesis in synthetic medium. Appl Microbiol Biotechnol 37:539–543

Specka U, Mayer F, Antranikian G (1991) Purification and properties of thermoactive glucoamylase from *Clostridium thermosaccharolyticum*. Appl Environ Microbiol 57:2317–2323

Srivastava RAK (1984) Studies on extracellular and intracellular purified amylases from a thermophilic *Bacillus stearothermophilus*. Enzyme Microb Technol 6:422–426

Stevenson DE, Furneaux RH (1996) High yield synthesis of ethyl and 2-fluroethyl β-D-galactopyranosides using *Streptococcus thermophilus* β-galactosidase. Enzyme Microb Technol 18:513–518

Stevenson DE, Stanley RA, Furneaux RH (1993) Optimization of alkyl β-D-galactopyronoside synthesis from lactose using commercially available β-galactosidase. Biotechnol Bioeng 42:657–666

Stoffer B, Frandsen T, Busk P, Schneider P, Svendsen I, Svensson B (1993) Production, purification and characterization of the catalytic domain of glucoamylase from *Aspergillus niger*. Biochem J 292:197–202

Stoffer B, Aleshin AE, Firsov LM, Svensson B, Honzatko RB (1995) Refined structure for the complex of D-gluco-dihydroacarbose with glucoamylases from *Aspergillus awamori* var. X100 to 2.2 Å resolution dual conformation for extended inhibitors bound to the active site of glucoamylases. FEBS Lett 358:57–61

Suzuki Y, Suzuki K (1991) Enzymatic formation of 4 G-α-D-glucopyranosyl rutin. Agric Biol Chem 55:181–187

Suzuki Y, Uchida K (1994) Enzymatic formation of a new derivative of thiamin, β-galactosylthiamin. Biosci Biotech Biochem 58:1273–1276

Svensson B, Larsen K, Svendsen I, Boel E (1983) The complete amino acid sequence of the glycoprotein glucoamylase G1 from *Aspergillus niger*. Carlsberg Res Commun 48:529–544

Svensson B, Clarke AJ, Svendsen I, Moller H (1990) Identification of carboxylic acid residues in glucoamylase G2 from *Aspergillus niger* that participate in the catalysis and substrate binding. Eur J Biochem 18:29–38

Takahashi T, Kato K, Ikegami Y, Irie M (1985) Different behavior towards raw starch of three forms of glucoamylase from a *Rhizopus* sp. J Biochem 98:663–671

Tanaka Y, Ashikari T, Nakamura N, Kiuchi N, Shibano Y, Amachi T, Yoshizumi H (1986) Comparison of amino acid sequences of three glucoamylases and their structure-function relationships. Agric Biol Chem 50:965–969

Tapavicza SV, Bell D, Kopp-Holtwiesche B (2000) Plant growth enhancement against phytopathogenic fungi and/or soil borne pests. Patent WO 0002451 (CA 132 60488)

Taylor PM, Napier EJ, Fleming ID (1978) Some properties of a glucoamylase produced by the thermophilic fungus *Humicola lanuginose*. Carbohydr Res 16:301–308

Trincone A, Pagnotta E, Giordano A, Perugino G, Rossi M, Moracci M (2003) Enzymatic synthesis of 2-deoxyglycosides using the β-glycosidase of the archaeon *Sulfolobus solfataricus*. Biocatal Biotransform 21:17–24

Uchida K, Suzuki Y (1998) Enzymatic synthesis of a new derivative of thiamin, *O*-α-glucosylthiamin. Biosci Biotech Biochem 62(2):221–224

Verdoucq L, Czjzek M, Moriniere J, Beven DR, Esen A (2003) Mutational and structural analysis of aglycone specificity in maize and sorghum β-glucosidase. J Biol Chem 278:25055–25062

Vic G, Crout DHG (1995) Synthesis of allyl and benzyl β-D-glucopyranosides and allyl β-D-galactopyranoside from D-glucose or D-galactose and the corresponding alcohol using almond β-D-glucosidase. Carbohydr Res 279:315–319

Vic G, Thomas D (1992) Enzyme-catalyzed synthesis of alkyl-β-D-glucosides in organic media. Tetrahedron Lett 33:4567–4570

Vic G, Biton J, Beller DL, Michel JM, Thomas D (1995) Enzymatic glycosylation of hydrolytic alcohols in organic medium by the reverse hydrolysis reaction using almond β-D-glucosidase. Biotechnol Bioeng 46:109–116

Vic G, Thomas D, Crout DHG (1997) Solvent effect on enzyme-catalyzed synthesis of β-D-glucosides using the reverse hydrolysis method application to the preparative-scale synthesis of 2-hydroxybenzyl and octyl β-D-glucopyronosides. Enzyme Microb Technol 20:597–603

Vijayakumar GR (2007) Enzymatic synthesis of selected glycosides. PhD thesis, University of Mysore

Voorhorst WGB, Eggen RIK, Luesink EJ, De Vos WM (1995) Characterization of the Cel B gene coding for β-glucosidase from the hyperthermophilic archean *Pyroccus furiosus* and its expression and site directed mutation in *Escherichia coli*. J Bacteriol 177:7105–7111

Vulfson EN, Patel R, Beecher JE, Andrews AT, Law BA (1990) Glycosidases in organic solvents I. Alkyl-β-glucoside synthesis in a water-organic two-phase system. Enzyme Microb Technol 12:950–954

Williamson G, Belshaw NJ, Williamson MP (1992) O-Glycosylation in *Aspergillus* glucoamylase. Confirmation and role in binding. Biochem J 282:423–428

Withers SG, Street IP (1988) Identification of a covalent α-D-glucopyranosyl enzyme intermediate formed on a β-glucosidase. J Am Chem Soc 110:8551–8553

Withers GG, Warren RAJ, Street IP, Rupitz K, Kempton JB, Abersold R (1990) Unequivocal demonstration of the involvement in the mechanism of a retaining glycosidase. J Am Chem Soc 112:5887–5889

Yoon SH, Fulton DB, Robyt JF (2004) Enzymatic synthesis of two salicin analogues by reaction of salicyl alcohol with *Bacillus macerans* cyclomaltodextrin glucanyltransferase and *Leuconostoc mesenteroides* B-742CB dextransucrase. Carbohydr Res 339:1517–1529

Yu RC, Hang YD (1991) Purification and characterization of a glucoamylase from *Rhizopus oryzae*. Food Chem 40:301–308

Zechel DL, Withers SG (2001) Dissection of nucleophilic and acid–base catalysis in glycosidases. Curr Opin Chem Biol 5:643–649

Zhou JH (2000) Herbal sweetening and preservative composition comprising licorice extract and mogrosides obtained from plants belonging to cucurbitaceae and/or momordica. Patent US 6103240 (CA 133 168393)

3 Lipases

Abstract

Among the hydrolytic enzymes, transformations mediated by lipases have been extensively studied. This chapter begins with a brief description on the three lipases – porcine pancreas lipase, *Rhizomucor* lipase and *Candida rugosa* lipase – employed extensively in the transformation work described in this book. This is followed by specificity of lipases and the reactions – both hydrolytic and esterification – catalysed by them with respect to their mechanism. After a brief mention about the advantages of using lipases in organic solvents over chemical reactions, this chapter concludes with a table furnished with literature data on mainly diverse esterification reactions carried out in the a past three decades using lipases in nonaqueous media.

3.1 Lipases

Lipases (EC 3.1.1.3) catalyse hydrolysis of triglycerides at the oil/water interface (Lehninger 1977). Under reverse hydrolytic conditions, lipases exhibit their ability to catalyse various other types of reactions like esterification, transesterification, polymerisation and lactonisation. High selectivity and mild conditions associated with lipase-mediated transformations have made them very attractive for the synthesis of a wide range of natural products, pharmaceuticals, fine chemicals, food ingredients and bio-lubricants (Schreier 1997; Dorm et al. 2004; Gill and Valivety 1997; Belarbi et al. 2000; Sharma et al. 2001). The main reason for the use of lipases is the growing interest and demand for the production of products through environmentally compatible natural means. Lipases are regarded as enzymes with high commercial potential due to their versatility in application. Lipase-catalysed esterification in organic solvents offers synthetic challenges, which if dealt with successfully can result in the generation of several useful compounds.

Lipases are hydrolases which catalyse the hydrolysis of triacylglycerols to glycerol and free fatty acids. In eukaryotes, lipases are involved in various stages of lipid metabolism including fat digestion, absorption, reconstitution and lipoprotein metabolism. In plants, lipases are found in energy reserve tissues. Lipases contain a hydrophobic oligopeptide lid, covering the entrance of its active site which requires interfacial activation

S. Divakar, *Enzymatic Transformation*,
DOI 10.1007/978-81-322-0873-0_3, © Springer India 2013

Table 3.1 Biotechnological applications of lipases (Vulfson 1994)

Industry	Action	Product or application
Detergents	Hydrolysis of fats	Removal of oil stains from fabrics
Dairy	Hydrolysis of milk fat, cheese ripening, modification of butter fat	Development of flavouring agents in milk, cheese and butter
Bakery	Flavour improvement and shelf life elongation	Bakery products
Beverages	Improved aroma	Beverages
Food	Quality improvement, transesterification	Mayonnaise, dressings and whippings, health foods
Meat and fish	Flavour development and removal of fats	Meat and fish products
Fats and oils	Transesterification, hydrolysis	Cocoa butter, margarine, fatty acids, glycerol, mono- and diglycerides
Chemicals	Enantioselectivity, synthesis	Chiral building blocks and chemicals
Cosmetics	Synthesis	Emulsifiers, moisturisers
Leather	Hydrolysis	Leather products
Paper	Hydrolysis	Paper with improved quality by removing wax
Cleaning	Synthesis and hydrolysis	Removal of cleaning agents like surfactants
Food dressing	Quality improvement	Mayonnaise, dressing and whipping
Pharmaceuticals	Transesterification, hydrolysis	Speciality lipids, digestive aids
Health food	Transesterification	Health foods

at lipid–water interface (Martinelle et al. 1995). This lid opens up in presence of hydrophobic interfaces, due to conformational change of lipases acquiring an open structure in which the active site residues become accessible to substrates. However, in the absence of interfaces, the lid covers the active site, making it inaccessible to substrates (Brzozowski et al. 1991). The catalysis by lipase encompasses a linear substrate concentration gradient at the interface, amenable orientation of a scissile ester bond, reduction in the micro-aqueous phase around the substrate ester molecules and the conformational change of the enzyme (Derewenda and Sharp 1993). Since lipases tolerate organic solvents in the reaction mixture, lipases find promising position in organic chemical processing (Kiran and Divakar 2001; Kiran et al. 2001; Therisod and Klibanov 1986; Berglund and Hutt 2000; Harikrishna and Karanth 2001), detergent formulations (Jaeger and Reetz 1998), synthesis of biosurfactants (Plou et al. 1999; Sarney and Vulfson 1995; Sarney et al. 1996), oleochemical industry (Bornscheuer 2000; Undurraga et al. 2001), dairy industry (Vulfson 1994), paper manufacture (Jaeger and Reetz 1998), nutrition (Pabai et al. 1995a, b; Undurraga et al. 2001) and cosmetics and pharmaceutical processing (Berglund and Hutt 2000; Liese et al. 2000). Other biotechnological applications of lipases are shown in Table 3.1.

3.1.1 Porcine Pancreas Lipase (PPL)

Pancreatic lipases exhibit high molecular activity with one molecule of lipase capable of cleaving nearly 7,000 ester bonds per second (Scharpe et al. 1997) under optimal conditions. Porcine pancreas has been the source of lipase for most of the work on pancreatic lipase. It is also the richest source for pancreatic lipases, and it is the first purified lipase (Peschke 1991).

Pancreatic lipase, a serine hydrolase, contains a single-chain glycoprotein of about 48-KDa molecular weight (Winkler and Gubernator 1994). From pH titration and photooxidation studies, among the catalytic triad, Ser-152, His-263 and Asp-176, found to be responsible for the catalytic activity, histidine

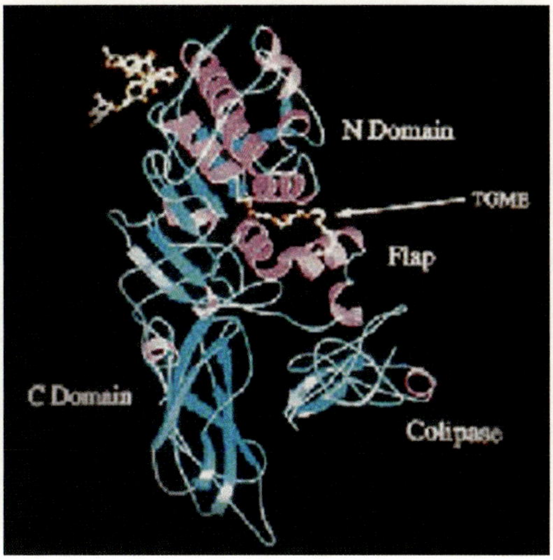

Fig. 3.1 Schematic ribbon diagram of the porcine lipase–colipase structure. The glycan chain, connected to the lipase N-terminal domain (N domain), is drawn as a stick model. One tetraethylene glycol monooctyl ether inhibitor molecule, located in the open active site, is represented by *balls* and *sticks*. Colipase interacts with the lipase C-terminal domain (C domain) and with the flap (Adopted from Hermoso et al. 1996)

residue was the most important (Winkler et al. 1990). Donner (1976) purified and measured some of the physical parameters: molecular weight 52 KDa, sedimentation coefficient (s degrees 20, w) 4.0 S, diffusion coefficient (D degrees 20, w) 6.7×10^{-7} cm^2 s^{-1}, Stokes radius (*r*) 30.3 Å, partial specific volume (*v*) 0.72 cm^3 g^{-1}, frictional ratio (f/f_0) 1.23 and isoelectric point (pI) 5.18. It consists of a central parallel β-sheet having a helical link (van Tilbeurgh et al. 1992; Hermoso et al. 1996) bound by prolipase at its edge with the plane of the prolipase roughly perpendicular to the C-terminal β-sheet domain of the lipase molecule (Fig. 3.1). Prolipase is a flattened molecule with dimensions of about 33 Å × 24 Å × 16 Å consisting of mainly three finger-shaped regions constituted by residues 26–39, 47–64 and 67–87 held together by disulphide bonds. Majority of hydrophobic amino acids are found in the region opposite of the lipase binding site. The catalytic active site also contains a surface helix amphipathic lid consisting of residues 248–258 covering the active site (van Tilbeurgh et al. 1992).

3.1.2 *Rhizomucor miehei* Lipase (RML)

Crystallographic study of *Rhizomucor miehei* lipase (RML) has shown a single polypeptide chain with 269 amino acid residues, and the molecular weight is 29.4 KDa (Brady et al. 1990; Brzozowski et al. 1991). Brady et al. (1990) identified that RML is an α-/β-type protein with three disulphide bonds responsible for stabilising the two terminal strands. The catalytic triad contains Ser-144, His-257 and Asp-203, buried under a single 17-residue lid (82–96 residues) that occludes the active site in the native structure (Derewenda et al. 1992). Brzozowski et al. (1991), studying an atomic model of inhibitor–RML complex, showed a direct covalent bond formation between nucleophilic O_γ of Ser-144 and substrate. Carbonyl oxygen of the substrate may be stabilised by the interaction of amide nitrogen, and the hydroxyl of Ser-82 (Leu-145 may also be involved in amide interaction) through hydrogen bonding and thus Ser-82 exhibiting a favourable conformation for the oxyanion interaction (Fig. 3.2a, b).

The pentapeptide sequence – Gly^{142}-His^{143}-Ser^{144}-Leu^{145}-Gly^{146} – corresponds to a tight turn between the fourth strand of the central β-sheet and a buried α-helix (Fig. 3.3). The catalytically active Ser-144 found in the middle of the turn is in the rare ε-conformation ($\varphi = 62$; $\phi = 121°$). This structural motif consisting of a β-strand followed by a tight turn containing the active Ser and an α-helix was called as β-εSer-α-motif (Fig. 3.3; Derewenda and Sharp 1993).

The inhibition study of lipase by serine protease inhibitors like di-isopropyl phosphorofluoridate indicated the role of a Ser residue in the active site. It has been found that all known amino acid sequence of neutral lipases share a consensus pentapeptide GX_1SX_2G (where X represents any amino acid, G represents glycine and S represents serine) which contain an essential Ser residue (Derewenda and Sharp 1993).

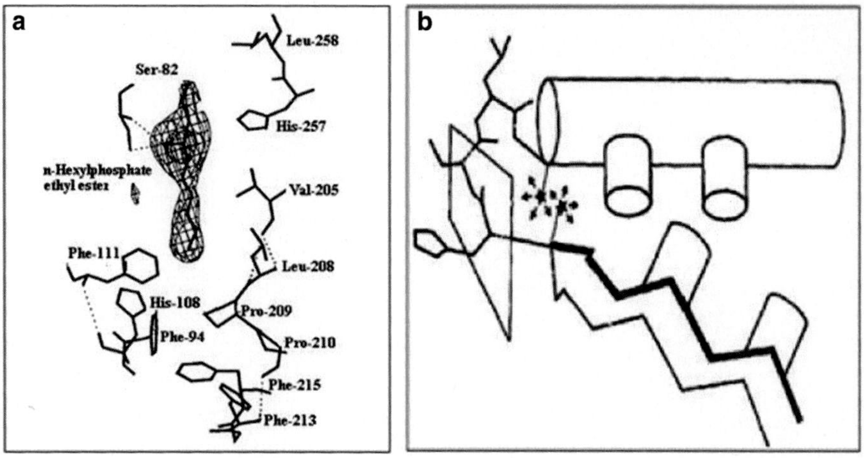

Fig. 3.2 (**a**) The hydrogen-bonding network in the active site of the *Rhizomucor miehei* lipase. The crystal structure (at 3-Å resolution) of a complex of RML lipase with n-hexylphosphonate ethyl ester in which the enzyme's active site is exposed by the movement of the helical lid. The catalytic Ser-144 is immediately beneath the phosphorus atom of the inhibitor. His-257 is clearly displaced towards the ethyl oxygen consistent with the proposed orientation of the substrate and the mechanism of hydrolysis. Hydrogen bond contacts between O_γ atom of Ser-82, and its amide NH are indicated (Adopted from Brzozowski et al. 1990, 1991) (**b**) A schematic drawing showing the packing within the β-εSer-α motif. The helix and strand pack against each other with four amino acids (tinted) forming the interface. The residues nearer the turn are in closer contact, and therefore, their side chains are restricted to those of smaller hydrophobic amino acids. The plane of the central peptide of the turn is perpendicular to the axis of the motif, which forces the catalytic Ser to adopt a strained ε conformation. The two stars show the positions that β-carbons of amino acids other than Gly would occupy if the two invariant Gly residues of the GX_1SX_2G pentapeptide were mutated (Adopted from Derewenda and Sharp 1993)

3.1.3 *Candida rugosa* Lipase (CRL)

Candida rugosa lipase is the first example of a native interface-activatable lipase in 'open' form (Fig 3.3). CRL is a single-domain protein belonging to the family of α-/β-hydrolase proteins consisting of a central hydrophobic eight-stranded β-sheet packed between two layers of amphiphilic α-helices (Fig. 3.3). CRL is made up of a single polypeptide chain with 534 amino acid residues with a molecular weight of 57 kDa. CRL appears in five isoforms, which have been cloned and sequenced (Kawaguchi et al. 1989; Longhi et al. 1992).

In CRL, the active site triad consists of Ser-209, Glu-341 and His-449 proximate to three surface loops (62–92, 122–129 and 294–305) very important for catalytic activity (Fig. 3.3). A characteristic super secondary structure motif, found in all lipases, houses an embedded Ser-209. In the open conformation of CRL, the lid extends nearly perpendicular to the protein surface, forming a large depression that surround the active site. Uncharged polar residues constitute the hydrophilic area. The hydrophobic face of the flap facing the active site is mainly composed of aliphatic side chain amino acid residues, and the flap facing opposite the active site is hydrophilic in character. The geometry of loops 13 and 4 together with the active site of CRL suggests that the oxyanion hole O_γ is formed by the backbone amide of Gly-123, Gly-124 and Ala-210, involved in hydrogen bond formation with the substrates (Fig. 3.3). Presence of two acyl binding 'pockets' in the active site of CRL depicts the substrate specificity for carbon chain lengths: a small pocket for short-chain acids and a bigger

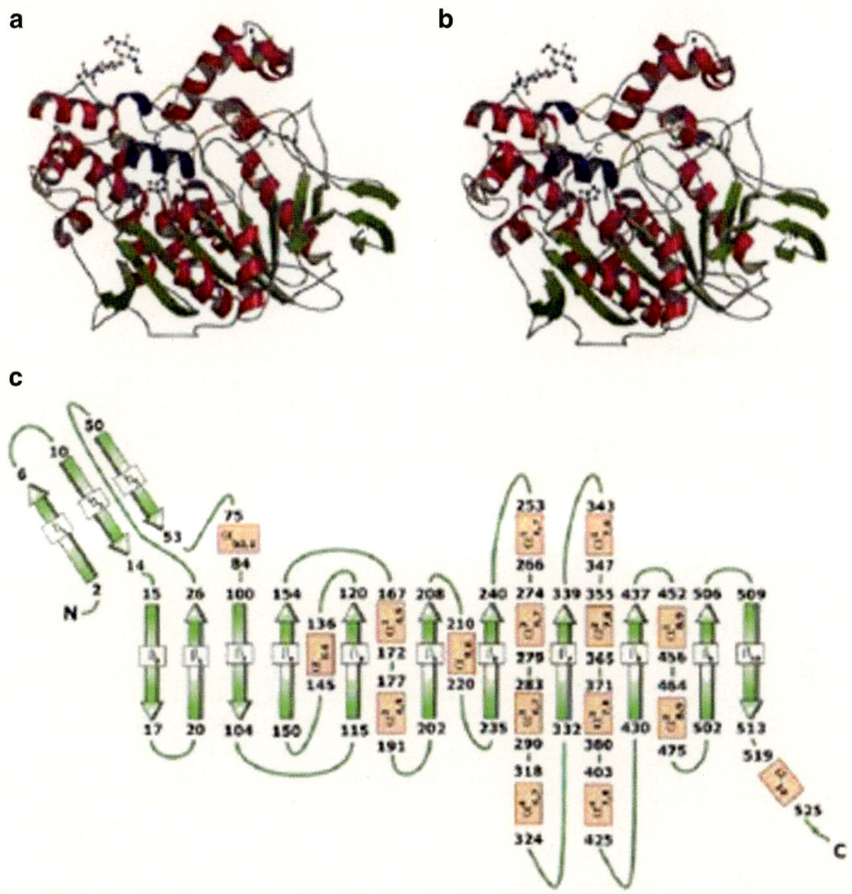

Fig. 3.3 Overall structure of *Candida rugosa* lipase. (**a**) Ribbon representation with a-helices, β-strands and coils coloured in red, green and grey, respectively. The helical and coil segments forming the flap region are shown in dark blue and orange, respectively. The catalytic triad residues (Ser-209, Glu-341 and His-449), the disulphide bridges and the Asn-attached N-acetylglucosamine moieties are shown in ball-and-stick representation. (**b**) A representation of the lipase 2 topology with the secondary structure elements identified (β, strands; α, helix) (Adopted from Mancheno et al. 2003)

pocket for binding longer-chain acids (Parida and Dordick 1993).

3.2 Lipase Specificity

Enzymes possesses extraordinary ability to exhibit regioselectivity and stereospecificity in reactions catalysed by them. Based on positional specificity, lipases can be divided into five different classes (Camp et al. 1998).

Lipases of first group do not exhibit regioselectivity in that they catalyse hydrolysis of fatty acyl triglycerides independent of their type or position. Examples of this class are lipases from *Candida cylindracea*, *Corynebacterium acnes* and *Staphylococcus aureus* (Camp et al. 1998).

Lipases from *Aspergillus niger*, *Rhizopus delemar*, *Rhizomucor miehei*, *Candida rugosa* and *porcine pancreas* catalyse reaction at *sn*-1 and *sn*-3 positions of triacylglycerides

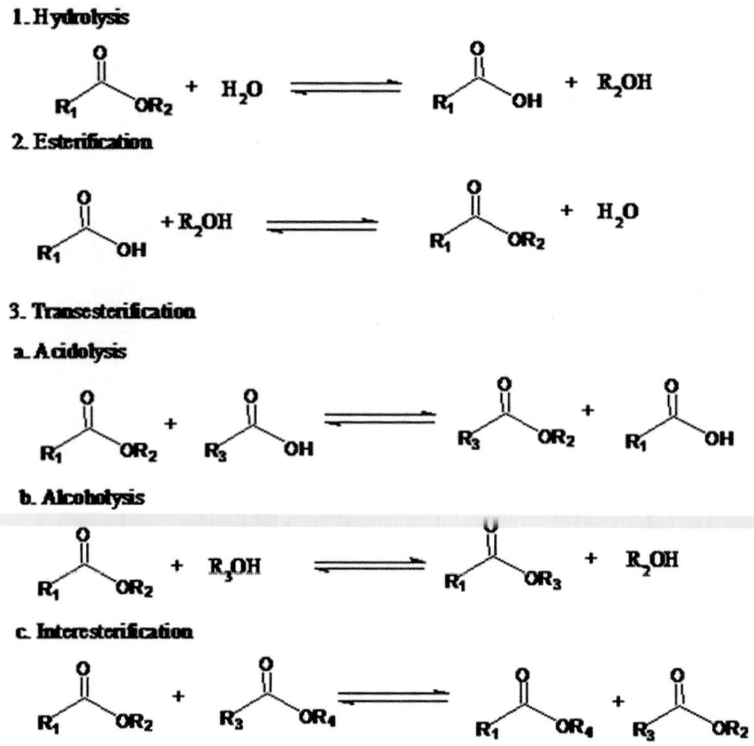

Scheme 3.1 Types of reactions catalysed by lipases

(Macrea 1985), hence known as 1,3-specific lipases.

The third group of lipases covers lipases with different rates of hydrolysis of monoacyl, diacyl and triacylglycerides. Some of these lipases are located in the tissues of rats and humans.

The fourth group of lipases catalyses the exchange of specific type of fatty acids exemplified by the extracellular lipases from the fungus *Geotrichum candidum* which preferentially releases unsaturated *cis-n*-9 fatty acid groups (Macrea 1985).

The fifth group of lipases shows stereospecificity, namely, a faster rate of hydrolysis of fatty acids placed at the *sn*-1 position than the *sn*-3 position or vice versa. Examples of this group are lipoprotein lipases from milk, adipose tissues and postheparin plasma which preferentially cleave the ester bond in *sn*-1 and human and rat lingual lipases which react preferentially with the fatty acids at *sn*-3 position (Jensen et al. 1983).

3.3 Reactions Catalysed by Lipases

The ranges of substrates with which lipases react and also the range of reactions they catalyse are probably far more than any other enzymes studied till date.

Lipases catalyse three types of reactions (Scheme 3.1):

3.3.1 Hydrolysis

Ester hydrolysis is the dominant reaction in aqueous media, when there is large excess of water.

3.3.2 Esterification

Esterification is achieved under low water conditions such as in nearly anhydrous solvents, and if the water content of the medium is controlled, relatively better product yields can be obtained.

3.3.3 Transesterification

The acid moiety of an ester is exchanged with another one. If the acyl donor is a free acid, the reaction is called *acidolysis*, and if the acyl donor is an ester, the reaction is called *interesterification*. In *alcoholysis*, the nucleophile alcohol acts as an acyl acceptor.

3.4 Mechanism of Lipase-Catalysed Esterification in Organic Solvents

Lipases show lipid splitting nature and the mechanism is same as that of serine proteases (Pleiss et al. 1998). Catalytic triad in lipases contains Ser, His and Asp/Glu residues. The serine residue in active centre is activated by histidine and aspartic acid/glutamic acid residues. The substrate acid forms a tetrahedral acyl–enzyme intermediate by reaction with the OH group of the catalytic serine residue. The resulting excess of negative charge that develops on the carbonyl oxygen atom is stabilised by the oxyanion hole (Brzozowski et al. 1991). The tetrahedral intermediate I forms a serinate ester with elimination of water molecule. Subsequent nucleophilic attack of alcohol to the acyl–enzyme intermediate leads to tetrahedral intermediate II. Finally, the product ester is released, and enzyme is free for the next molecule to attack. Grochulski et al. (1994), Cygler et al. (1994) and Schrag and Cygler (1997) proposed this mechanism for the ester formation in case of RML (Scheme 3.2).

Step I. Acylation Step

Initially, serine hydroxyl group forms a tetrahedral intermediate complex I with acyl donor; the negative charge that is formed in the tetrahedral intermediate is stabilised by hydrogen bonding with the acid given which are responsible for the oxyanion hole formation.

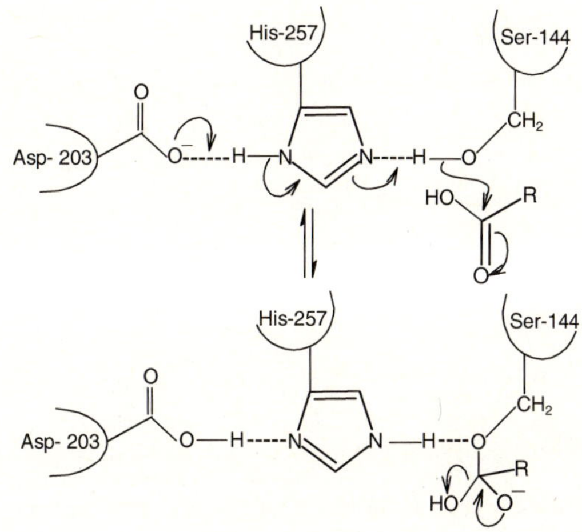

Tetrahedral complex I

Step II. Formation of Acyl–Enzyme Complex

After the formation of the tetrahedral intermediate I, an acyl–enzyme complex is formed through covalent bond with Ser residue by losing one molecule of water.

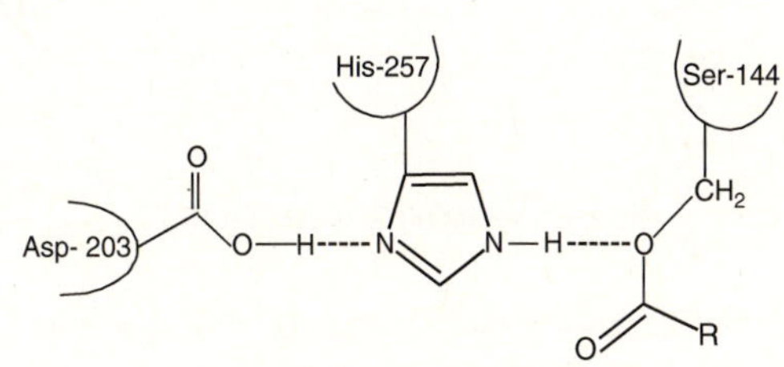

Acyl enzyme complex

Step III. Nucleophilic Attack by Alcohol (Carbohydrate)

Nucleophile alcohol attacks the carbonyl centre of the tetrahedral intermediate forming a tetrahedral complex II, forming an enzyme–acid–alcohol complex.

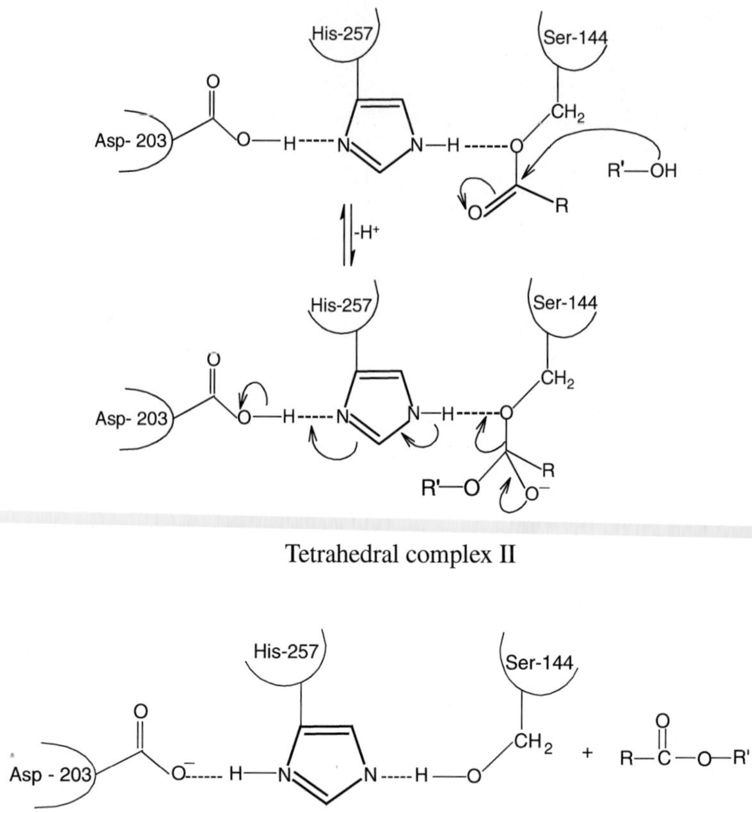

Scheme 3.2 Lipase mediated esterification (Grochulski et al. 1993; Cygler et al. 1994; Schrag and Cygler 1997)

Step IV. Release of Ester (L-Amino Acyl Ester of Carbohydrate)

Finally, the ester is released, and the enzyme will be ready for the next molecule to attack.

3.5 Esterification Reactions

Table 3.2 lists some of the commercially important flavour, fragrance, surfactant and sweetener esters prepared through lipase catalysis.

3.6 Advantages of Lipase Catalysis over Chemical Catalysis

There are many advantages of using lipases as biocatalysts (Lohith 2007; Somashekar 2009):

1. Specificity of the reaction.
2. Milder reaction conditions under which the lipolytic process can be operated.
3. Non-generation of by-products associated with the use of several chemical procedures.
4. Improved product yield and better product quality.
5. Exploitation of the stereo- and regiospecificities shown by lipases to produce high-value chiral synthons.
6. Success in immobilisation techniques that have enabled the reuse of lipases leading to economically viable processes.
7. Good conversion yields.
8. Lipases are highly thermostable, exhibiting activity at 100°C.
9. Use of nonpolar solvents, which impart stability to lipase rather than in water, renders insolubility

Table 3.2 Commercially important esters synthesised by lipase mediated catalysis

Compound	Use	Lipase	References
Flavour esters			
Isoamyl acetate	Banana flavour	*Pseudomonas fluorescence*	Takahashi et al. (1988)
		Candida antarctica	Langrand et al. (1990)
		Rhizomucor miehei	Rizzi et al. (1992), Chulalaksananukul et al. (1993)
		Rhizomucor miehei	Razafindralambo et al. (1994)
		Candida cylindracea, PPL	Divakar et al. (1999), Harikrishna et al. (2001a, b)
		Aspergillus niger	Welsh and Williams (1990), Welsh et al. (1990)
		Novozyme 435	
		Rhizomucor miehei, Candida antarctica	Gubicza et al. (2000)
		Pseudomonas pseudomallei	Guvenc et al. (2002)
		Porcine liver lipase	Romero et al. (2005)
			Kanwar and Goswami (2002), Ngrek (1974), Kumar et al. (2005)
Isoamyl butyrate	Banana flavour	*Candida antarctica*	Langrand et al. (1988, 1990)
		Rhizomucor miehei	Mestri and Pai (1994a, b)
		Candida cylindracea, PPL, *Aspergillus niger*	Welsh and Williams (1990), Welsh et al. (1990)
		Candida antarctica,	Gubicza et al. (2000)
		Geotrichum sp. and *Rhizopus* sp.	Macedo et al. (2004)
Isoamyl propionate	Banana flavour	*Candida antarctica*	Langrand et al. (1988)
Isoamyl isovalerate	Apple flavour	*Rhizomucor miehei*	Chowdary et al. (2002)
Isobutyl isobutyrate	Pineapple flavour	*Rhizomucor miehei*	Hamsaveni et al. (2001)
Methyl propionate	Fruity flavour	*Rhizomucor miehei*	Perraud and Laboret (1989)
Ethyl butyrate	Pineapple flavour	*Candida cylindracea*	Gubicza et al. (2000), Gillies et al. (1987)
Butyl isobutyrate	Sweet fruity odour	*Candida cylindracea*, PPL and *Aspergillus niger*	Yadav and Lathi (2003), Welsh and Williams (1990), Welsh et al. (1990)
Protocatechuic aldehyde		*Rhizomucor miehei*, PPL	Divakar (2003)
Short-chain alcohol esters of C_2–C_{18} acids	Fruity odour	*Staphylococcus warneri*	Talon et al. (1996)
Short-chain fatty acid esters	Fruity odour	*Staphylococcus xylosus*	Mestri and Pai (1994a), Macedo et al. (2003), Xu et al. (2002)
Long-chain alcoholic esters of lactic acids	Flavour	*Candida antarctica*	From et al. (1997), Torres and Otero (1999), Parida and Dordick (1991)
		Novozyme 435 *Rhizomucor miehei*	Bousquet et al. (1999)
Methyl benzoate	Exotic fruit and berries flavour	*Candida rugosa*	Leszczak and Tran-Minh (1998)
Tetrahydrofurfuryl butyrate	Fruity favour	Novozyme 435	Yadav and Devi (2004)

(continued)

Table 3.2 (continued)

Compound	Use	Lipase	References
Cis-3-hexen-1-yl acetate	Fruity odour	*Rhizomucor miehei*	Chang et al. (2003)
Fragrance esters			
Tolyl esters	Honey note	*Rhizomucor miehei*, PPL	Burdock (1994), Suresh-Babu et al. (2002), Manohar and Divakar (2002)
Anthranilic acid esters of C_2–C_{18} alcohols	Flowery odour of jasmine	*Candida cylindracea*, PPL	Kittleson and Pantaleone (1994), Suresh-Babu and Divakar (2001), Manohar and Divakar (2004a)
4-t-Butylcyclohexyl acetate	Woody and intense flowery notes	PPL	Manohar and Divakar (2004b)
Geranyl methacrylate	Floral fruity odour	Rhizomucor miehei, PPL, *Pseudomonas cepacia*	Athawale et al. (2002)
Citronellyl acetate Citronellyl propionate Citronellyl valerate	Fruity rose odour	*Candida antarctica* SP435	Claon and Akoh (1994b)
		Pseudomonas fragi	Mishio et al. (1987)
			Marlot et al. (1985)
Geranyl butyrate Geranyl propionate Geranyl valerate Farnesol butyrate Farnesol propionate Farnesol valerate Phytol butyrate Phytol propionate Phytol valerate	Fruity odour	*Candida rugosa*	Akoh et al. (1992), Shieh et al. (1995)
Citronellyl laurate	Fruity, characteristic lavender and bergamot-like fragrance	Novozyme SP	Yadav and Lathi (2004)
α-Terpinyl acetate α-Terpinyl propionate		*Rhizomucor miehei*	Rao and Divakar (2002)
α-Terpinyl esters of fatty acids α-Terpinyl esters of short-chain acids		*Rhizomucor miehei*	Rao and Divakar (2001)
Terpinyl esters of triglycerols		*Aspergillus niger, Rhizopus delemar, Geotrichum candidum,*	Iwai et al. (1980)
		Penicillium cyclopium	Claon and Akoh (1994a)
Surfactant esters			
Oleic acid esters of short-chain alcohols	Surfactants	Novozyme 435	Dorm et al. (2004)
Butyl oleate	Surfactants	*Rhizomucor miehei*	Knez et al. (1990)
2-O-Alkanoyl lactic acid esters of C_2–C_{18} alcohols	Surfactants	*Rhizomucor miehei*, PPL	Kiran and Divakar (2001), Kiran et al. (1998)
Surfactants and sweetener esters			
N-Acetyl-L-leucyl–D-glucose N-Acetyl-L-methionyl–D-glucose N-Acetyl-L-tyrosinyl–D-glucose N-Acetyl-L-tryptophanyl–D-glucose	Surfactants	*Mucor javanicus, Pseudomonas cepacia,* subtilisin	Maruyama et al. (2002)

(continued)

Table 3.2 (continued)

Compound	Use	Lipase	References
N-Acetyl-L-phenylalanyl–D-glucose	Surfactant	Subtilisin	Maruyama et al. (2002), Riva et al. (1988)
N-Acetyl-L-phenylalanyl–D-galactose			
N-Acetyl-L-phenylalanyl–fructose			
N-Acetyl-L-phenylalanyl–mannose			
N-Acetyl-L-phenylalanyl–lactose			
N-t-Boc-L-phenylalanyl–glucose			
N-Acetyl-L-methionyl–methyl-β-galactopyranoside	Surfactants	Optimase M-440, Proleather, APG 380	Park et al. (1996)
N-t-Boc-L-phenylalanyl–galactose			
N-t-Boc-L-phenylalanyl–fructose			
N-t-Boc-L-phenylalanyl–methyl α-D-glucopyranoside			
N-t-Boc-L-phenylalanyl–sorbitol	Surfactants	Optimase M-440	Park et al. (1999)
N-t-Boc-L-phenylalanyl–sucrose			
N-t-Boc-L-phenylalanyl–cellobiose			
N-t-Boc-L-phenylalanyl–raffinose			
N-t-Boc-L-phenylalanyl–trehalose			
N-t-Boc-L-phenylalanyl–maltose			
N-t-Boc-L-phenylalanyl–lactose			
N-t-Boc-L-leucyl–sucrose	Surfactants	Optimase M-440	Jeon et al. (2001)
N-t-Boc-L-tyrosinyl–sucrose			
N-t-Boc-L-methionyl–sucrose			
N-t-Boc-L-aspartyl–sucrose			
Di-N-t-Boc-L-lysyl–sucrose			
N-t-Boc-L-phenylalanyl–xylitol			
N-t-Boc-L-phenylalanyl–arabitol			
N-t-Boc-L-phenylalanyl–mannitol			
N-t-Boc-L-phenylalanyl–N-acetyl–D-glucosamine			
L-Prolyl, L-phenylalanyl, L-tryptophanyl and L-histidyl esters of carbohydrates	Surfactants, sweeteners	*Rhizomucor miehei*, PPL, *Candida rugosa*	Vijayakumar et al. (2004), Somashekar and Divakar (2007), Lohith et al. (2003), Lohith and Divakar (2005)
1-*O*-ester, 2-*O*-ester, 3-*O*-ester, 4-*O*-ester, 5-*O*-ester, 6-*O*-ester, 6′-*O*-ester			
2,5-di-*O*- ester, 3,5-di-*O*-ester			
2, 6-di-*O*-ester, 3, 6-di-*O*-ester			
6,6′-di-*O*-ester			
N-Acetyl-L-alanyl-methyl-β-D-galactopyranoside	Surfactants, sweeteners	Subtilisin	Riva et al. (1988)
2-*O*-ester, 3-*O*-ester, 4-*O*-ester, 6-*O*-ester		*Rhodotorula lactosa*	Suzuki et al. (1991)
6-*O*-Butyl glucose	Surfactants	PPL	Therisod and Klibanov (1986)
		Subtilisin	Kirk et al. (1992)
			Zaks and Dodds (1997)

(continued)

Table 3.2 (continued)

Compound	Use	Lipase	References
6-*O*-Acetyl glucose			
6-*O*-Capryloyl glucose			
6-*O*-Acetyl galactose			Klibanov (1986)
6-*O*-Acetyl maltose			Dordick (1989)
6-*O*-Acetyl fructose			Schlotterbeck et al. (1993)
1-*O*-Acetyl fructose			Boyer et al. (2001)
Fructose oleate	Surfactant	Lipozyme, *Rhizomucor miehei*	Khaled et al. (1991)
Fatty acid esters of glycosides		*Candida antarctica*	Adlerhorst et al. (1990)
Butyl oleate		*Candida rugosa*	Zaidi et al. (2002)
Oleyl butyrate			
Oleyl oleate			
6-*O*-Lauroyl sucrose	Surfactants	*Humicola lanuginosa*	Ferrer et al. (1999)
6-*O*-Lauroyl glucose		*Candida antarctica* B	Ferrer et al. (2005)
6-*O*-Lauroyl maltose			
6-*O*-Palmitoyl maltose			
1,6-di-*O*-Lauroyl sucrose			
6,6-di-*O*-Lauroyl sucrose			
6′-*O*-Palmitoyl maltose			
β-Methylglucoside methacrylate/acrylate	Surfactants	*Candida antarctica*	Kim et al. (2004)

of the enzyme, solubility of substrates and products in organic solvents resulting in homogenous reaction conditions, easy product workout procedures and easy removal of water formed as a by-product.

References

Adlerhorst K, Björking F, Godtfredsen SE, Kirk O (1990) Enzyme catalyzed preparation of 6-*O*-acylglucopyranosides. Synthesis 1:112–115

Akoh CC, Copper C, Nwosu CV (1992) Lipase-G catalyzed synthesis of monoglycerides in organic solvent and analysis by HPLC. J Am Oil Chem Soc 69:257–260

Athawale V, Manjrekar N, Athawale M (2002) Lipase-catalyzed synthesis of geranyl methacrylate by transesterification study of reaction parameters. Tetrahedron Lett 43:4797–4800

Belarbi EH, Molina E, Chisti Y (2000) A process for high yield and scaleable recovery of high purity eicosapentaenoic acid esters from micro algae and fish oil. Enzyme Microb Technol 26:516–529

Berglund P, Hutt K (2000) Biocatalytic synthesis of enantiopure compounds using lipases. In: Patel RN (ed) Stereoselective biocatalysis. Marcel Dekker, New York

Bornscheuer UT (2000) Enzymes in lipid modification. Wiley-VCH, Weinheim/Berlin

Bousquet M, Willemot R, Monsan P, Boures E (1999) Enzymatic synthesis of unsaturated fatty acid glucoside esters for dermo-cosmetic applications. Biotechnol Bioeng 63:730–736

Boyer V, Stanchev M, Fairbanks AJ, Davis BG (2001) Ready protease catalysed synthesis of carbohydrate-amino acid conjugates. Chem Commun 19:1908–1909

Brady L, Brzozowski AM, Derewenda U, Derewenda ZS, Dodson GG, Tolley S, Turkenburg JP, Christiansen L, Huge-Jensen B, Nashkov L, Thim L, Menge U (1990) A serine protease triad forms the catalytic center of triglycerol lipase. Nature 343:767–770

Brzozowski AM, Derewenda U, Derewenda ZS (1990) A model for interfacial activation in lipases from the structure of a fungal lipase-inhibitor complex. Nature 351:491–494

Brzozowski AM, Derewenda U, Derewenda ZS, Dodson GG, Lawson DM, Turkenburg JP, Bjorkling F, Huge-Jensen B, Patkar SA, Thim L (1991) A model for interfacial activation in lipases from the structure of a fungal lipase-inhibitor complex. Nature 351:491–494

Burdock GA (1994) In Fenaroli's Handbook of flavor ingredients, vol II, 3rd edn. CRC Press, Boca Raton

Camp JV, Huyghebaert A, Goeman P (1998) In: Christope AB (ed) Structural modified food fats synthesis biochemistry and use. AOCS Press, Champaign

Chang SW, Shaw JF, Shieh CJ (2003) Optimization of enzymatically prepared hexyl butyrate. Food Technol Biotechnol 41:237–242

Chowdary GV, Divakar S, Prafulla SG (2002) Modeling on isoamyl isovalerate synthesis from *Rhizomucor miehei* lipase in organic media optimization studies. World J Microbiol Biotechnol 18:179–185

Chulalaksananukul W, Condoret JS, Combes D (1993) Geranyl acetate synthesis by lipase catalysed transesterification in supercritical carbon dioxide. Enzyme Microb Technol 15:691–698

Claon PA, Akoh CC (1994a) Lipase catalyzed synthesis of terpene esters by trans esterification in *n*-hexane. Biotechnol Lett 16:235–240

Claon PA, Akoh CC (1994b) Effect of reaction parameters on SP435 lipase-catalyzed synthesis of citronellyl acetate in organic solvent. Enzyme Microb Technol 16:835–838

Cygler M, Grochulski P, Kazlauskas RS, Schrag JD, Bouthillier F, Rubin B, Serreqi AN, Gupta AK (1994) A structural basis for the chiral preferences of lipases. J Am Chem Soc 116:3180–3186

Derewenda ZS, Sharp AM (1993) News from the interface the molecular structure of triacyl glyceride lipases. Trends Biochem Sci 18:20–25

Derewenda U, Brzozwski AM, Lawson DM, Derewenda ZS (1992) Catalysis at the interface The anatomy of a conformational change in a triglyceride lipase. Biochemistry 31:1532–1541

Divakar S, Kiran KR, Harikrishna S, Karanth NG (1999) An improved process for the preparation of esters of organic acids and alcohols. Indian Patent, 1243/ DEL/99 No. 191078

Divakar S (2003) Lipase catalysed regioselective esterification of protocatechuic aldehyde. Indian J Chem Sect B 42B:1119–1122

Donner M (1976) Preparation of porcine pancreatic lipase free of co-lipase activity. Acta Chem Scand B 30:430–434

Dordick JS (1989) Enzymatic catalysis in monophasic organic solvents. Enzyme Microb Technol 11:194–211

Dorm N, Belafi-Bak K, Bartha L, Ehrenstein U, Gubicza L (2004) Manufacture of an environmental-safe biolubricant from fusel oil by enzymatic esterification in solvent-free system. Biochem Eng J 21:229–234

Ferrer M, Cruces MA, Bernable M, Ballesteros A, Plou FJ (1999) Lipase catalysed regioselective acylation of sucrose in two solvent mixtures. Biotechnol Bioeng 65:10–16

Ferrer M, Soliveri J, Plou FJ, Cortes NL, Duarte DR, Christensenc M, Patinob JLC, Ballesterosa A (2005) Synthesis of sugar esters in solvent mixtures by lipases from *Thermomyces lanuginosus* and *Candida antarctica* B, and their antimicrobial properties. Enzyme Microb Tech 36:391–398

From M, Adlercreutz P, Mattiasson B (1997) Lipase catalyzed esterification of lactic acid. Biotechnol Lett 19:315–317

Gill I, Valivety R (1997) Polyunsaturated fatty acids Part 1. Occurrence, biological activities and applications. Trends Biotechnol 15:401–409

Gillies B, Yamazaki H, Armstrong DW (1987) Production of flavor esters by immobilized lipase. Biotechnol Lett 9:709–714

Grochulski P, Li Y, Schrag JD, Bouthillier F, Smith P, Harrison D, Rubin B, Cygler M (1993) Insight into interfacial activation from an open structure of *Candida rugosa* lipase. J Biol Chem 268:12843–12847

Grochulski P, Bouthillier F, Kazlauskas RJ, Serreqi AN, Schrag JD, Ziomek E, Cygler M (1994) Analogs of reaction intermediates identify a unique substrate binding site in *Candida rugosa* lipase. Biochemistry 33:3494–3500

Gubicza L, Kabiri-Badr A, Keoves E, Belafi-Bako K (2000) Large-scale enzymatic production of natural flavour esters in organic solvent with continuous water removal. J Biotechnol 84:193–196

Guvenc A, Kapucu N, Mehmetoglu I (2002) The production of isoamyl acetate using immobilized lipases in a solvent-free system. Process Biochem 38:379–386

Hamsaveni DR, Prafulla SG, Divakar S (2001) Optimization of isobutyl butyrate synthesis using central composite rotatable design. Process Biochem 36:1103–1109

Harikrishna S, Karanth NG (2001) Lipase-catalyzed synthesis of isoamyl butyrate. A kinetic study. Biochim Biophys Acta 1547:262–267

Harikrishna S, Manohar B, Divakar S, Prapulla SG, Karanth NG (2001a) Optimization of isoamyl acetate production using immobilized lipase from *Mucor miehei* by response surface methodology. Enzyme Microb Technol 26:132–138

Harikrishna S, Divakar S, Karanth NG (2001b) Enzymatic synthesis of isoamyl acetate using immobilized lipase from *Rhizomucor miehei*. J Biotechnol 87:193–201

Hermoso J, Pignol D, Kerfelec B, Crenon I, Chapus C, Fontecilla-camps JC (1996) Lipase activation by nonionic detergents. The crystal structure of the porcine lipase co-lipase-tetraethylene glycol monooctyl ether complex. J Biol Chem 270:18007–18016

Iwai M, Okumura S, Tsujisaka Y (1980) Synthesis of terpene alcohol esters by lipase. Agri Biol Chem 44:2731–2732

Jaeger KE, Reetz TM (1998) Microbial lipases from versatile tools for biotechnology. Trends Biotechnol 16:396–403

Jensen RG, de-Jong FA, Clark RM (1983) Determination of lipase specificity. Lipid 18:239–252

Jeon GJ, Park OJ, Hur BK, Yang JW (2001) Enzymatic synthesis of amino acid-sugar alcohol conjugates in organic media. Biotechnol Lett 23:929–934

Kanwar L, Goswami P (2002) Isolation of a *Pseudomonas* lipase produced in pure hydrocarbon substrate and its application in the synthesis of isoamyl acetate using membrane-immobilised lipase. Enzyme Microb Technol 31:727–735

Kawaguchi V, Honda H, Toniguchi-Morimura J, Iwasaki S (1989) The codon CUG is read as serine in an asporogenic yeast *Candida cylindracea*. Nature 341:164–166

Khaled N, Montet D, Pina M, Graille J (1991) Fructose oleate synthesis in a fixed catalyst bed reactor. Biotechnol Lett 13:167–172

Kim J, Haam S, Park DW, Ahn IS, Lee TG, Kim HS, Kim WS (2004) Biocatalytic esterification of methylglucoside for synthesis of biocompatible sugar-containing vinyl esters. Chem Eng J 99:15–22

Kiran KR, Karanth NG, Divakar S (1998) An improved enzymatic process for the preparation of fatty acid hydroxyacid ester. Indian Patent, 1978/DEL/98 187313

Kiran KR, Divakar S (2001) Lipase catalysed esterification of organic acids with lactic acid. J Biotechnol 87:109–121

Kiran KR, Manohar B, Divakar S (2001) A central composite rotatable design analysis of lipase catalysed synthesis of lauroyl lactic acid at bench-scale level. Enzyme Microb Technol 29:122–128

Kirk O, Bjorkling F, Godfredsen SE, Larsen TS (1992) Fatty acid specificity in lipase catalysed synthesis of glucoside esters. Biocatalysis 6:127–134

Kittleson JR, Pantaleone (1994) Enzymic biphasic process for the synthesis of aromatic esters flavoring agents from corresponding carboxylic acid and alcohol by esterification mediated by a lipase from *Candida cyclindracea*. US. Patent, 5,437,991

Klibanov AM (1986) Enzymes that work in organic solvents. Chem Technol 16:354–359

Knez Z, Leitgeb M, Zavrsnik D, Lavrie B (1990) Synthesis of oleic acid esters with immobilized lipase. Fat Sci Technol 4:169–172

Kumar R, Modak J, Madras G (2005) Effect of the chain length of the acid on the enzymatic synthesis of flavors in supercritical carbon dioxide. Biochem Eng J 23:199–202

Langrand G, Triantaphylides C, Baratti J (1988) Lipase catalysed formation of flavor esters. Biotechnol Lett 10:549–554

Langrand G, Rondot N, Triantaphylides C, Baratti J (1990) Short-chain flavor esters synthesis by microbial lipases. Biotechnol Lett 12:581–586

Lehninger LA (1977) Biochemistry, 3rd edn. Worth Publishers Inc., New York, pp 355–357

Leszczak JP, Tran-Minh C (1998) Optimized enzymatic synthesis of methyl benzoate in organic medium. Operating conditions and impact of different factors on kinetics. Biotechnol Bioeng 60:556–561

Liese A, Seelbach K, Wandrey C (2000) Industrial biotransformations. Wiley-VCH, Weinheim

Lohith K, Vijaya-kumar GR, Manohar B, Divakar S (2003) An improved enzymatic process for the preparation of amino acyl esters of mono and disaccharides. Indian Patent, NF-492/03, PCT/03/00466

Lohith K, Divakar S (2005) Lipase catalysed synthesis of L-phenylalanine esters of D-glucose. J Biotechnol 117:49–56

Lohith K (2007) Enzymatic synthesis of selected amino acid esters of sugars. PhD thesis, University of Mysore

Longhi S, Fasetti F, Grandori R, Lotti M, Vanoni M, Alberghina L (1992) Cloning and nucleotide sequences of two lipase genes from *Candida cylindracea*. Biochim Biophys Acta 1131:227–232

Macedo GA, Lozano MMS, Pastore GM (2003) Enzymatic synthesis of short chain citronellyl esters by a new lipase from *Rhizopus* sp. Electron J Biotechnol 6(1):0717–3458

Macedo GA, Pastore GM, Rodrigues MI (2004) Optimising the synthesis of isoamyl butyrate using *Rhizopus* sp. lipase with a central composite rotatable design. Process Biochem 39:687–692

Macrea AR (1985) In: Tramper J, Vander-Plas HC, Linko P (eds) Biocatalyst in organic synthesis. Elsevier, Amsterdam, pp 195–208

Mancheno JM, Pernas MA, Martinez MJ, Ochoa B, Ruo ML, Hermosa JA (2003) Structural insights Into the lipase/esterase behavior in the *Candida rugosa* family crystal structure of the lipase 2 isoenzyme at 1.97 A resolution. J Mol Biol 332:1059–1069

Manohar B, Divakar S (2002) Application of central composite rotatable design to lipase catalyzed syntheses of *m*-cresyl acetate. World J Microbiol Biotechnol 18:745–751

Manohar B, Divakar S (2004a) Application of surface plots and statistical designs to selected lipase catalysed esterification reactions. Process Biochem 39:847–851

Manohar B, Divakar S (2004b) Porcine pancreas lipase acetylation of beta-cyclodextrin anchored 4-t-butylcyclohexanol. Indian J Chem Sect B 43B:2661–2665

Marlot C, Langrand G, Triantaphylides C, Baratti J (1985) Ester synthesis in organic solvent catalyzed by lipase immobilized on hydrophilic supports. Biotechnol Lett 7:647–650

Martinelle M, Holmquist M, Hult K (1995) On the interfacial activation of *Candida antarctica* lipase A and B as compared with *Humicola lanuginosa* lipase. Biochim Biophys Acta 1258:272–276

Maruyama T, Nagasawa SI, Goto M (2002) Enzymatic synthesis of sugar esters in organic solvents. J Biosci Bioeng 94:357–361

Mestri S, Pai JS (1994a) Effect of moisture on lipase catalyzed esterification of geraniol palmarosa oil in nonaqueous system. Biotechnol Lett 17:459–461

Mestri S, Pai JS (1994b) Synthesis of isoamyl butyrate by lipase by lipase of *Mucor miehei*. PAFAI J 2:24–26

Mishio T, Takahashi K, Yoshimoto T, Kodera Y, Saito Y, Inada Y (1987) Terpene alcohol ester synthesis by polyethylene glycol modified lipase in benzene. Biotechnol Lett 9:187–190

Ngrek S (1974) Synthesizing power of liver lipase. Acta Biol Expt 14:157–174

Pabai F, Kermasha S, Morin A (1995a) Interesterification of butter fat by partially purified extracellular lipases from *Pseudomonas putida*, *Aspergillus niger* and *Rhizopus oryzae*. World J Microbiol Biotechnol 11:669–677

Pabai F, Kermasha S, Morin A (1995b) Lipase from *Pseudomonas fragi* CRDA 323 partial purification, characterization and interesterification of butter fat. Appl Microbiol Biotechnol 43:42–51

Parida S, Dordick JS (1991) Substrate structure and solvent hydrophobicity control lipase catalysis and enantioselectivity in organic media. J Am Chem Soc 113:2253–2259

Parida S, Dordick JS (1993) Tailoring lipase specificity by solvent substrate chemistries. J Org Chem 58:3238–3244

Park OJ, Park HG, Yang JW (1996) Enzymatic transesterification of monosaccharides and amino acid esters in organic solvents. Biotechnol Lett 18:473–478

Park OJ, Jeon GJ, Yang JW (1999) Protease catalysed synthesis of disaccharide amino acid esters in organic media. Enzyme Microb Technol 25:455–462

Perraud R, Laboret F (1989) Optimization of methyl propionate production catalyzed by *Mucor miehei* lipase. Appl Microbiol Biotechnol 44:321–326

Peschke G (1991) Active components and galenic aspects of enzyme preparations. In: Lankisch PG (ed) Pancreatic enzymes in health and disease. Springer, Berlin, pp 55–64

Pleiss J, Fisher M, Schimid RD (1998) Anatomy of lipase binding site. Chem Phys Lipids 93:67–80

Plou FJ, Cruces MA, Pastor E, Ferrer M, Bernabe M, Ballesterose A (1999) Acylation of sucrose with vinyl esters using immobilized hydrolysis demonstration that chemical catalysis may interfere with enzymatic catalysis. Biotechnol Lett 21:635–639

Rao P, Divakar S (2001) Lipase catalysed esterification of α-terpineol with various organic acids application of the Plackett-Burman design. Process Biochem 36:1125–1128

Rao P, Divakar S (2002) Response surface methodological approach for the *Rhizomucor miehei* lipase-mediated esterification of α-terpineol with propionic acid and acetic anhydride. World J Microbiol Biotechnol 18:341–345

Razafindralambo H, Blecker C, Lognoy G, Marlier M, Wathlet JP, Severin M (1994) Improvement of enzymatic synthesis yields of flavor acetates the example of isoamyl acetate. Biotechnol Lett 16:247–250

Riva S, Chopineau J, Kieboom APG, Klibanov AM (1988) Protease catalysed regioselective esterification of sugars and related compounds in anhydrous dimethylformamide. J Am Chem Soc 110:584–589

Rizzi M, Stylos P, Riek A, Reuss M (1992) A kinetic study of immobilized lipase catalyzing the synthesis of isoamyl acetate by transesterification in n-hexane. Enzyme Microb Technol 14:709–714

Romero MD, Calvo L, Alba C, Habulin M, Primozic M, Knez Z (2005) Enzymatic synthesis of isoamyl acetate with immobilized Candida antarctica lipase in supercritical carbon dioxide. J Supercrit Fluids 33:77–84

Sarney DB, Vulfson EN (1995) Application of enzymes to the synthesis of surfactants. Trends Biotechnol 13:164–172

Sarney DB, Barnard MJ, MacManus DA, Vulfson EN (1996) Application of lipases to the regioselective synthesis of sucrose fatty acid monoesters. J Am Oil Chem Soc 73:1481–1487

Scharpe S, Uyttenbroeck W, Samyn N (1997) Pancreatic enzyme replacement. In: Lauwers A, Scharpe S (eds) Pharmaceutical enzymes. Marcel Dekker, New York, pp 187–221

Schlotterbeck A, Lang S, Wray V, Wagner F (1993) Lipase catalyzed monoacylation of fructose. Biotechnol Lett 15:61–64

Schrag JD, Cygler M (1997) Lipase and α/β hydrolase fold. Methods Enzymol 284:85–107

Schreier P (1997) In: Berger RG (ed) Biotechnology of aroma compounds. Adv Biochem Eng Biotechnol 17:52

Sharma R, Chisti Y, Banerjee UC (2001) Production, purification, characterization, and applications of lipases. Biotechnol Adv 19:627–662

Shieh CJ, Akoh CC, Koehler PE (1995) Four-factor response surface optimization of the enzymatic modification of triolein to structured lipids. J Am Oil Chem Soc 72:619–623

Somashekar BR (2009) Lipase catalysed esterification of sugars with alkyl side chain containing amino acids. PhD thesis, University of Mysore

Somashekar BR, Divakar S (2007) Synthesis of L-valyl, L-leucyl and L-isoleucyl esters of carbohydrates using lipase as biocatalyst. Indian J Chem Sect B 46B:2026–2044

Suresh-Babu CV, Divakar S (2001) Selection of alcohols through Plackett-Burman design in lipase catalyzed synthesis of anthranilic acid. J Am Oil Chem Soc 78:49–52

Suresh-Babu CV, Karanth NG, Divakar S (2002) Lipase catalysed esterification of cresols. Ind J Chem Sect B 41B:1068–1071

Suzuki Y, Shimizu T, Takeda H, Kanda K (1991) Fermentative or enzymatic manufacture of sugar amino acid esters. Japan Patent, 03216194 A2

Takahashi K, Saito Y, Inada Y (1988) Lipases made active in hydrophobic media. J Am Oil Chem Soc 65: 911–916

Talon R, Montel MC, Berdague JL (1996) Production of flavor esters by lipases of *Staphylococcus warneri* and *Staphylococcus xylosus*. Enzyme Microb Technol 19:620–622

Therisod M, Klibanov AM (1986) Facile enzymatic preparation of mono acylated sugars in pyridine. J Am Chem Soc 108:5638–5640

Torres C, Otero C (1999) Part I Enzymatic synthesis of lactate and glycolate esters of fatty alcohols. Enzyme Microb Technol 25:745–752

Undurraga D, Markovits A, Erazo S (2001) Cocoa butter equivalent through enzymic interesterification of palm oil mid-fraction. Process Biochem 36:933–939

van Tilbeurgh H, Sarda L, Verger R, Cambillau C (1992) Structure of the pancreatic lipase-procolipase complex. Nature 359:159–162

Vijayakumar GR, Lohith K, Somashekar BR, Divakar S (2004) Lipase catalysed synthesis of L-alanyl, L-leucyl and L-phenylalanyl esters of D-glucose using unprotected amino acids. Biotechnol Lett 26:1323–1328

Vulfson EN (1994) Industrial applications of lipases. In: Wolley P, Petersen SB (eds) Lipases their structure biochemistry and application. Cambridge University Press, New York

Welsh FW, Williams RE (1990) Lipase-mediated production of ethyl butyrate and butyl butyrate in nonaqueous systems. Enzyme Microb Technol 12:743–748

Welsh FW, Williams RE, Dawson KH (1990) Lipase-mediated synthesis of low molecular weight flavor esters. J Food Sci 55:1679–1682

Winkler F, D'Arcy A, Hunziker W (1990) Structure of human pancreatic lipase. Nature 343:771–775

Winkler FK, Gubernator K (1994) Structure and mechanism of human pancreatic lipase. In: Wooley P, Peterson SB (eds) Lipase: their structure, biochemistry and applications. Cambridge University Press, Cambridge, pp 139–157

Xu Y, Wang D, Qing-Mu X, Ao-Zhao G, Zhang KC (2002) Biosynthesis of ethyl esters of short chain fatty acids using whole-cell lipase from *Rhizopus Chinensis* CCTCCM 201021 in non aqueous phase. J Mol Catal B Enzyme 18:29–37

Yadav GD, Devi KM (2004) Immobilized lipase-catalysed esterification and transesterification reactions in non-aqueous media for the synthesis of tetrahydrofurfuryl butyrate comparison and kinetic modeling. Chem Eng Sci 59:373–383

Yadav GD, Lathi PS (2003) Kinetics and mechanism of synthesis of butyl isobutyrate over immobilised lipases. Biochem Eng J 16:245–252

Yadav GD, Lathi PS (2004) Synthesis of citronellol laurate in organic media catalyzed by immobilized lipases kinetic studies. J Mol Catal B Enzyme 27:113–119

Zaidi A, Gainer JL, Carta G, Mrani A, Kadiri T, Belarbi Y, Mir A (2002) Esterification of fatty acids using nylon-immobilized lipase in *n*-hexane kinetic parameters and chain length effects. J Biotechnol 93: 209–216

Zaks A, Dodds DR (1997) Applications of biocatalysts and biotransformations to the synthesis of pharmaceuticals. Drug Dev Today 2:513–531

Enzymatic Esterification of Compounds Possessing Multifunctional Hydroxyl and Carboxyl Groups

4

Abstract

This entire chapter is devoted to the esterification reactions involving compounds containing multifunctional groups like OH, COOH, CH_3 and CHO. The description involves optimisation of reaction conditions for use of lipases in nonpolar solvents under low water activity by both conventional and response surface methodological conditions. It attempts to bring out the superiority of lipase catalysis over chemical synthesis in the few reactions discussed. Esterification of the OH group of lactic acid with the COOH group of few long-chain fatty acids to synthesise 2-O-alkanoyl acids is discussed in detail. This is followed by the work on the preparation of tolyl esters, protocatechuic aldehyde esters, 4-t-butylcyclohexyl acetate and acetylation of β-cyclodextrin. State of the lipases employed under nonaqueous solvents in the above-mentioned reactions is best brought by studies on thermostability and scanning electron microscopy. The role of water, constituting the micro-aqueous phase around the enzyme during its employment under nonpolar conditions, attempts to explain the integrity of the enzyme under such conditions, in terms of developing a theoretical model on the micro-aqueous pH and various equilibria occurring at and associated with the micro-aqueous phase.

4.1 Introduction

Besides catalysing the hydrolysis of triglycerides at the oil/water interface, lipases (triacyl glycerol hydrolases (E.C.3.1.1.3)) are also known to catalyse various other types of reactions like esterification, transesterification, polymerisation and lactonisation.

4.2 2-O-Acyl Esters of Lactic Acid

Multifunctional hydroxy compounds contain one or more functional groups along with a hydroxyl group. Esters of multifunctional hydroxy compounds like polyhydric alcohols and hydroxy acids play an important role in the area of food, since they comprise a wide range of compounds from biodegradable plastics to emulsifiers.

S. Divakar, *Enzymatic Transformation*,
DOI 10.1007/978-81-322-0873-0_4,

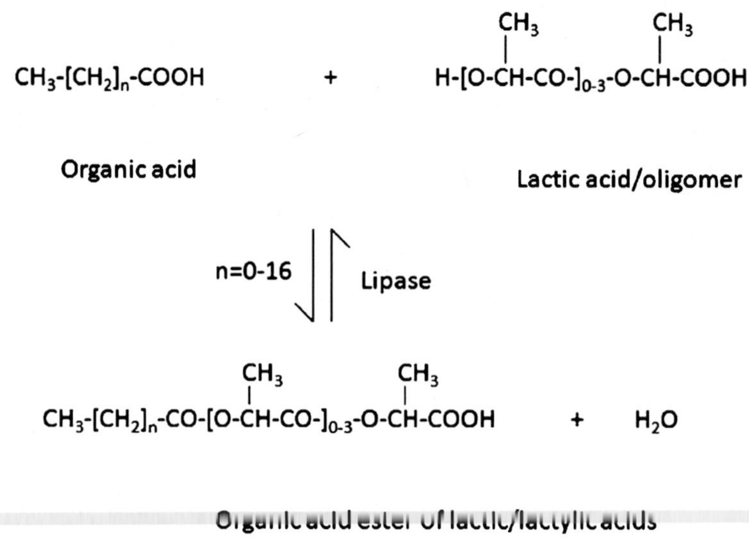

Scheme 4.1 Lipase-catalysed esterification of organic acids with lactic/lactylic acids

Fatty acid esters of lactic acid like 2-O-palmitoyl lactic acid and 2-O-stearoyl lactic acid are commercially important, biocompatible, ionic esters which find wide applications in food, pharmaceutical and cosmetic industry as surfactants.

The reaction involves the hydroxyl group of lactic acid at the 2- position and the carboxyl group of the fatty acid. Till date only chemical routes for the synthesis of fatty acid esters of lactic acid are known. Enzymatic synthesis of these esters is not known till date. Fatty acid esters of lactic acid like 2-O-palmitoyl lactic acid, 2-O-stearoyl lactic acid, 2-O-lauroyl lactic acid and other 2-O-acyl esters of lactic acid have been synthesised using immobilised lipases from *Rhizomucor miehei* (RML) and porcine pancreas (PPL), Scheme 4.1. For methodologies, the readers are recommended to refer Kiran and Divakar (2001).

4.2.1 Lactic Acid

Lactic acid invariably contained 12% water, which hindered the esterification reaction as it facilitated transesterification reactions involving water of reaction present as well as that accumulated during the reaction. Crystalline lactic acid, prepared by removing water by azeotropic reflux using benzene, gave better yields in enzymatic reactions.

Due to its highly hydrophilic nature, lactic acid was found to be insoluble in many nonpolar solvents. Polar solvents like dioxane and ethylmethyl ketone were found to dissolve both lactic acid and fatty acids. Ethylmethyl ketone was found to be a better solvent than polar solvents like dioxane which gave only 4% esterification only with RML as they strip the water off the enzyme rendering catalytically inactive.

4.2.2 2-O-Stearoyl Lactic Acid

The enzymatic reaction was controlled by several factors. Optimisation of the reaction by varying enzyme/substrate (E/S) ratios showed a maximum yield of 99% at an E/S ratio of 22.6 AU $mmol^{-1}$ (Table 4.1, Kiran and Divakar 2001). For a lactic acid/fatty acid concentration of 0.06 M, an E/S ratio of 125.0 gL $mole^{-1}$ gave maximum esterification in case of RML (50 and 40%, respectively, for 2-O-palmitoyl and 2-O-stearoyl lactic acids). In case of PPL, it was 40 and 30%, respectively, at an E/S ratio of 40 gL $mole^{-1}$.

Table 4.1 Preparation of 2-O-stearoyl lactic acid (bench-scale)[a]

Solvent	Enzyme/substrate ratio (E/S)[b] (PPL)	Time (h)	Ester yield (%)	Initial rate (mmol h^{-1})
$CHCl_3$/hexane	11.3	64.5	43.1	0.32
$CHCl_3$	14.7	70	63.2	0.40
$CHCl_3$/hexane	18.6	68	34.5	0.31
$CHCl_3$	20.5	89.5	95.8	0.36
$CHCl_3$	22.6	68	99.1	0.32
$CHCl_3$	23.6	86	90.4	0.40
$CHCl_3$	27.0	67	34.5	0.32
$CHCl_3$	29.5	63	40.2	0.33

[a]Lactic acid and stearic acid concentrations: 0.025 mole
[b]Activity in hydrolytic units

Table 4.2 Preparation of 2-O-palmitoyl lactic acid by using RML and PPL (bench-scale)

Palmitic acid conc. (mole)	Lactic acid conc. (mole)	Solvent	Enzyme/substrate ratio (E/S)[a] (AU $mmol^{-1}$)	% yield and time (h)	Initial rate (mmol h^{-1})
0.11	0.11	Ethylmethyl ketone	RML (7.2)	0 (45)	-
0.11	0.11	Hexane	RML (7.2)	17.9 (44)	0.29
0.055	0.055	$CHCl_3$	PPL (10.3)	18.2 (39)	0.32
0.025	0.025	$CHCl_3$	PPL (22.6)	76 (40)	0.59
0.025	0.025	$CHCl_3$	PPL (22.6)	72 (112)	0.21
0.028	0.025	$CHCl_3$/hexane	PPL (10.1)	48.1 (65)	0.32
0.025	0.025	$CHCl_3$	PPL (14.8)	64.1 (70)	0.35
0.025	0.025	$CHCl_3$	PPL (18.1)	75.6 (94)	0.29
0.025	0.025	$CHCl_3$/hexane	PPL (18.2)	45.9 (68)	0.30
0.025	0.025	$CHCl_3$	PPL (21.3)	85.1 (92)	0.21
0.025	0.028	$CHCl_3$	PPL (27.1)	51.7 (68)	0.28
0.025	0.025	$CHCl_3$	PPL (29.4)	40.2 (36)	0.14

[a]Activity in hydrolytic units

4.2.3 2-O-Palmitoyl Lactic Acid

When hexane was used as the solvent, though lactic acid remained insoluble, 18% esterification was observed after 48 h (Kiran and Divakar 2001). When chloroform was employed as the solvent, 18.2 and 76% ester formation were observed with RML and PPL, respectively (Table 4.2).

A maximum esterification of 85.1% was obtained with chloroform. A chloroform to hexane (30:70) mixture at (Table 4.2) an enzyme/

Table 4.3 Comparison of ester yields as determined by titration, HPLC and ^1H NMR

Ester	Ester yield (%) Titration[a]	HPLC[b]	^1H NMR[c]
2-O-stearoyl lactic acid	83.6	38.2	34.1
2-O-palmitoyl lactic acid	61.6	24.9	22.2
2-O-lauroyl lactic acid	43.7	21.7	24.3

Error in measurements will be:
[a]±5–10%
[b]±5%
[c]±5%

Table 4.4 Yield of organic acid esters of lactic acid as deduced from ^{1}H NMR[a]

Sample	Ester yield (%) (potentiometry)	Free lactic acid (%)	Lactylate esters[b] (%)	Organic acid esters of lactic/lactylic acids (%)
Commercial lactic acid		64.1	35.9	
Crystalline lactic acid		28.1	71.9	
Stearoyl lactic acid	83.6	40.9	25.0	34.1
Stearoyl lactic acid–water layer	83.6	85.5	2.72	10.8
Stearoyl lactic acid–bicarbonate layer	83.6	–	–	–
Stearoyl lactic acid–chloroform layer	83.6	8.2	21.3	70.5
Palmitoyl lactic acid	61.6	44.4	33.3	22.2
Lauroyl lactic acid	43.7	28.2	45.6	24.3
Palmitoyl methyl lactate	57.0	69.3	18.4	12.3
Palmitoyl butyl lactate	44.0	80.7	14.5	2.8

[a]Percentage of organic acid esters of lactic/lactylic acids was determined from increase in the area of –CH–O– signals at 4.9–5.20 ppm region in the spectra of samples with respect to that of crystalline lactic acid at 4.05 ppm; error in NMR measurements will be ±5%

[b]Oligomers of lactic acid

substrate ratio of 18–21 AU mmol^{-1} of PPL gave a yield of about 85% after 96 h.

The ester proportions determined by NMR and HPLC showed good correspondence (Table 4.3).

In ^{1}H NMR the signals in the region 4.0–5.3 ppm corresponding to –CH–O– from free and ester samples were quite informative (Table 4.4). In ^{13}C NMR the signals in the region 60–70 ppm and 170–190 ppm corresponding to –CH–O– and –CO– region, respectively, were informative. A detailed analysis of the signals in the product samples showed unreacted fatty acid, lactic acid and lactylic acids along with esterified fatty acid, lactic acid and lactylic acids. The yields of the esters were also determined. However, the yields were lesser than those detected by titrimetry.

4.2.4 Optimisation of Reaction Parameters Using Response Surface Methodology

Response surface methodological analyses of the esterification reaction between palmitic acid (Kiran et al. 2000) and stearic acid (Kiran et al. 1999) with lactic acid in the presence of immobilised lipases from *Rhizomucor miehei* and porcine pancreas lipase were attempted.

4.2.4.1 2-O-Stearoyl Lactic Acid

Design: Box–Behnken design, 27 experiments, 3 variables at 3 levels

Variables: Enzyme/substrate ratio, incubation period, lactic acid/stearic acid

Equation:

$$\text{RML}: Y = 23.95 - 3.96X_e + 2.96X_t - 8.28X_eX_e + 5.19X_eX_s$$

$$\text{PPL}: Y = 10.43 + 1.65X_s + 5.42X_t - 3.14\,X_sX_s + 4.26\,X_sX_t$$

where

Y = percentage of esterification

X_e = enzyme/substrate ratio (E/S), g mole^{-1}

X_t = incubation period, h

X_s = lactic acid (LA)/stearic acid (SA), mole L^{-1}

The maximum yields predicted by the theoretical equations for both the lipases matched well with the observed experimental values. In case of RML, 2-O-stearoyl lactic acid formation was found to increase with incubation period and

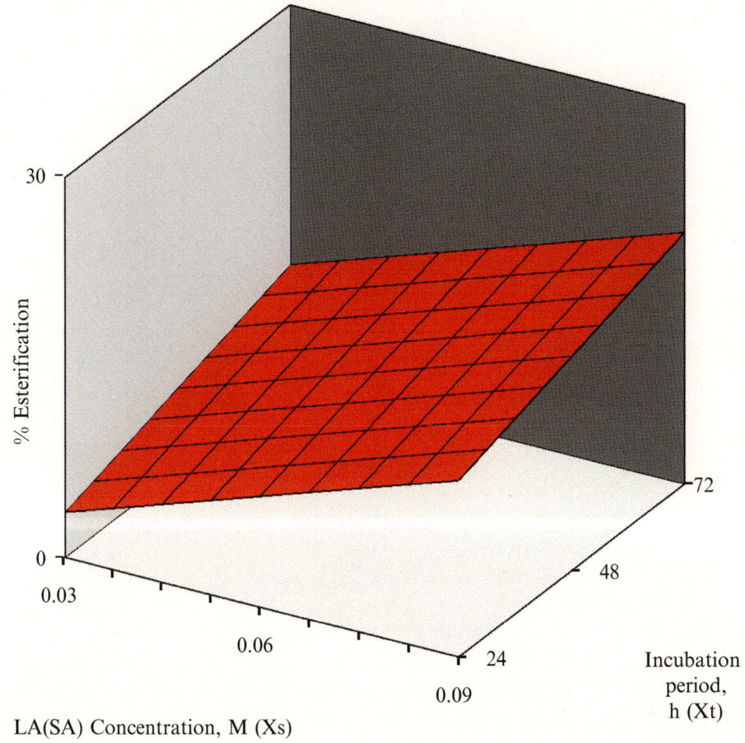

Fig. 4.1 Response surface plot predicting esterification percentage of lactic acid for its reaction with stearic acid using Lipozyme IM20 at a fixed E/S ratio of 166.7 g mol^{-1}

lactic acid/stearic acid concentrations with maximum esterification at an E/S ratio of 125 gL $mole^{-1}$ (Fig. 4.1). In case of PPL, esterification showed steady increase with increase in incubation period and lactic acid/stearic acid concentrations independent of the enzyme/substrate ratios employed (Fig. 4.2).

The optimum conditions predicted for esterification showed good correspondence with experimental values.

$$\text{RML: At } X_e = 102.1 \text{ gL / mole}^{-1}; X_s = 0.03 \text{ M}; X_t = 72 \text{ h}$$
$$\text{Predicted Yield: } 29.43\,\% \quad \text{Experimental Yield: } 24.10\,\%.$$

$$\text{PPL: At } X_e = 40 \text{ gL / mole}^{-1}; X_s = 0.0882 \text{ M}; X_t = 72 \text{ h}$$
$$\text{Predicted Yield: } 18.86\% \quad \text{Experimental Yield: } 17.85\,\%.$$

4.2.4.2 2-O-Palmitoyl Lactic Acid

Design: Box–Behnken design, 27 experiments, 3 variables at 3 levels

Variables: Enzyme/substrate ratio, incubation period, lactic acid/palmitic acid

Equation:

$$\text{RML}: Y = 17.02 + 4.85X_e + 4.32\,X_s + 4.39\,X_t - 10.72X_s\,X_s + 3.15\,X_sX_t$$
$$\text{PPL}: Y = 4.90 + 2.27\,X_e + 1.37\,X_s + 2.99\,X_t + 3.57X_eX_s$$

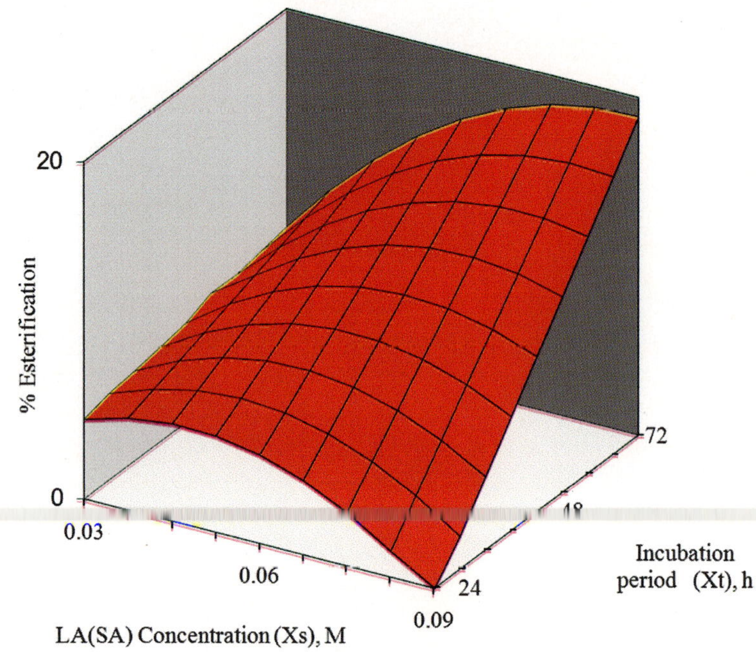

Fig. 4.2 Response surface plot predicting esterification of lactic acid with stearic acid using PPL as a function of LA(SA) concentrations at different incubation periods at all E/S ratios

where

Y = percentage of esterification

X_e = enzyme/substrate ratio (E/S), AU mmole^{-1}

X_t = incubation period, h

X_s = lactic acid (LA)/palmitic acid (SA), mM

In case of RML, palmitoyl lactic acid formation increased with increase in LA(PA) concentration up to 60 mM above which the esterification decreased. Esterification increased with increase in E/S ratios and incubation periods. In case of PPL, the three-dimensional plots showed almost linear relationships. Esterification increased with increase in E/S ratios and incubation periods. PPL was found to give lower yields of the ester when compared to RML. The correspondence between experimental and predicted conditions for optimum esterification was found to be good.

4.2.4.3 2-O-Lauroyl Lactic Acid

Design: Central composite rotatable design (CCRD), 32 experiments, 5 variables at 5 levels (Kiran et al. 2001a)

Variables: Enzyme/substrate ratio (0.09–1.14 AU mmol^{-1}), lactic acid concentration (5–25 mmol), incubation time (6–54 h), buffer volume (0–0.2 mL) and buffer pH values (4.0–8.0)

Equation:

$$Y = 1.562 + 0.338\,X_t - 0.013\,X_e + 0.135\,X_a - 0.081\,X_b + 0.062\,X_p - 0.031\,X_tX_t + 0.178\,X_eX_e + 0.105\,X_aX_a + 0.028\,X_bX_b - 0.044\,X_pX_p - 0.059\,X_tX_e + 0.082X_tX_a + 0.150\,X_tX_b + 0.038\,X_tX_p - 0.335\,X_eX_a - 0.102\,X_eX_b - 0.081\,X_eX_p - 0.036\,X_aX_b - 0.007\,X_aX_p + 0.044\,X_bX_p$$

where

Y = ester yield (mmol)

X_t = incubation period

X_e = E/S ratio (AU mmol^{-1})

X_a = lactic acid concentration (mmol)

X_b = buffer volume (mL)

X_p = buffer pH

A typical three-dimensional plot is given in Fig. 4.3 which shows the effect of E/S ratios and lactic acid concentrations at a fixed buffer volume of 0 mL and an incubation period of 54 h on ester yield. Highest ester yield of 4.8 mmol was observed at the lowest E/S ratio of 0.09 AU mmol^{-1}. Lower E/S ratios gave higher yields, and higher E/S ratios gave lesser yields. This behaviour clearly explained the competitive nature of binding between lauric and lactic acids for

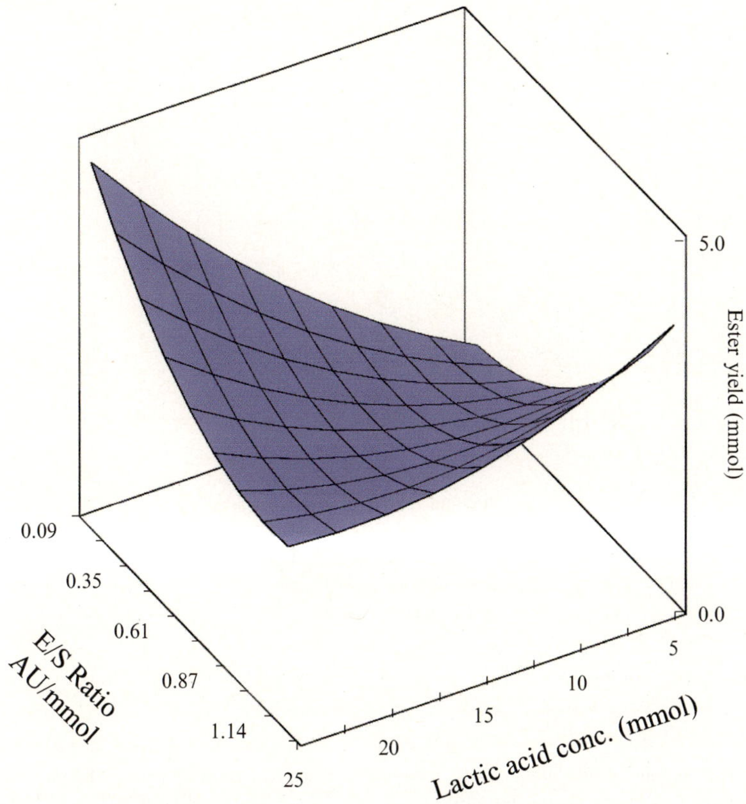

Fig. 4.3 Response surface plot showing variation in predicted yield of lauroyl lactic acid as a function of E/S ratios and lactic acid concentrations at an incubation period of 54 h. Since the buffer volume is 0 mL, terms connected with buffer volume and pH values were ignored

the same binding site on the enzyme. Addition of buffer in terms of both volume and pH did not have a profound effect on increase in ester yield. Predicted yields showed good validation with experimental yields when experiments corresponding to selected points on the contour plots were carried out.

4.2.5 Effect of Acid Carbon Chain Length on Esterification with Lactic Acid

It was observed that increase in carbon chain of the organic acid increased the ester yield (Fig. 4.4, Table 4.5).

The initial rates measured also showed a similar pattern. Only exception to this rule was in the case of propionic acid which showed as esterification of 3.4%, whereas acetic acid showed an esterification of 19.3%. Highest yield of 99% was observed with stearic acid. The low molecular weight stronger acids (lesser K_a than long-chain fatty acids) undergo more dissociation resulting in increased hydrogen ion concentration at the micro-aqueous interphase. Lower carbon chain length organic acids also are more soluble at the interphase than the long-chain ones, thereby enhancing the dissociation, leading to lower observed yields (Kiran and Divakar 2001).

4.2.6 Reusability of Porcine Pancreas Lipase

The ester yield decreased with each cycle from 75.1 to 14.6% after 5th recycle (Table 4.6 and Fig. 4.5). Due to strenuous conditions like heat and incubation in nonpolar solvents for longer periods,

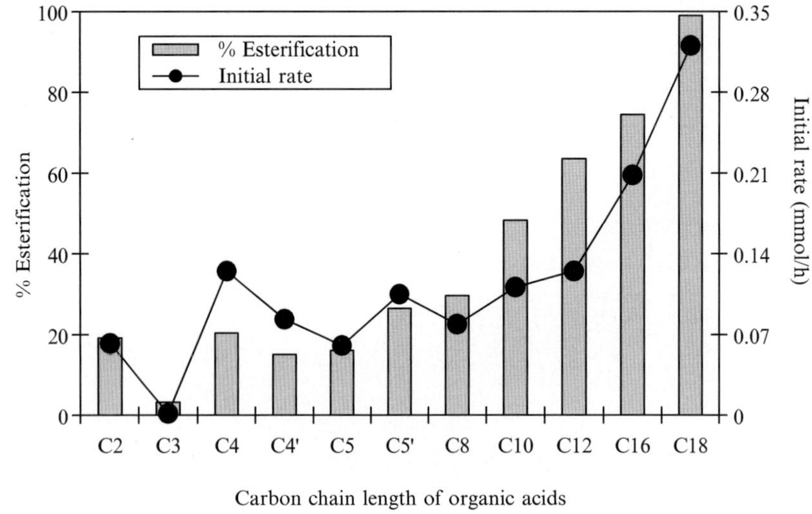

Fig. 4.4 Effect of acid carbon chain length on extent of esterification of lactic acid and initial rate of the reactions (bench-scale). Reactions conditions: lactic and organic acids: 0.025 mol; solvent: chloroform; E/S ratio: PPL, 22.6 AU $mmol^{-1}$ (activity units were with respect to hydrolytic activity units); incubation period: 96 h

Table 4.5 Effect of organic acid carbon chain length 2-O-acyl ester of lactic acid preparation using PPL[a]

Acid employed	Maximum esterification (%)	Initial rate (mmol h^{-1})	Ester yield (mmol)
Acetic C_2	19.2	0.06	4.8
Propionic C_3	3.8	0.01	0.9
Butyric C_4	15.1	0.08	3.8
Isobutyric C_4	20.4	0.13	5.1
Valeric C_5	24.5	0.11	3.4
Isovaleric C_5	14.1	0.06	4.4
Octanoic C_8	29.6	0.08	7.7
Decanoic C_{10}	48.4	0.11	9.4
Lauric C_{12}	63.6	0.13	15.9
Palmitic C_{16}	74.5	0.21	18.7
Stearic C_{18}	90.4	0.32	24.8

AU $mmol^{-1}$ – E/S ratio is with respect to hydrolytic activity units; solvent – chloroform; incubation period – 96 h
[a]Reaction conditions: lactic acid and organic acids – 0.025 mole; E/S ratio – 22.6

catalytic activity of the enzyme was affected due to probable structural changes. The initial rates of the reactions also showed a gradual decrease which was in correspondence with decrease in esterification.

Table 4.6 Reusability of PPL for the bench-scale synthesis of 2-O-stearoyl lactic acid[a]

Recycle NO[b]	Maximum esterification percentage	Initial rate (mmol h^{-1})
1	75.1	0.28
2	40.5	0.17
3	25.6	0.09
4	28.4	0.05
5	14.6	0.05
6	14.9	0.04

[a]Lactic acid and stearic acid concentrations : 0.025 moles E/S ratio: 22.6 AU $mmol^{-1}$
[b]Incubation period: 72 h, solvent: chloroform
Activity in hydrolytic units

4.2.7 Food Chemical Codex Specifications for Enzymatically Synthesised 2-O-Acyl Esters of Lactic Acid

Enzymatically synthesised 2-O-acyl esters of lactic acid in comparison with the commercially available sodium-2-O-stearoyl lactylate (from Enzyme India Ltd. Chennai) showed good correspondence

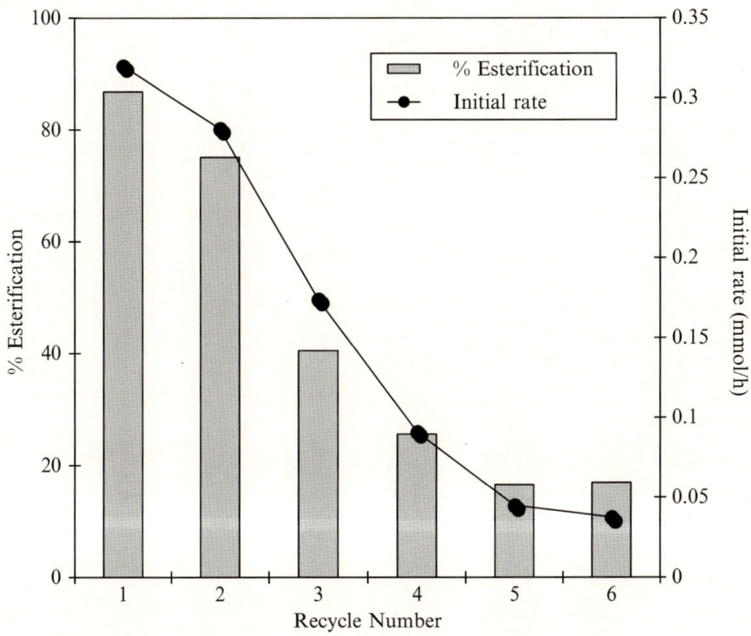

Fig. 4.5 Effect of enzyme reusability on extent of esterification of lactic acid with stearic acid (bench-scale). Reaction conditions: lactic and stearic acids: 0.025 mol; solvent: chloroform; E/S ratio: 22.6 AU $mmol^{-1}$ (activity units were with respect to hydrolytic activity units); incubation period: 72 h

to food chemical codex (FCC specifications) in terms of acid value, ester value, sodium content and lactic acid contents (Table 4.7).

Enzymatically prepared stearoyl lactic acid shows an acid value of 69.4, whereas commercial preparation shows 110.4. Those of palmitoyl and lauroyl lactic acids showed acid values of 79.5 and 91.5, respectively. In case of ester values also, all the three enzymatically prepared esters showed better values than the commercial sample. Free lactic acid and sodium contents of enzymatically prepared ones were also within the specified range. Unlike chemically prepared sample, the enzymatic ones are neat without any side products. These characteristics showed that the enzymatically prepared ester samples are better than the commercial preparation and will be more suitable for applications in food formulations.

4.3 Tolyl Esters

Phenolic esters of organic acids especially those of cresols are good flavour compounds as they possess a combination of sweet, floral and fruity odours which are very much desired in food and cosmetic industry (Burdock 1994). Enzymatic esterification of fatty acids with primary alcoholic groups of functionalised phenols and interesterification reactions of peracetylated polyphenolic compounds have been reported (Habulin and Krmelj 1996). However, before this report, direct synthesis of phenolic esters involving phenolic OH groups mediated by lipases was practically nil (Scheme 4.2).

Porcine pancreas lipase and *Rhizomucor miehei* lipase showed very little esterification in the presence of solvents under shake-flask conditions, for the preparation of esters of m- and

Table 4.7 Comparative data on commercially available and enzymatically prepared 2-O-acyl esters of lactic acids

Sample	Acid value	Ester value	Sodium content (%w/w)	Lactic acid Content (%w/w)	Arsenic (ppm)	Heavy metals (ppm)
Food chemical codex specifications	60–80	150–190	3.5–5.0	31–34	<3	<10
Commercial 2-O-stearoyl lactate	110.4	149.9	ND	24.4	ND	ND
Enzymatic samples						
2-O-stearoyl lactic acid						
Sodium 2-O-stearoyl lactic acid	69.6	169.8	3.2	22.2	ND	ND
2-O-stearoyl lactic acid	144.7	203.6	0	22.5	ND	ND
2-O-stearoyl lactic acid(free lactic acid removed)	109.1	159.2	0	24.1	ND	ND
2-O-palmitoyl lactic acid						
Sodium 2-O-palmitoyl lactic acid	79.5	172.9	3.3	21.8	ND	ND
2-O-palmitoyl lactic acid	162.3	120.3	0	32.3	ND	ND
2-O-stearoyl lactic acid(free lactic acid removed)	147.2	148.3	0	24.2	ND	ND
2-O-lauroyl lactic acid						
Sodium 2-O-lauroyl lactic acid	91.3	153.3	3.5	24.8	ND	ND
2-O-lauroyl lactic acid	169.8	183.7	0	28.2	ND	ND
2-O-lauroyl lactic acid(free lactic acid removed)	152.6	141.2	0	23.9	ND	ND

ND not determined

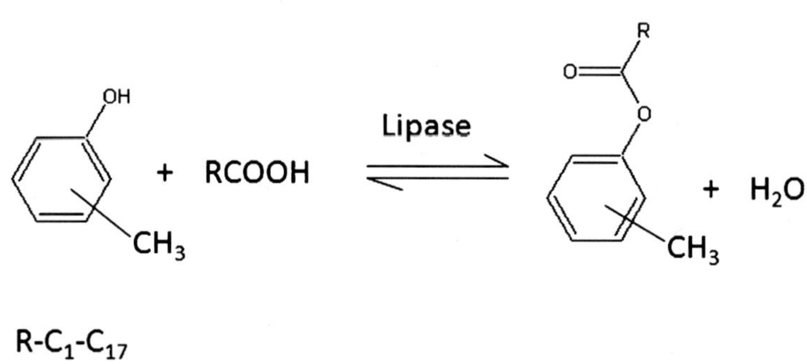

Scheme 4.2 Lipase-catalysed esterification of cresols

p-cresols using organic acids of carbon chain lengths C_2–C_{18} (Table 4.8). Benchscale conditions employed a specially devised experimental set-up (Kiran et al. 1998; Suresh-Babu et al. 2002) which maintained azeotropic conditions where the solvent system $CHCl_3$ to petroleum ether (30:70) not only maintained a temperature at 56–60°C (when the reaction mixture was under reflux) but also aided effective removal of water of reaction from the reaction mixture, thereby facilitating further esterification. The reaction gave better yields of ester when 0.05–0.2 mL of 0.1 M phosphate buffer of pH 7.0 was present in the reaction mixture. In the absence of the buffer, ester yields were very poor.

Effect of increasing amounts of RML addition showed enhanced ester yields from 51.6% for

Table 4.8 Porcine pancreas lipase-catalysed esterification of m- and p-cresols

	Shake-flask[a]	Bench-scale[b]
Esters	% Yield	% Yield
m-cresyl acetate	18.6 (51)	44.1 (95)
m-cresyl propionate	31.4 (70)	19.7 (43)
m-cresyl butyrate	38.4 (44)	52.6 (72)
m-cresyl isobutyrate	18.9 (69)	10.5 (64)
m-cresyl valerate	20.8 (45)	1.3 (72)
m-cresyl isovalerate	32.9 (45)	8.6 (72)
m-cresyl octanoate	12.8 (30)	10.0 (72)
m-cresyl decanoate	0.0 (72)	3.1 (72)
m-cresyl laurate	10.1 (30)	64.6 (46)
m-cresyl palmitate	0.0 (72)	7.4 (72)
m-cresyl stearate	15.5 (44)	0.0 (72)
p-cresyl acetate	64.9 (66)	67.9 (88)
p-cresyl propionate	31.4 (69)	33.7 (74)
p-cresyl butyrate	8.13 (21)	12.2 (46)
p-cresyl isobutyrate	15.8 (48)	8.1 (44)
p-cresyl valerate	11.3 (70)	4.5 (63)
p-cresyl isovalerate	14.7 (21)	10.4 (23)
p-cresyl octanoate	22.1 (53)	0.0 (89)
p-cresyl decanoate	28.7 (28)	8.3 (44)
p-cresyl laurate	17.2 (40)	12.5 (40)
p-cresyl palmitate	23.8 (50)	8.6 (72)
p-cresyl stearate	7.9 (20)	0.0 (70)

[a]Equimolar concentrations of cresols and organic acids (0.0023 mole) with 60 mg PPL
[b]Equimolar concentrations of cresols and organic acids (0.01 mole) with 400 mg PPL and 0.1 mL 0.1 M phosphate buffer pH 7.0
Figures in parentheses indicate period of incubation in hours to achieve maximum yield

450 mg to 74% for 1 g of RML. A maximum esterification of 52.9% was observed with 100 mg RML at pH 7.0 at an optimum buffer volume of 0.1 mL. Under identical reaction conditions, m- and p-cresol gave conversion yields above 30% for organic acids up to butyric acid. With increase in organic acid carbon chain length, the conversion yields decreased.

It was generally observed that larger amounts of enzymes are required for better conversions when p-cresol was employed. A detailed analysis of m- and p-cresyl ester synthesis using lipases has shown that cresols (m- and p-) generally act as inhibitors for these enzymes. This may be true for other phenols also.

4.3.1 Application of Central Composite Rotatable Design to Lipase-Catalysed Synthesis of m-Cresyl Acetate

Design: Central composite rotatable design (CCRD), 32 experiments, 5 variables at 5 levels (Manohar and Divakar 2002)

Variables: *m*-Cresol concentrations (0.005–0.025 mole), enzyme/substrate ratios (0.18–1.22 AU mmole^{-1}), incubation periods (6–54 h), pH (4–8) and buffer volumes (0–0.2 mL)

Equation: A second-order polynomial equation was developed.

$$\begin{aligned}Y = {} & 1.5692 + 0.59067X_t + 0.0532X_e - 0.096X_a \\ & + 0.2917X_b + 0.2X_p + 0.142X_tX_t + 0.202X_eX_e \\ & + 0.289X_aX_a + 0.2216X_bX_b + 0.05583X_pX_p \\ & + 0.115X_tX_e - 0.169X_tX_a - 0.026X_tX_b + 0.0134X_tX_p \\ & - 0.1092X_eX_a + 0.246X_eX_b - 0.283X_eX_p + 0.036X_aX_b \\ & + 0.0089X_aX_p + 0.0282X_bX_p\end{aligned}$$

where

X_a = *m*-cresol concentration (moles)
X_e = enzyme/substrate ratio (AU mmoles^{-1})
X_t = incubation period (h)
X_b = buffer volume (mL)
X_p = buffer pH

The predictability of the model is at 74% confidence level. Three-dimensional plots and contour plots were obtained using the above equation.

Lipase-catalysed reaction between *m*-cresol and acetic acid did not proceed in the absence of the buffer. It did not also occur in the absence of the enzyme. The reaction required presence of buffer for better conversions, and a minimum amount of 0.1 mL buffer was found necessary for this reaction. Increase in buffer volumes at all the pH values studied (in the range pH 4.0–8.0) showed decrease in ester yield up to a buffer volume of 0.1 mL. Above 0.1 mL, increase in buffer volume increased ester yield up to 0.2 mL (Fig. 4.6). Buffer pH values around 4.0 and below appear to favour better esterification than those at pH values in the range 4.0–8.0. The methodology projected conditions for higher yields up to 7.2 mmoles. Validation experiments carried out under these predicted conditions gave a highest

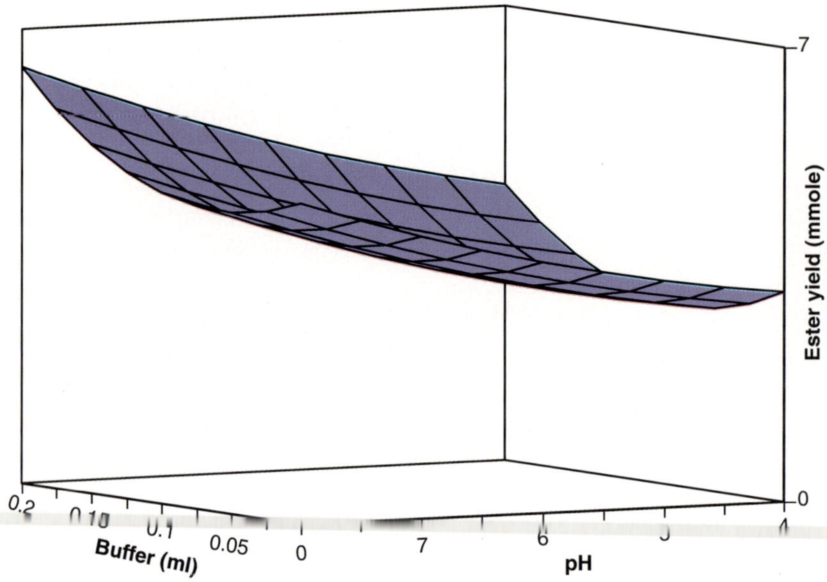

Fig. 4.6 Three-dimensional plot showing the effect of pH and buffer volume on esterification for an E/S ratio of 0.7 AU mmol^{-1}, 0.25 mole m-cresol concentration and 72 h incubation period

experimental yield of 4.3 mmoles. The CCRD treatment clearly showed inhibitory nature of *m*-cresol in the esterification process.

4.4 Application of Plackett–Burman Design for Lipase-Catalysed Esterification of Anthranilic Acid

In order to determine the effectiveness of esterification of anthranilic acid using alcohols of carbon chain length C_1–C_{18}, Plackett–Burman (PB) design allowed the selection of better alcohols for esterification of anthranilic acid (Scheme 4.3).

Of the alcohols employed, methanol, decanol, cetyl alcohol and stearoyl alcohols showed 99.9% significance. Esterification of anthranilic acid with methanol showed the highest yield of 45.6% (Suresh-Babu and Divakar 2001).

The experimental and predicted yields with experimental design are shown in Table 4.9. Generally it was observed that alcohols with 99.9% significance levels showed better yields, with highest yield from methanol followed by stearyl alcohol and decanol.

The reaction between anthranilic acid and methanol showed that an optimum enzyme amount of 250 mg of PPL was required for the reaction. Higher amounts of enzyme did not increase the yield. Experiments were conducted at different concentrations of methanol (0.01, 0.02, 0.05, 0.1 and 0.2 mole). At lower concentrations of methanol, esterification was slow. This may be due to evaporation of methanol. But at higher methanol concentrations (0.05 and 0.1 mole), conversion yields were good. A maximum yield of 45.6% was obtained when 0.05 mole of methanol was employed.

4.5 Kinetic Study of Porcine Pancreas Lipase Inhibition by *p*-Cresol (*p*-Cresyl acetate) and Lactic Acid (2-O-Stearoyl Lactate)

The enzymatic reactions between p-cresol and acetic acid and lactic acid and stearic acid required larger concentrations of enzymes for lesser concentrations of substrates indicating the inhibitory nature of p-cresol and lactic acid towards porcine pancreas lipase (Kiran and Divakar 2002). The results from these kinetics investigations are described below.

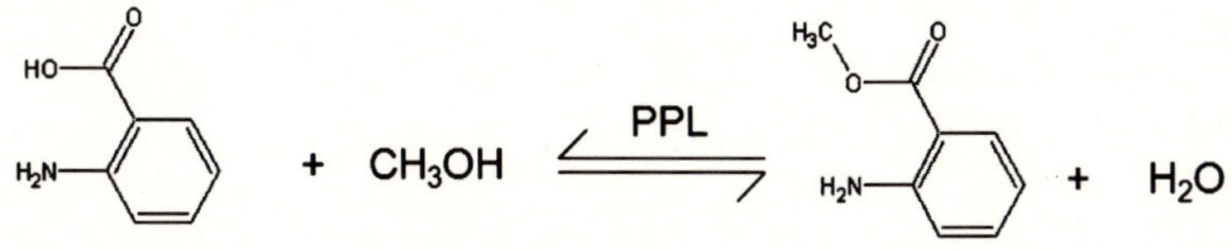

Scheme 4.3 Lipase-catalysed esterification of anthranilic acid

Table 4.9 Experimental design with experimental and predicted yields

Number of alcohol carbon atoms												
C_1	C_2	C_3	C_4	C_5	C_6	C_8	C_{10}	C_{12}	C_{16}	C_{18}	Experimental yield (mmole)	Predicted yield (mmole)
1	−1	1	−1	−1	−1	1	1	1	−1	1	4.8	4.8
1	1	−1	1	−1	−1	−1	1	1	1	−1	4.4	4.7
−1	1	1	−1	1	−1	−1	−1	1	1	1	3.7	3.9
1	−1	1	1	−1	1	−1	−1	−1	1	1	3.8	3.8
1	1	−1	1	1	−1	1	−1	−1	−1	1	2.4	2.7
1	1	1	−1	1	1	−1	1	−1	−1	−1	3.1	3.4
−1	1	1	1	−1	1	1	−1	1	−1	−1	0.0	0.2
−1	−1	1	1	1	−1	1	1	−1	1	−1	3.9	3.3
−1	−1	−1	1	1	1	−1	1	1	−1	1	2.8	1.2
1	−1	−1	−1	1	1	1	−1	1	1	−1	2.7	3.0
−1	1	−1	−1	−1	1	1	1	−1	1	1	4.8	4.5
−1	−1	−1	−1	−1	−1	−1	−1	−1	−1	−1	1.1	1.4

Initial velocities of the reaction measured and double reciprocal plots constructed by plotting 1/v versus 1/[acetic acid] or 1/[stearic acid] showed that the lines obtained at different p-cresol (lactic acid) concentrations intersected on the *Y*-axis, giving the maximum velocity (V_{max}). The various kinetic parameters determined, namely, $K_{M(p\text{-cresol})}$ ($K_{M(PC)}$), $K_{M(acetic\ acid)}$ ($K_{M(AA)}$), $K_{M(lactic\ acid)}$ ($K_{M(LA)}$), $K_{M(Stearic\ acid)}$ ($K_{M(SA)}$), $K_{i(p\text{-cresol})}$ ($K_{i(PC)}$) and $K_{i(lactic\ acid)}$ ($K_{i(LA)}$), from experimental data are shown in Table 4.10.

Scheme 4.4 indicates the mechanism of inhibition of lipase activity by p-cresol and lactic acid, respectively, due to the formation of dead-end enzyme–p-cresol or enzyme–lactic acid complexes. In this process, the acyl–enzyme intermediate formed between acetic acid and lipase (stearic acid–lipase) resulted in the products by attacking the nucleophile (p-cresol or lactic acid as the case may be). During this process, the formation of two products, namely, release of water molecule due to acyl–enzyme complex formation and release of product due to lipase–ester dissociation, resulted from two substrates.

The rate equation for the mechanism described is given as

$$v = \frac{V_{max}}{K_{M[Y]} \times [X] \times \{1 + [X]/K_i\} + K_{M[X]} \times [Y] + [X] \times [Y]}$$

where

[X] = p-cresol or lactic acid concentration

[Y] = acetic acid or stearic acid concentration

$K_{M[X]}$ = Michaelis-Menten constant of p-cresol or lactic acid

$K_{M[Y]}$ = Michaelis-Menten constant of acetic acid or stearic acid

K_i = inhibition constant of p-cresol or lactic acid

By suitably modifying the equation, theoretical values for kinetic parameters were developed.

Table 4.10 lists the theoretical values which were in close agreement to the experimental ones. This indicates that Ping-Pong Bi-Bi mechanism is applicable in both cases and clearly illustrates the inhibitory nature of p-cresol and lactic acid in the respective reactions.

Table 4.10 Data on kinetic parameters

	p-cresyl acetate			Stearoyl lactic acid	
Kinetic parameters	Experimental	Theoretical	Kinetic parameters	Experimental	Theoretical
V_{max} (mole h^{-1})	0.000143	0.0002	V_{max} (M h^{-1})	0.00154	0.00132
$K_{M(AA)}$ (mole)	0.00311	0.0226	$K_{M(SA)}$ (M)	0.0548	0.0647
$K_{M(PC)}$ (mole)	0.000035	0.007	$K_{M(LA)}$ (M)	0.0098	0.0286
$K_{i(PC)}$ (mole)	0.00966	0.0355	$K_{i(LA)}$ (M)	0.0951	0.2364

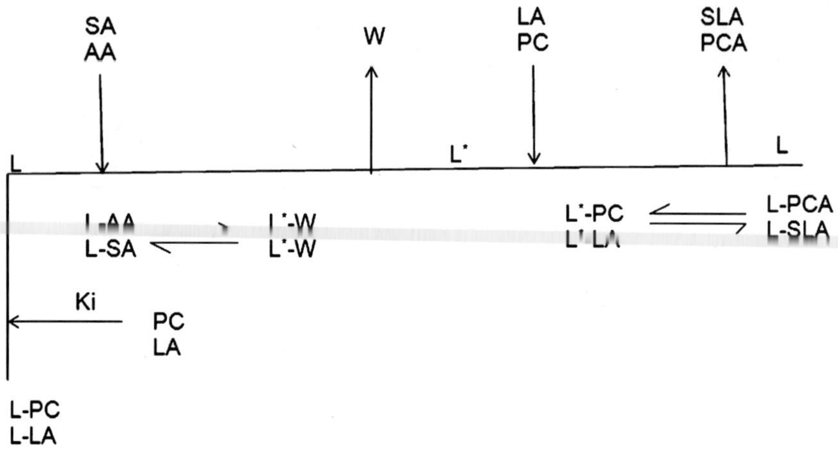

Scheme 4.4 Schematic representation of Ping-Pong Bi-Bi mechanism of esterification of p-cresol with acetic acid and stearic acid with lactic acid L – lipase; PC – p-cresol; AA – acetic acid; L–AA – lipase–acetic acid complex; PCA – p-cresyl acetate; LA – lactic acid; SA – stearic acid; L–PC and L–LA – dead-end enzyme–inhibitor complex; L–SA – lipase–stearic acid complex; SLA – stearoyl lactic acid; L*- acyl enzyme; W – water (Segel 1993)

4.6 Thermostability of Porcine Pancreas Lipase

In order to understand the behaviour of PPL under different temperatures for longer periods of incubations, which is essential for employment of lipase in esterification reactions which are time consuming and requires higher temperatures and nonpolar solvents, thermostability of porcine pancreas lipase was studied through response surface analysis (Kiran et al. 2001b). Native enzyme showed loss in activity at 60°C which could be due to a conformation with a greater unfolding at 60°C than at 40°C and 80°C (Fig. 4.7). Longer periods of incubation of PPL at especially 80°C did not affect the active conformation of PPL even after incubation for a period up to 10 days. However, presence of small amounts of buffer stabilised the enzyme at 60°C contrary to what was observed with the native enzyme. In the presence of 0.2 M lactic acid, there was a general loss in active conformation and hence activity, at all temperatures and periods studied.

The equation obtained to study the thermal stability in terms of activity measurements of native PPL is as follows:

$$\text{Percentage activity} = 408.17 - 12.73\, X_t + 0.219\, X_p + 0.108\, X_t^2 - 0.005\, X_p^{\,2} - 0.001\, X_t X_p$$

where

X_t = incubation temperature
X_p = incubation period

The results from activity measurements were supported by ultraviolet (UV) spectroscopic data of the same in every respect indicating that

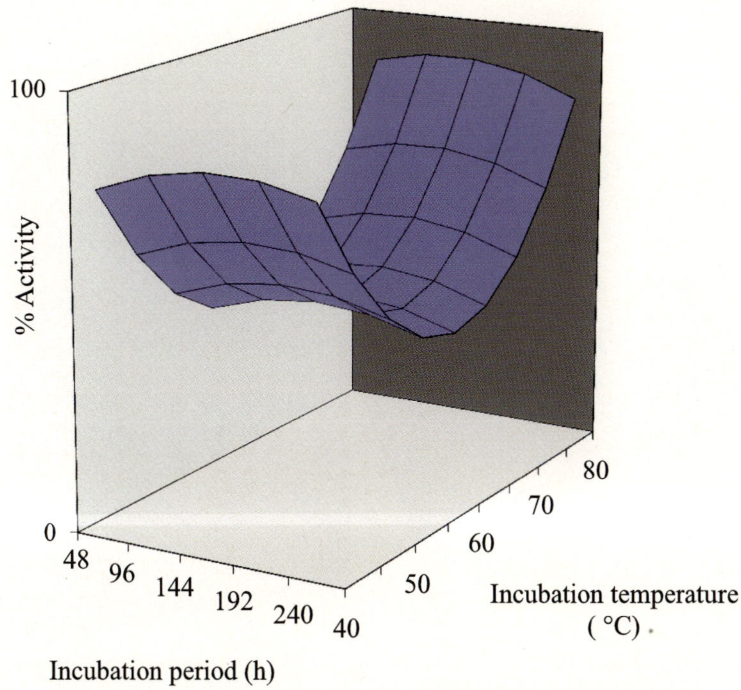

Fig. 4.7 Three-dimensional plot showing the effect of incubation period and temperature on percentage activity. Reaction conditions: 10 mg PPL was incubated in 2 mL MIBK at 40°C, 60°C and 80°C for 48 h, 144 h and 240 h, respectively. Percentage activity is with respect to esterification activity of native PPL

variation in conformational changes is responsible for loss in activity (Fig. 4.8).

The following equation was obtained from UV measurements:

$$A_{290} = 1.154 - 0.038\,X_t - 0.003\,X_p - 0.299\,X_t^2 - 0.001\,X_p^2 - 0.032\,X_tX_p$$

where

X_t = incubation temperature

X_p = incubation period

Table 4.11 shows the results of regression analysis with experimental and predicted percentage activity and absorbance values at 290 nm.

4.7 Scanning Electron Microscopy

The state of three lipase preparations, two from *Rhizomucor miehei* and one from porcine pancreas, employed in the esterification reactions of *p*-cresol and lactic acid in nonpolar solvents at 50–60°C investigated by scanning electron microscopy (SEM) showed the compact nature of adsorption and distribution of lipase on matrix in case of *Rhizomucor miehei* and carrier in case of porcine pancreas (Suresh-Babu et al. 2001). All the three lipase preparations subjected to high temperatures and nonpolar solvents for a prolonged period of incubation of 72–120 h showed decrease in the 'compactness' when compared with unused lipase. Presence of buffer preserved the activity and compactness, and the absence of the same, in some cases, reduced the amount of enzyme per unit area on the support. *Rhizomucor miehei*

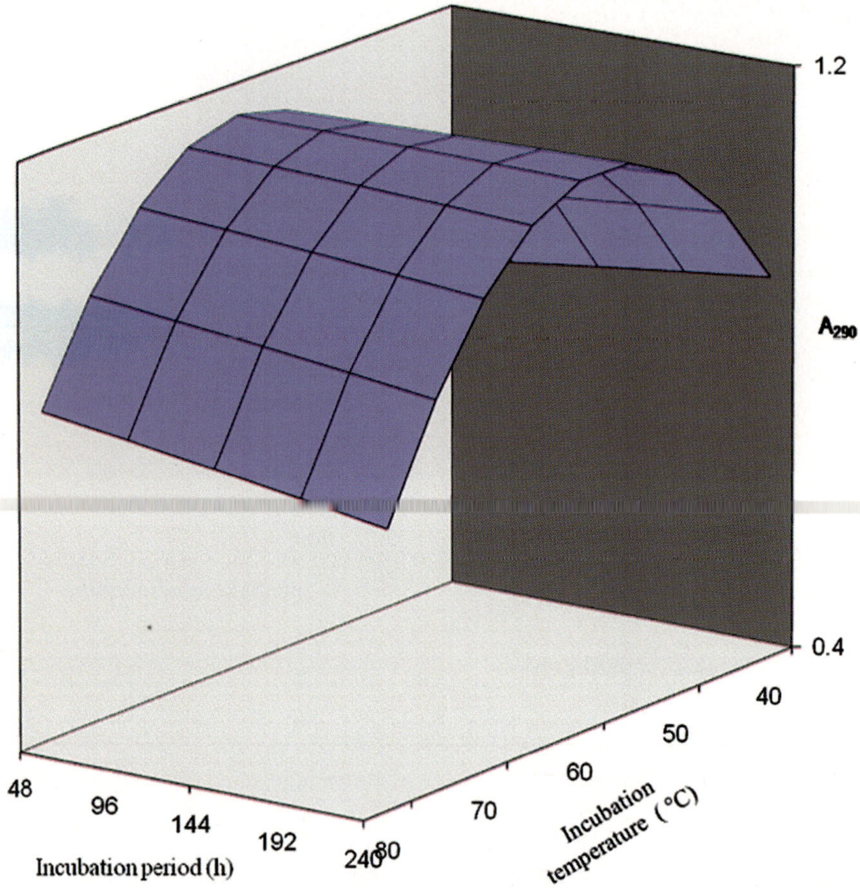

Fig. 4.8 Three-dimensional plot showing the effect of incubation period and temperature on absorbance at 490 nm. Reaction conditions: 10 mg PPL was incubated in 2 mL MIBK at 40°C, 60°C and 80°C for 48 h, 144 h and 240 h, respectively

Table 4.11 Predicted and experimental yields in case of incubation of PPL in MIBK

	X_t (°C)		X_p (h)		Percentage activity		Absorbance at 290 nm	
Exp no.	Coded	Actual	Coded	Actual	Experimental	Predicted	Experimental	Predicted
1	−1	40	−1	48	72.8	79.3	0.953	0.862
2	−1	40	0	144	90.5	84.2	0.851	0.894
3	−1	40	1	240	85.4	83.2	0.876	0.923
4	0	60	−1	48	50.5	40.1	0.974	1.156
5	0	60	0	144	41.6	44.9	1.243	1.154
6	0	60	1	240	32.8	39.8	1.244	1.151
7	1	80	−1	48	83.7	87.6	0.942	0.851
8	1	80	0	144	89.3	90.3	0.771	0.816
9	1	80	1	240	87.	83.1	0.734	0.780

lipase samples subjected to reaction in the presence of 0.1 mL and 0.1 M buffer at different pH values (4.0–9.0) showed decrease in 'compactness' of the enzyme on the surface which correlated to increase in esterification activity of the same. Increase in volume of buffer (0.05–0.75 mL) in the reaction mixture at pH 7.0 showed a decrease in compactness and also reduction in activity. The studies indicated that a compromise between extent of adsorption, distribution and activity could be achieved by varying the pH and volume of buffer in the reactions, for maximum conversions.

4.8 Hydrogen Ions in Micro-aqueous Phase During Lipase-Catalysed Esterification in Nonaqueous Media

Water plays a crucial role in both hydrolysis and esterification in lipase catalysis (Hahn-Hagerdal 1986). While a critical amount of water is essential for maintaining the active conformation of the enzyme, excess water facilitates hydrolysis (Halling 1989; Zaks and Klibanov 1988). Though the former aspect has been extensively studied as water activity, information on the latter is very scanty. While exhaustive stripping of water from the enzyme resulted in lowering of reaction rate, presence of a critical amount of water favoured esterification due to enhancement of enzyme activity (Valivety et al. 1992).

In a nonpolar solvent, excess water adds to the already existing hydration shell of the enzyme constituting a micro-aqueous enzyme–water–solvent phase (Fig. 4.9). Micro-aqueous phase, although being not a well-defined thermodynamic structure, holds the key for problems associated with many equilibrium processes of the esterification reaction that can be only complementary to the associated enzymatic factors like solvent polarity and desiccants which affect the micro-aqueous phase thereby the extent of esterification as well. Micro-aqueous phase provides the appropriate solvent characteristics like dielectric constant and polarity that maintain the active conformation in a nonpolar solvent.

The kinetic aspects of lipase-catalysed esterification reactions have been thoroughly investigated (Chulalaksanaukul et al. 1992; Lee et al. 1995; Janssen et al. 1999). However, this is not the case for non-enzymatic thermodynamic equilibrium processes associated with the same. Molecular surface area of the enzyme, volume/thickness of the micro-aqueous phase, total [H^+] in the micro-aqueous phase, polarity of the solvent, diffusion behaviour of the acid/alcohol/ester and partitioning of water/[H^+]/alcohol/acid between the organic solvent and the micro-aqueous phase all play a crucial role in determining the extent of esterification for a given amount of the enzyme. Among these, the major equilibrium processes are dissolution and dissociation of the organic acid in the micro-aqueous phase and partitioning of water/[H^+]/acid/alcohol/ester between the micro-aqueous phase and the organic solvent (Fig. 4.9). Dissociation of water soluble organic acids and their dissociation in the micro-aqueous phase lead to the formation of the conjugate acid, H_3O^+. In case of weak organic acids, H_3O^+ is stronger than the organic acid itself and hence, facilitates hydrolysis, thereby limiting the extent of esterification. Many features like the micro-aqueous phase are difficult to measure. Attempts have also been made by a few researchers to measure [H^+] at the micro-aqueous phase (Brown et al. 1997).

A theoretical model was developed based on mass transfer equations for arriving at the concentration of [H+] in the micro-aqueous phase and other factors which regulate esterification (Kiran et al. 2002). The reaction chosen was the esterification equilibria of 2-O-stearoyl lactic acid synthesis using immobilised lipase from Rhizomucor miehei and porcine pancreas.

Table 4.12 lists the extent of esterification observed for different concentrations of substrate and enzyme for both RML and PPL in ethylmethyl ketone. The water of reaction formed is less than that required to cause aggregation of the enzyme, but just sufficient to form a thin film on the enzyme to regulate its activity.

A quantity termed hydrogen ion number (HIN) in the micro-aqueous phase was determined using the relation

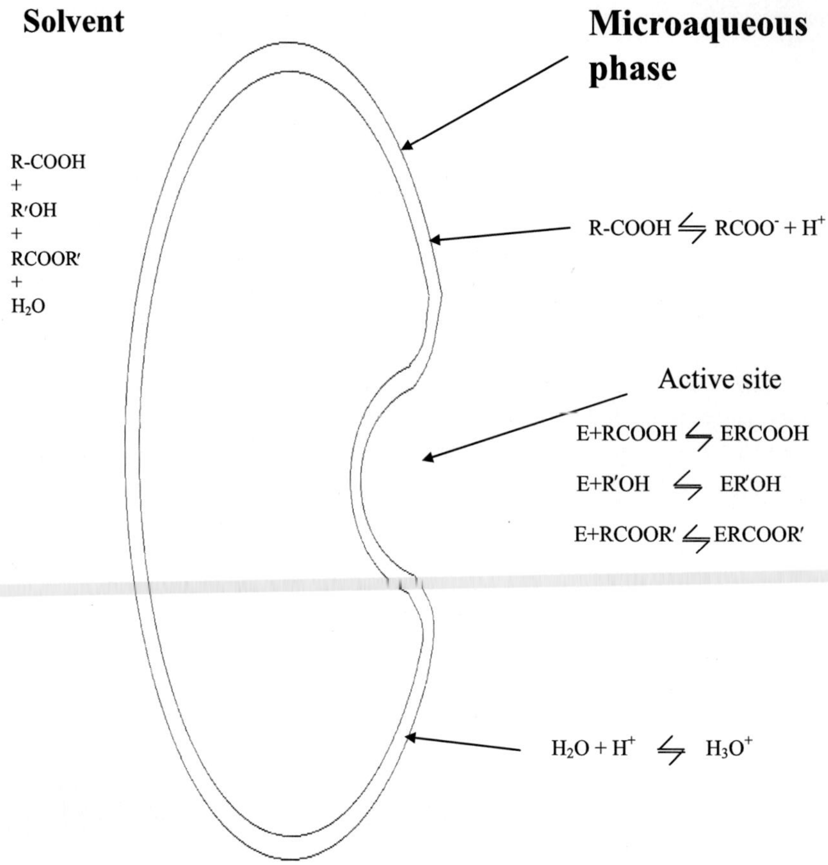

Fig. 4.9 Schematic representation of micro-aqueous phase around lipase

Table 4.12 Theoretical HIN, pH and water for the enzymatic preparation of 2-O-steroyl lactic acid

Substrate M	Enzyme g L^{-1}	Ester M	Yield %	Water of reaction mL L^{-1}	pH	HIN ($\times 10^{-7}$)	Log HIN
	Lipozyme						
0.03	3.8	0.008	26.7	0.144	2.86	1.20	17.08
0.03	3.8	0.010	33.3	0.179	2.88	1.43	17.16
0.03	3.8	0.012	40.0	0.216	2.90	1.63	17.51
0.03	2.5	0.008	26.7	0.144	2.86	1.20	17.08
0.06	5.0	0.014	23.3	0.252	2.70	3.03	17.48
0.06	7.5	0.018	30.0	0.324	2.72	3.73	17.57
0.06	10.0	0.008	13.3	0.144	2.67	1.84	17.26
0.09	7.5	0.020	22.2	0.360	2.61	5.34	17.73
0.09	11.3	0.020	22.2	0.360	2.61	5.34	17.73
0.09	15.0	0.022	24.4	0.396	2.62	5.79	17.76
	PPL						
0.03	0.8	0.006	20.0	0.108	2.84	0.94	16.97
0.03	1.2	0.004	13.3	0.072	2.82	0.65	16.81
0.06	0.6	0.006	10.0	0.108	2.66	1.41	17.15
0.06	1.5	0.010	16.7	0.180	2.68	2.26	17.35
0.06	2.4	0.018	30.0	0.324	2.72	3.73	17.57
0.09	2.3	0.002	2.5	0.040	2.60	6.74	16.83
0.09	3.6	0.021	23.0	0.369	2.61	5.51	17.74
0.09	3.6	0.018	20.0	0.320	2.60	4.88	17.69

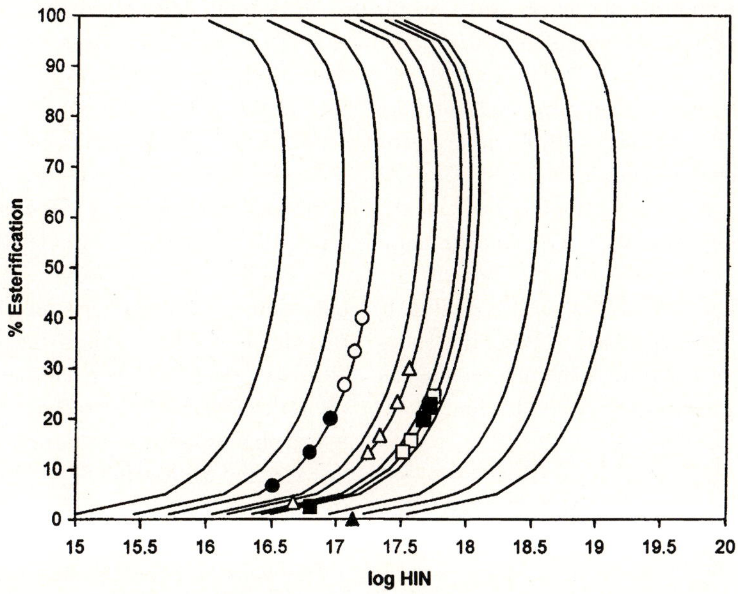

Fig. 4.10 Theoretical plot log HIN versus percentage of esterification

$$\text{HIN} = \left[\text{H}^{+}\right]_{\text{T}} \times 1.008 \times 4.023 \times 10^{23} \times \text{Xml} / 1{,}000$$

where 1.008 = atomic weight of H^+.

The total [H^+] in the water of reaction constituting the micro-aqueous phase was arrived at (Atkins 1987) using the relation:

$$\left[\text{H}^{+}\right]_{\text{T}} = \left(\text{K}_{\text{LA}} \times \text{EC}_{\text{LA}}\right)^{0.5} + \left(\text{K}_{\text{SA}} \times \text{EC}_{\text{SA}}\right)0.5$$

$$\text{KSA} = 1.76 \times 10^{-6} \text{ and KLA} = 1.37 \times 10^{-4}\,\text{M}^{-1}$$

From the partitioning coefficient values determined for stearic and lactic acids in ethylmethyl ketone (0.058 and 1.46, respectively), the effective concentration of the acid (EC_{LA} and EC_{SA}) at the micro-aqueous phase constituted by the water of reaction was determined.

Theoretical plots of log HIN versus % esterification from 0 to 100% were constructed for acid concentrations 0.01–1.0 M. Parenthesis-shaped curves were obtained (Fig. 4.10). Irrespective of the acid concentration, during the course of reaction, log HIN varied with extent of esterification in an explicable manner.

At the initial stage up to 10% esterification, dissociation of the more unreacted acid in the lesser water of reaction contributed to the micro-aqueous phase registering higher log HIN. Further increase in esterification up to 50% consumed more free acid and registered a very slight net increase in log HIN, due to decrease in free acid concentration although water of reaction was larger. Beyond 55% esterification, decrease in free acid in larger volume of water of reaction registered marginal decrease in log HIN which became progressive as the extent of conversion reached 99%. Experimentally derived data were also fitted on to the theoretical plots, and they cover a very narrow range of esterification (Fig. 4.10). The values outside this region could be obtained with higher enzyme concentrations. Higher yields are not possible due to hydrolysis of the ester formed by [H^+] build-up.

At any stage of esterification, an equilibrium concentration of [H^+], i.e. a net proton build-up, is always maintained at the micro-aqueous phase. Such a build-up could affect esterification by effecting hydrolysis. The question arises as what happens to such a large number of [H+] in the micro-aqueous phase? In real-time systems, log HIN values will be low and that this model clearly brings out the role of enzymes and buffers employed to neutralise this excess [H^+] at the micro-aqueous phase. The depletion could occur by (1) ionisable groups from the amino acid residues on the surface of the enzyme molecule like Asp, Glu, Tyr, Lys and Arg at different levels of ionisation capable of 'consuming' these ions and (2) partitioning of [H^+], stearic and lactic acids, ester and water between the micro-aqueous phase and the surrounding bulk organic solvent that reduces the effective load of all these constituents in the micro-aqueous phase. Thus, this study explained the processes at the micro-aqueous phase dictating the extent of esterification.

4.9 Acetylation of Protocatechuic Aldehyde

Selectivity in enzymatic acetylation of the two phenolic OH groups at positions 3 and 4 of protocatechuic aldehyde (PA) using lipases from *Rhizomucor miehei* (RML and Chirazyme) and porcine pancreas lipase (Scheme 4.5) in methyl isobutyl ketone/benzene (15:85) was investigated (Divakar 2003). Chemical esterification using acetic acid and acetic anhydride did not produce any product as no catalyst was employed. Similarly under identical conditions, acetic acid did not react even in the presence of the three lipases (Table 4.13).

However, acetic anhydride in the presence of all the three lipases showed esterification (Table 4.13). Product formation of mono-acetoxy and diacetoxy products was monitored by the H-5 signal of the reactant and the esters. Yields of the order of 41.0–73.5% were obtained with good selectivity (Table 4.14) with the highest selectivity of 97% for the PPL-catalysed reaction.

In order to enhance the selectivity and yield, enzymatic acetylation was also carried out by anchoring PA inside β-cyclodextrin cavity. β-Cyclodextrin (βCD) is a cyclic oligosaccharide consisting of 7 glucose units joined in the form of a ring by virtue of it, possessing an hydrophobic cavity of 6–8 A in size capable of accommodating a plethora of organic molecules that fit into this size stabilised by H bonding, hydrophobic and other interactions. Earlier structural studies carried out by us have shown that protocatechuic aldehyde could be anchored inside β-cyclodextrin (βCD) cavity in a 1:1 stoichiometric proportion with a binding constant value of $6{,}700 \pm 770\ M-1$ in water and $5{,}100 \pm 580\ M-1$ in $NaHCO_3$ buffer (pH 10.5) (Maheswaran and Divakar 1997).

β-Cyclodextrin was peracetylated to prevent acetic anhydride being consumed by the OH groups of βCD. Reaction between PA and acetic anhydride in the presence of peracetylated βCD was carried out using the three lipases (Table 4.13) successfully. All the three lipases showed more than 30% yield. However, the acetylation with RML showed >99% selectivity by yield 3-acetoxy derivative exclusively.

4.10 4-t-Butylcyclohexyl Acetate

4-t-Butylcyclohexanol exists as *cis* and *trans* isomers with different flavour notes. The *trans* esters possess a woody note, whereas the *cis* esters exhibit a strong, intense, flowery note. The esters are used in soaps, detergents, creams, lotions and perfumes and also as a solubility-reducing agent for surfactants. At room temperature, the *trans*/*cis* ratio of the alcohol is 3:1 (Karger et al. 1968).

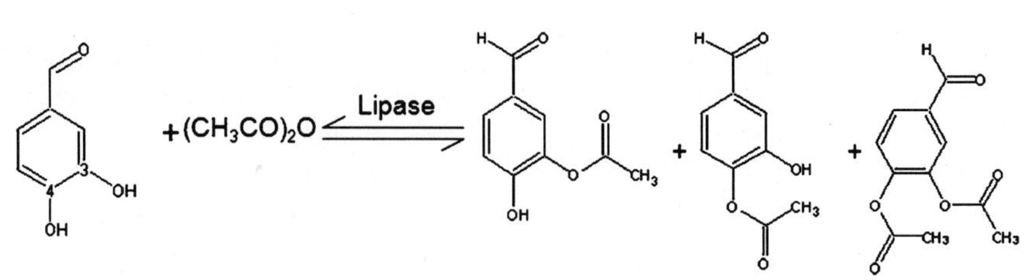

Scheme 4.5 Lipase-catalysed regioselective esterification of protocatechuic aldehyde

Table 4.13 Lipase-catalysed esterification data on protocatechuic aldehyde[a]

Reactions	Ester yield				
			3-acetoxy	3,4-diacetoxy	
(Moles)	Titrationb	NMRb	Derivative	Derivative	Selectivity
PA + acetic anhydride (0.009 + 0.0044)	No rean	No rean	–	–	–
PA + acetic acid + PPL (0.01+0.01)	63.1	No rean			
PA + acetic acid +chirazyme (0.01+0.01)	54.9	No rean			
PA + acetic acid + RML (0.01+0.01)	60.1	No rean			
PA + acetic anh.+PPL (0.01+0.005)	54.8	41.0	39.8	1.2	97.0
PA + acetic anh.+ chirazyme (0.01+0.005)	51.2	54.5	52.2	4.3	92.4
PA + acetic anh + RML (0.01+0.005)	52.9	73.5	64.2	7.3	90.0
PA + acetic anh+βCDA+PPL (0.01+0.005+0.001)	35.2	60.5	58.2	2.3	94.2
PA + acetic anh+βCDA+Chirazyme (0.01+0.005+0.001)	31.8	50.0	44.2	3.8	92.4
PA + acetic anh+βCDA+_RML (0.01 + 0.005 + 0.001)	35.3	48.1	44.4	3.7	92.3
PA + acetic anh + βCDA + PPL (0.01+0.01+0.001)	54.4	61.5	57.7	3.8	93.8
PA + acetic anh + βCDA + Chirazyme (0.01 + 0.01 + 0.001)	51.7	79.4	64.7	14.7	81.5
PA + acetic anh + βCDA + RML (0.01 + 0.01 + 0.001)	55.1	17.6	17.6	0	>99

[a]PA – protocatechuic aldehyde; +βCDA –βCD acetate
[b]Monitoring by titration and ^{1}HNMR

Chemical synthesis of the esters of 4-t-butylcyclohexanol was carried out with organic acids from acetic to heptanoic acids (Karger et al. 1968). The *trans*/*cis* ratio of acetate, propionate, n-butyrate, valerate, caproate, heptylate, isobutyrate and pivalate remained at approximately 3.5 in all the cases. Acetylation of 4-t-butylcyclohexanol anchored inside βCD was found to give a *trans*/*cis* ratio of 5.5 (Pattekhan and Divakar 2001, 2002).

Enzymatic esterification of 4-t-butylcyclohexanol was carried out by using porcine pancreas lipase in order to obtain better yield and selectivity (Scheme 4.6). The yields were low in nonpolar solvents (Manohar and Divakar 2004). However, when it was anchored inside β-cyclodextrin (βCD) cavity, the esterification yields were higher (Table 4.14).

The low-yield 4-t-butylcyclohexyl acetate in nonpolar solvents was attributed to the formation of aggregate of 4-t-butylcyclohexanol with the buried alcohol OH groups, thereby preventing esterification. Hence, βCD was employed for breaking down the aggregates, by anchoring 4-t-butylcyclohexanol inside βCD cavity. Peracetylated βCD was used and in the presence of the enzyme to promote interesterification between the acetate groups of βCD and 4-t-butylcyclohexanol even when acetic anhydride was employed. With increase in incubation period, the esterification was found to increase

Table 4.14 Conversion yield and *trans*/*cis* ratios of 4-t-butylcyclohexyl acetate

Acetic anhyd. mole	βCD acetate mole	PPL mg	Time hr	Conversion yield %	*Trans*/*cis* ratios
0.0016	0.00025	275	48	18.8	2
0.0039	0.00025	275	48	43.8	2.95
0.0016	0.00075	275	48	29.0	2.6
0.0039	0.00075	275	48	34.2	2.4
0.0016	0.00025	625	48	18.1	4.34
0.0039	0.00025	625	48	39.4	2.1
0.0016	0.00075	625	48	7.6	2.83
0.0039	0.00075	625	48	66.6	2.29
0.0016	0.00025	275	96	47.1	2.37
0.0039	0.00025	275	96	56.3	2.07
0.0016	0.00075	275	96	44.9	2.55
0.0039	0.00075	275	96	55.8	2.35
0.0016	0.00025	625	96	39.5	2.8
0.0039	0.00025	625	96	73.4	2.71
0.0016	0.00075	625	96	21.2	2.7
0.0039	0.00075	625	96	57.1	2.9
0.0005	0.0005	450	72	10.4	4.01
0.005	0.0005	450	72	86.6	2.38
0.0028	0	450	72	12.1	2.1
0.0028	0.001	450	72	36.5	3.5
0.0028	0.0005	100	72	41.5	2.9
0.0028	0.0005	800	72	45.7	2.61
0.0028	0.0005	450	24	25.7	3.02
0.0028	0.0005	450	120	68.6	2.51
0.0028	0.0005	450	72	57.1	2.45
0.0028	0.0005	450	72	47.3	2.02
0.0028	0.0005	450	72	55.0	2.35
0.0028	0.0005	450	72	58.8	2.26
0.0028	0.0005	450	72	59.9	2.35
0.0028	0.0005	450	72	57.2	2.37

steadily from 18.5% after 12 h to a maximum of 65.9% after 120 h. From the initial

slope value, the rate of acetylation was determined to be 4.36 μmole h^{-1}. Further, an increase in βCD–acetate concentration from 0 to 0.005 mole (0.5 M equivalent to 4-t-butylcyclohexanol) resulted in increase in esterification up to a maximum of 55.9%, and it decreased thereafter up to 0.001 mole (1 M equivalent of 4-t-butylcyclohexanol). Esterification also increased with increase in PPL concentration reaching a maximum of 55.7% at 450 mg and decreased thereafter up to 1 g (Table 4.14).

Gas chromatographic analyses of the reaction mixtures showed that *trans*/*cis* ratios vary from 2.1 to 4.4 which were higher than that of the alcohol 2.47. However, no general trend was observed between the reaction conditions employed and *trans*/*cis* ratios. Inclusion of 4-t-butylcyclohexanol inside the peracetylated βCD cavity preferentially facilitates the esterification of the *trans* isomer of the alcohol as the *cis* isomer is sterically hindered from reacting with the incoming acyl group due to the presence of bulky acetyl groups on peracetylated βCD.

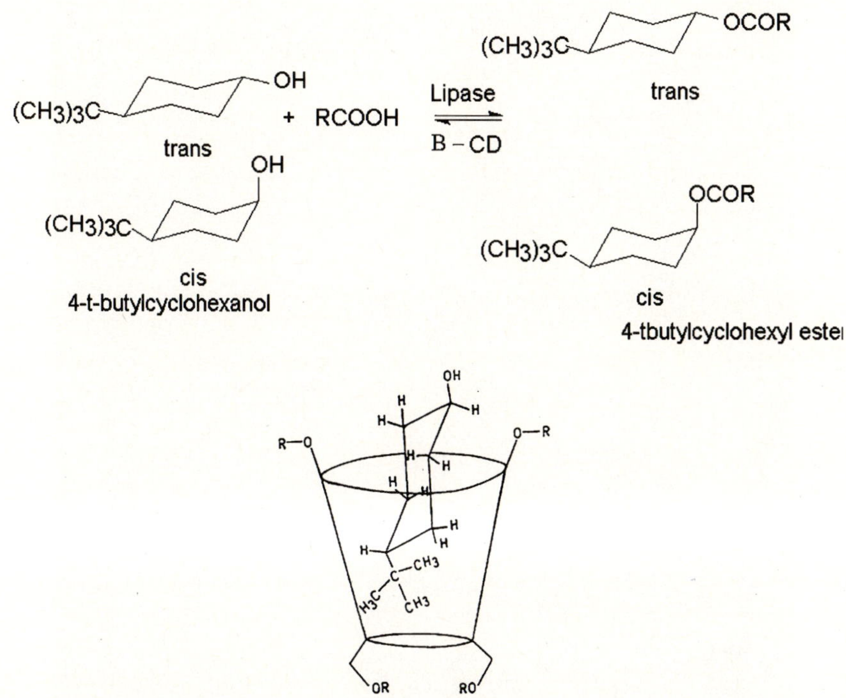

Scheme 4.6 Lipase-catalysed esterification of 4-t-butylcyclohexanol

4.11 Esterification of β-Cyclodextrin

β-Cyclodextrin (βCD) is soluble in water only to the extent of 1.8 g L^{-1} in water, and its solubility in organic solvents other than DMSO, DMF and pyridine is negligible or low. Poor solubility of cyclodextrin in many solvents can be reduced by derivatisation. Hence, the hydroxyl groups of βCD (7 hydroxyl groups from 2-OH, 7 hydroxyl from 3-OH and 7 hydroxyl groups from 6-CH_2OH) were esterified (Pattekhan and Divakar 2002) using lipases (Scheme 4.7).

Among the organic acids employed – acetic, propionic, butyric, isobutyric, valeric, isovaleric, lauric, octanoic, palmitic and stearic acids – only acetic, propionic and isobutyric acids were found to undergo esterification with 58.8, 25.8 and 17.4% yields, respectively (Table 4.15). Derivatised βCD like DM βCD (heptakis-2,6-di-O-methyl βCD) and HP βCD (hydroxypropyl βCD) with acetic acid gave 32.8 and 55.8% esters, respectively (Table 4.15).

All these esters were soluble in methanol. βCD was able to react with 4.78 molecules, DM βCD with 2.45 molecules and HP βCD with 2 molecules of acetic acid. βCD reacted with 0.38 molecules of propionic acid and 2.43 molecules of isobutyric acid.

Out 7 primary hydroxyl groups and 14 secondary hydroxyl groups of βCD from 2-OH and 3-OH, very few hydroxyl groups were converted to the esters. While primary hydroxyl groups can react with ease, the secondary hydroxyl groups at 2-OH reacted with difficulty, and 3-OH groups may not react at all being buried inside the cavity. Very low substitution of acetic, propionic and isobutyric groups is due to steric hindrance of the incoming acyl groups by the βCD cavity and the disposition of free hydroxyl groups within it.

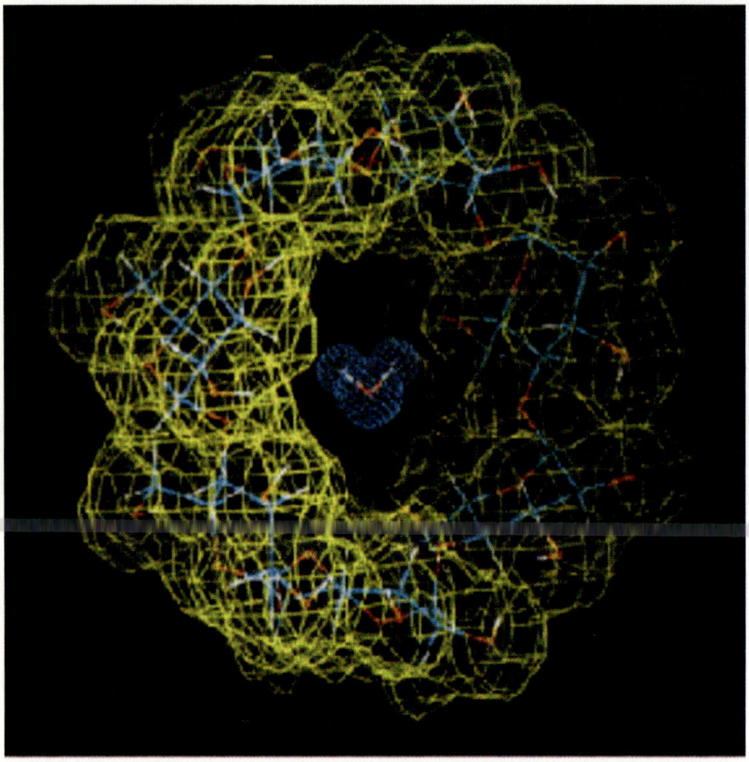

Scheme 4.7 Three-dimensional structure of β-cyclodextrin (Roquett 1991)

Table 4.15 Esterification of βCD and its derivatives catalysed by *Rhizomucor miehei* lipase[a]

System	Ratio	Percentage esterification by titration	Percentage esterification by ^{1}H NMR	Degree of substitution by ^{1}H NMR[b]
βCD–acetic acid	0.1:5	58.8	48.4	1:6.78
βCD–propionic acid	0.1:1	25.8	2.7	1:0.38
βCD–isobutyric acid	0.1:1	17.4	17.4	1:2.43
DMβCD–acetic acid	0.1:5	32.8	17.5	1:2.45
HPβCD–acetic acid	0.1:5	55.8	14.3	1:2

[a]Degree of substitution – ratio of peak area for CH_3 protons of acid to the number of H-1 protons of glucose units of βCD and its derivatives

[b]Percentage esterification by ^{1}HNMR – number of hydroxyl groups of βCD esterified out of 14 per molecule of βCD or its derivatives. Error in ^{1}HNMR measurements ±5%

References

Atkins PW (1987) Physical chemistry, 2nd edn. Oxford ELBS, Oxford Press, Oxford, p 280

Brown JR, Guther MLS, Field RA, Ferguson MAJ (1997) Hydrophobic mannosides act as acceptors for trypanosome α-mannosyltransferases. Glycobiology 7:549–558

Burdock GA (1994) In Fenaroli's handbook of flavor ingredients, vol II, 3rd edn. CRC Press, Boca Raton

Chulalaksanaukul W, Condort JS, Combes D (1992) Kinetics of geranyl acetate synthesis by lipase catalyzed transesterification in *n*-hexane. Enzyme Microb Technol 14:293–298

Divakar S (2003) Lipase catalysed regioselective esterification of protocatechuic aldehyde. Indian J Chem Sect B 42B:1119–1122

Food Chemical Codex specifications

Habulin M, Krmelj V (1996) Synthesis of olcic acid esters catalyzed by immobilized lipase. J Agric Food Chem 44(1):338–342

Hahn-Hagerdal B (1986) Water activity a possible external regulator in biotechnical processes. Enzyme Microb Technol 8:322–327

Halling PJ (1989) Organic liquids and biocatalysts theory and practice. Trends Biotechnol 7:50–52

Janssen AEM, Sjursnes BJ, Vakurov AV, Halling PJ (1999) Kinetics of lipase catalyzed esterification in organic media correct model and solvent effects on parameters. Enzyme Microb Technol 24:463–470

Karger BL, Stern RL, Zannucci JF (1968) Anal Chem 40(4):727

Kiran KR, Divakar S (2001) Lipase catalysed esterification of organic acids with lactic acid. J Biotechnol 87:109–121

Kiran KR, Divakar S (2002) Enzyme inhibition by *p*-cresol and lactic acid in lipase mediated syntheses of *p*-cresyl acetate and stearoyl lactic acid A kinetic study. World J Microbiol Biotechnol 18:707–712

Kiran KR, Karanth NG, Divakar S (1998) An improved enzymatic process for the preparation of fatty acid hydroxyacid ester. Indian Patent, 1978/DEL/98 187313

Kiran KR, Karanth NG, Divakar S (1999) Preparation of steroyl lactic acid catalysed by immobilized lipases from *Mucor miehei* and porcine pancreas optimization using response surface methodology. Appl Microbiol Technol 52:579–584

Kiran KR, Manohar B, Karanth NG, Divakar S (2000) Response Surface Methodological study of esterification of lactic acid with palmitic acid catalysed by immobilised lipases from Mucor miehei and porcine pancreas. Z Lebm Unt Fors 211:130–135

Kiran KR, Manohar B, Divakar S (2001a) A central composite rotatable design analysis of lipase catalysed synthesis of lauroyl lactic acid at bench-scale level. Enzyme Microb Technol 29:122–128

Kiran KR, Suresh-Babu CV, Divakar S (2001b) Thermostability of porcine pancreas lipase in non-aqueous media. Process Biochem 36:885–892

Kiran KR, Karanth NG, Divakar S (2002) Hydrogen ion concentration at the microaqueous phase in lipase catalysed esterification in non-aqueous organic media – steroyllactic acid. Ind J Biochem Biophys 39:101–105

Lee Y, Howard LR, Villalon B (1995) Flavonoids and antioxidant activity of fresh pepper (*C. annum*) cultivars. J Food Sci 60:473–476

Maheswaran MM, Divakar S (1997) Structural studies on inclusion compounds of β-cyclodextrin with some substituted phenols. J Incln Phenomenon 27:113–126

Manohar B, Divakar S (2002) Application of central composite rotatable design to lipase catalyzed syntheses of *m*-cresyl acetate. World J Microbiol Biotechnol 18:745–751

Manohar B, Divakar S (2004) Porcine pancreas lipase acetylation of beta-cyclodextrin anchored 4-t-butylcyclohexanol. Indian J Chem Sect B 43B:2661–2665

Pattekhan HH, Divakar S (2001) Regioselectivity in the preparation of 2-hydroxy-4-ethoxybenzaldehyde from resorcinol in presence of β- cyclodextrin and its derivatives. J Mol Catal A Chem 169(2001):185–191

Pattekhan HH, Divakar S (2002) Regioselective acetylation of 4-t-butylcyclohexanol in the presence of β-cyclodextrin and its derivatives. J Mol Catal A Chem 184:79–83

Roquett Catalog (1991) Kleptose-β-Cyclodextrin www.roquette-food.com

Suresh-Babu CV, Divakar S (2001) Selection of alcohols through Plackett-Burman design in lipase catalyzed syntheses of anthranilic acid. J Am Oil Chem Soc 78:49–52

Suresh-Babu CV, Kiran KR, Divakar S (2001) Scanning electron microscopic studies of lipase catalysed esterification catalysis for the synthesis of stearoyl lactate and p-cresyl laurate. World J Microbiol Biotechnol 17:659–665

Suresh-Babu CV, Karanth NG, Divakar S (2002) Lipase catalysed esterification of cresols. Ind J Chem Sect B 41B:1068–1071

Valivety RH, Halling PJ, Macrae AR (1992) *Rhizomucor miehei* lipase remains highly active at water activity below 0.001. FEBS Lett 301:258–260

Zaks A, Klibanov AM (1988) Enzyme catalysis in monophasic organic solvents. J Biol Chem 263:3194–3201

Enzymatic Polymerisation 5

Abstract

Potentiality of lipases to effect esterification has been explored to prepare polymers of monomer molecules which possess hydroxyl and carboxyl functions like lactic acid, ε-caprolactone, p-hydroxybenzoic acid, p-aminobenzoic acid, adipic acid and 1,6-hexanediol. Ring opening polymerisation of ε-caprolactone gave a polycaprolactone polymer of molecular weight 10,000, which was the best compared to the low molecular weight polymers of molecular weight 400–5,000 obtained with the other abovementioned monomers. Relative advantages and disadvantages of the use of lipases in such polymerisation reactions along with the film-forming properties of polylactic acid and polycaprolactone and its blends are presented in this chapter.

5.1 Introduction

The total world annual demand for synthetic polymers amounts to 1.3 million t. Due to their environmental hazard, the emphasis is shifting towards the production of biodegradable polymers which are eco-friendly in nature. Polyhydroxyalkanoates are one such class of biodegradable polymers. Hydroxy acids like glycolic acid, beta-hydroxypropionic acid, lactic acid, gamma-hydroxybutyric acid, delta-hydroxyvaleric acid and their lactones which are bifunctional in nature can be polymerised to give biodegradable polymers. There are many chemical methods available for the preparation of polylactic acid (Seiji and Masahiro 1992; Rika et al. 1996; Voss and Spielan 1958). Enzymatic production (in vitro) also leads to production of these polymers through milder reaction conditions, good yield, clean product and pollution-free process conditions (Gross et al. 2001; Kumar and Gross 2000).

Among the various polyhydroxyalkanoates studied, both low and high molecular weight polycaprolactone esters possess several desirable physicochemical properties. Ring opening polymerisation of lactones and macrolides of various sizes using lipases has been reported. *Pseudomonas fluorescens* lipase-catalysed polymerisation of α-methyl-β-propiolactone resulted in a polymer with a number average molecular weight of 600–2,900 (Svirkin et al. 1996). Copolymerisation of β-propiolactone and ε-caprolactone using *Pseudomonas fluorescens* lipase resulted in a random copolymer with a number average molecular weight of 520 (Namekawa et al. 1996). In these cases, the reaction rates and molecular

S. Divakar, *Enzymatic Transformation*,
DOI 10.1007/978-81-322-0873-0_5, © Springer India 2013

weight build-up were found to be slow. In order to enhance the reaction rates and reduce the incubation period, initiators like methanol and butanol were used. Use of methanol as initiator resulted in polycaprolactone with a degree of polymerisation of 35 at 40°C within 4 days (Knani et al. 1993). In the presence of butanol as an initiator, PPL-catalysed polymerisation of ε-caprolactone at 65°C in n-heptane resulted in an esterification of 33–100% with a number average molecular weight in the range 313–1,600. So far, the highest molecular weight achieved was 7,600 in case of polycaprolactone by PPL when butanol was used as initiator. Zhang et al. (1996) have used bacterial protease from *Bacillus subtilis* for the ring opening polymerisation of ε-caprolactone which resulted in a polymer with a number average molecular weight of 810. Many reports are also available on lipase-catalysed polymerisation of macrolides, namely, octanolide (Kobayashi et al. 1998), undecanolide dodecanolide, pentadecanolide (Uyama and Kobayashi 1996; Bisht et al. 1998) and hexadecanolide (Namekawa et al. 1996). Generally, it was observed that lipases from *Candida* and *Pseudomonas* species showed higher activity and better molecular weight build-up. The highest molecular weight of 25,000 was achieved with undecanolide (Uyama and Kobayashi 1996).

5.2 Polylactic Acid

Polylactic acid, a biodegradable polyhydroxyalkanoate (PHA), is one of promising replacements to synthetic polymers. The advantages of polylactic acid are its high strength, thermoplasticity, fabricability, biodegradability and bioenvironmental compatibility. The demand for polylactic acid amounts to 200 million pounds per year (Lipinsky and Sinclair 1986).

Both low and high molecular weight polylactic acids are known. High molecular weight polylactic acid possesses very good mechanical properties. Low molecular weight polylactic acids, when blended with other polymers, also exhibit characteristics of high molecular weight polymers. A detailed study on its preparation enzymatically was carried out (Scheme 5.1, Kiran and Divakar 2003).

5.2.1 Shake-Flask Level

Reactions carried out at an incubation temperature of 60°C by employing crystalline lactic acid dissolved in EMK in the presence of Lipozyme IM20, Chirazyme or PPL showed lesser yields, and the molecular weights were in the range 217–384 irrespective of divergent conditions employed. Transesterification reaction involving water of reaction was responsible for the reduced yields.

5.2.2 Bench-Scale Level

At bench-scale level, most of these drawbacks were overcome in terms of both extent of esterification and the molecular weight build-up (Table 5.1).

Lipozyme IM20: In the presence of Lipozyme IM20, higher E/S ratios of 4.39 and 9.21 AU mmol^{-1} resulted in 45.4 and 48.6% esterification with molecular weights of 411 and 459, respectively.

Chirazyme: The highest esterification of 66.9% at an E/S ratio of 0.79 AU mmol^{-1} with a molecular weight of 783 was obtained.

Porcine Pancreas Lipase: Increase in E/S ratio from 0.43 to 1.07 AU mmol^{-1} resulted in improvement of molecular weight from 809 to 1,128. Further increase in E/S ratio to 1.28 AU mmol^{-1} resulted in lower yield of 35.1% and lesser molecular weight (901).

A typical reaction profile of lactic acid polymerisation using PPL is shown in Fig. 5.1. The molecular weight build-up was found to follow the relation

$$E = 24.2 \log M - 2.74$$

where E = the percentage of esterification and M = the molecular weight. Figure 5.1 shows the theoretical plot for this equation and experimental points from Chirazyme and PPL-catalysed polymerisation reactions. These experimental points corresponded well with the theoretical line showing the validity of the prediction. During the initial stages of the reaction, decrease in acid content corresponded only to a small increase in

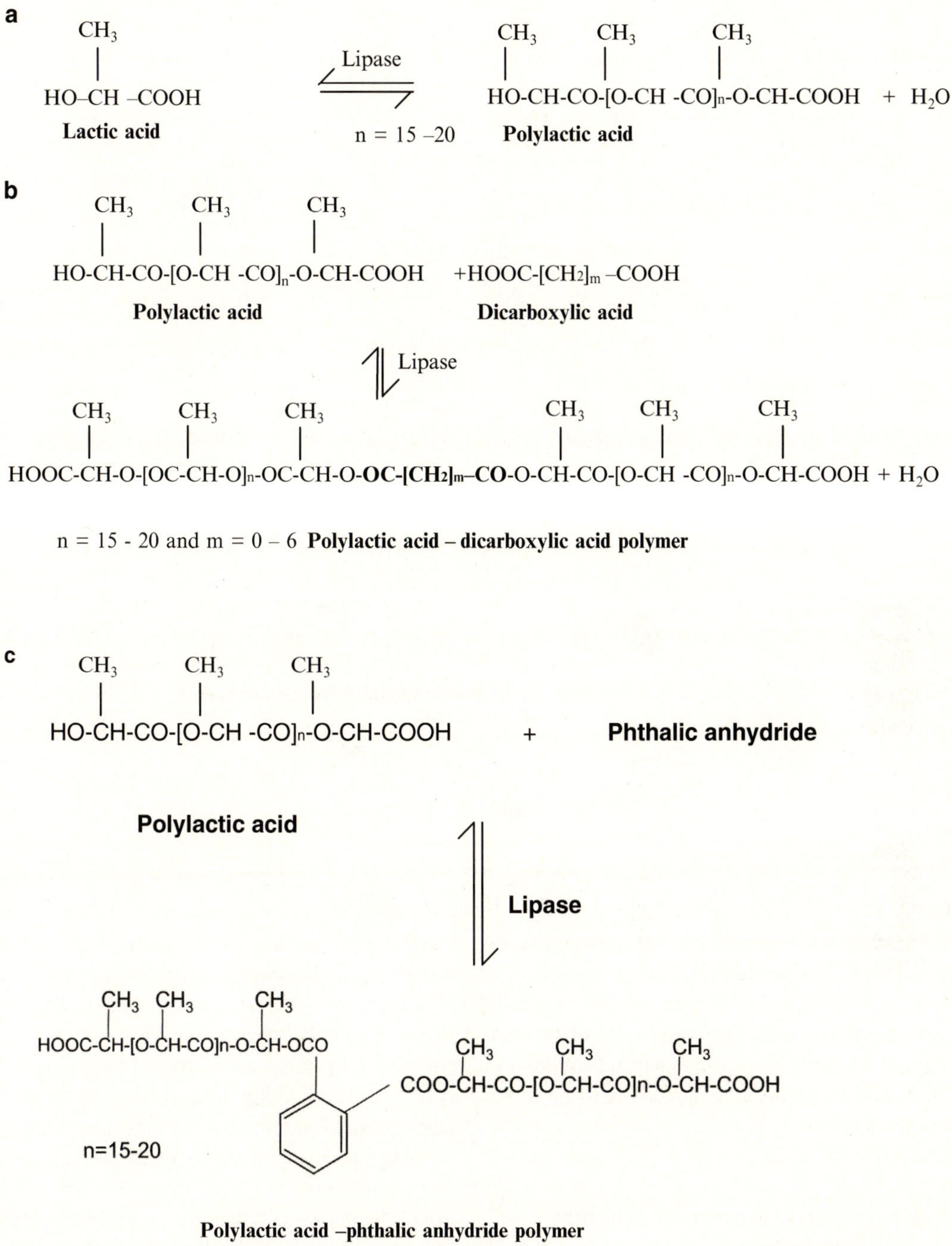

Scheme 5.1 Reactions depicting the strategy employed for the preparation of polylactic acid and its polymers with linker molecules

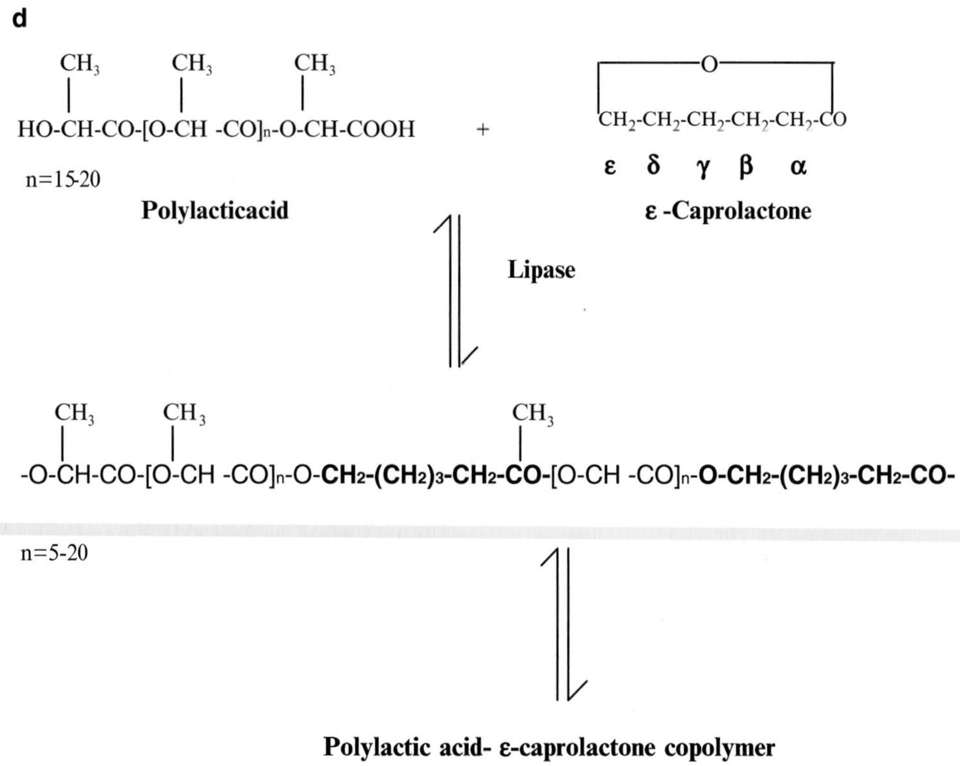

Scheme 5.1 (continued)

molecular weight due to predominant formation of dimers and trimers. However, above 85% esterification, a small increase in extent of esterification significantly improved the molecular weight since it corresponded to formation of an ester linkage between already existing oligomeric chains.

5.2.3 Further Molecular Weight Build-Up

A molecular weight of 1,423 probably might indicate the presence of 18 lactic acid monomers formed on an average from two polylactic acid chains of nine residues each. Lipases may not be able to bind longer chains, and hence, acyl transfer may not be possible beyond a nine-residue chain length leading to decrease in extent of esterification. In order to further enhance the molecular weight, polylactic acid chains were linked by other molecules. Several molecules containing hydroxyl and carboxyl functionalities could serve as linker molecules as polylactic acid chains contained a free hydroxyl and a carboxyl group at their terminal positions. Of them, diols could not be used, as they might inhibit the enzyme (Chaudhary and Qadri 1990). Dicarboxylic acids and anhydrides such as oxalic acid, succinic acid, succinic anhydride, adipic acid, malonic acid and phthalic anhydride at different molar equivalents in the range 0.1–1.0 (with respect to the polylactic acid concentration) were used along with enzymatically or chemically prepared polylactic acid of molecular weights 1,400–1,600 for the reactions.

Table 5.1 Data on polymerisation of lactic acid monomer by lipases at bench-scale level[a]

Lactic acid (mol)	E/S ratio[b] (AU mmol^{-1})	Max. esterification[c] (%) and incubation period (h)	Final esterification[c] (%) and incubation period (h)	Molecular weight[d]
	Lipozyme IM20			
0.056	1.76	35.4 (148)	35.4 (148)	417
0.056	4.39	45.4 (188)	45.4 (166)	411
0.056	9.21	48.6 (117)	48.6 (188)	459
	Chirazyme			
0.056	0.75	52.2 (145)	31.4 (240)	430
0.056	1.51	39.5 (115)	39.5 (115)	486
0.056	2.27	34.5 (125)	34.5 (125)	480
0.056[e]	0.82	46.5 (180)	46.5 (180)	447
0.055[f]	0.79	50.0 (165)	50.0 (165)	531
0.055[g]	0.79	66.9 (181)	62.5 (205)	783
0.055[g]	1.59	45.9 (162)	42.8 (186)	532
0.056[g]	2.25	59.3 (234)	22.2 (329)	833
0.056[h]	1.52	36.8 (138)	15.8 (207)	683
	PPL			
0.056	0.43	39.7 (141)	39.7 (165)	468
0.056[i]	0.43	15.6 (84)	15.6 (116)	705
0.056	0.9	46.2 (273)	41.9 (298)	799
0.056[g]	0.43	51.4 (211)	49.6 (234)	809
0.056[g]	0.86	74.4 (426)	65.8 (545)	1,040
0.056[g]	1.07	73.8 (402)	73.8 (470)	1,128
0.056[g]	1.28	35.1 (273)	35.1 (401)	901
0.056[j]	1.28	81.1 (387)	81.1 (423)	1,295
0.056[j]	0.43	79.9 (498)	79.9 (550)	1,300
0.556[j]	0.43	80.2 (507)	80.2 (545)	1,423

[a]Procedure described in Kiran and Divakar (2003)
[b]Enzyme/Substrate ratio
[c]Esterification at the end of reaction
[d]Esterification after product work out
[e]Methanol was added as initiator at 0.1 M equivalent
[f]Butanol was added as initiator at 0.1 M equivalent
[g]0.1-mL, 0.1-M pH 7 sodium phosphate buffer was added
[h]0.2-mL, 0.1-M pH 7 sodium phosphate buffer was added
[i]Reactions were carried out in benzene at 78°C
[j]Reactions were conducted in hexane–MIBK (7:1) mixture

5.2.3.1 Malonic Acid

The highest esterification of 36.2% and the highest molecular weight of 3,795 were observed with 1.0 M equivalent of malonic acid.

5.2.3.2 Succinic Acid

In case of succinic acid, maximum esterification observed was 66.7% at 0.1 M equivalent with a molecular weight of 2,685. Higher succinic acid equivalents gave less than 41% esterification molecular weights less than 1,830.

5.2.3.3 Adipic Acid

Use of 1.0 M equivalent of adipic acid resulted in the highest esterification of 19.9% with a molecular weight of 1,734.

5.2.3.4 Phthalic Anhydride

Polylactic acid of molecular weight 1,600 gave the maximum esterification of 62.3% with 0.1 M equivalent, the molecular weight being 2,685. Maximum molecular weight of 1,980 was observed with 20% esterification with 0.2 equivalent.

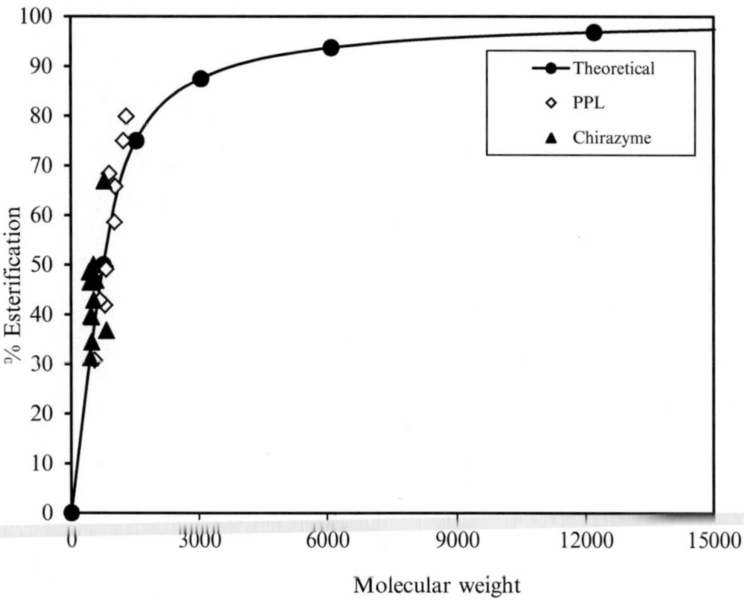

Fig. 5.1 Theoretical plot for prediction of molecular weight from extent of esterification along with experimental points. Percentage of esterification = 24.2 log M − 2.74

Table 5.2 Lipase-catalysed synthesis of polymers of lactic acid using linker succinic anhydride at bench-scale level[a]

Polylactic acid concentration	Max. esterification[b] (%) and incubation period (h)	Succinic anhydride (molar equivalents)	Molecular weight[c]
0.055[d]	79.8 (545)	0.1	3,102
0.055[e]	80.1 (544)	0.1	3,300
0.057[e]	42.7 (500)	0.2	3,086
0.055[e]	38.4 (552)	0.5	2,439
0.055[f]	2.2 (374)	0.1	-

[a]PPL was employed at an E/S ratio of 0.43 AU mmol^{-1}; activity units were with respect to esterification activity units; solvent: hexane, MIBK, chloroform mixture
[b]Error in extent of esterification was ±5–10%
[c]Molecular weight was determined by end-group analysis (Gowariker et al. 1992)
[d]Enzymatically prepared 1,300-molecular weight polylactic acid was employed
[e]Chemically prepared 1,600-molecular weight polylactic acid was employed
[f]Chemically prepared 4,200 polylactic acid was employed

5.2.3.5 Succinic Anhydride

Enzymatically prepared polylactic acid of molecular weight 1,300 used along with 0.1 M equivalent of succinic anhydride resulted in a maximum esterification of 79.8% after 545 h with a molecular weight of 3,102. A maximum yield of 80.1% was observed with 0.1 M equivalent of succinic anhydride with chemically prepared polylactic acid (1,600), giving a molecular weight of 3,300 (Table 5.2).

5.2.4 Nuclear Magnetic Resonance Spectroscopy

Polylactic acids prepared through lipase catalysis were characterised by ^{1}H and ^{13}C NMR.

Molecular weight of the polylactic acid was determined through ^{1}H NMR. The number of repeat units was determined from the ratios of areas of the signals of –CH-O- corresponding to the polymer

Table 5.3 Data on properties of polylactic acid films

Films	Tensile strength[a] (psi)	(%) Elongation	Optical properties (%) Transmittance	(%) Haze	Heat sealability
E-1400[b] 50%+PS[c] 50%	186	1.3	4.4	94.7	Yes
E-3300[d] 50%+PS[c] 50%	160	8.4	4.9	91.5	Yes
PS[c] 100%	341	0.8	84.0	5.7	Yes

[a]Tensile strength was measured at 20°C at an R_H of 55%
[b]Enzymatically prepared polylactic acid of molecular weight 1,400
[c]Polystyrene of molecular weight 2×10^5
[d]Enzymatically prepared polylactic acid–succinic anhydride polymer of molecular weight 3,300

ester chain (5.08–5.15 ppm) and those corresponding to the –CH-O- from the hydroxyl terminal of the polymer chain (4.35 ppm). The enzymatically prepared polylactic acid with molecular weight of 1,300 gave a molecular weight of 288 by ^{1}H NMR. Molecular weights determined by viscometry and ^{1}H NMR were found to show good agreement. Both viscometry and ^{1}H NMR gave weight average molecular weight, M_w, whereas end-group analysis gave number average molecular weight, M_n.

5.2.4.1 ^{13}C NMR

Three signals observed at 15.1, 20.4 and 20.5 ppm, respectively, corresponded to –CH_3 signals from polylactic acid chain, hydroxyl terminal of polylactic acid and from free lactic acid, respectively. Two downfield signals at 65.0 and 65.2 ppm corresponded to those of free lactic acid and hydroxyl terminal of the polylactic acid, respectively. Those at 69.6 and 70.3 ppm corresponded to ester –CH-O- of polylactic acid and carboxyl terminal of polylactic acid chain, respectively. The ester –CH-O- region at 69.6 ppm was a huge signal compared to the others, indicating larger extent of polymerisation. In the carbonyl carbon region, while 178.8- and 175.3-ppm signals corresponded to free lactic acid and hydroxyl terminal -CO- of polylactic acid respectively, those at 174.1, 173.8, 170.4 and 170.2 ppm corresponded to carbonyl carbon signals from various polymeric species.

5.2.5 Polylactic Acid Films

It was found that polylactic acid of molecular weights 400–3,300 could be cast into films only when blended with polystyrene of molecular weight 2×10^5. Various proportions of polystyrene were mixed with polylactic acid in chloroform and cast into films. For polystyrene content less than 50%, mixing with plasticisers up to 10% was required to cast a film with polylactic acid (1,600). Polylactic acids with molecular weight of 3,300 and above formed films with polystyrene without the addition of the plasticiser. However, addition of plasticiser slightly improved the mechanical properties of the film.

Polylactic acid films prepared were characterised in terms of their tensile strength, percentage elongation and optical properties like transmittance and haze (Table 5.3). Tensile strength of 1:1 blends of enzymatically prepared polylactic acids (1,400 and 3,300) with polystyrene was better (186 and 160 psi) than those of chemically prepared polylactic acids with molecular weights of 1,600 and 4,200 (157 and 128 psi, respectively). Compared to 100% polystyrene film, all the other blends of polylactic acid with polystyrene showed better elongation of around 1.5–2.0%. Higher elongation of 8.4% was observed with a blend of enzymatically prepared polylactic acid–succinic anhydride polymer (3,300). Due to non-transparent nature, all the blended films showed low transmittance (<5.0%) and high haze (>90%). Films were also tested for heat sealability. All the blended films were found to be heat sealable.

Of all the films tested, a 1:1 blend of enzymatically prepared polylactic acid–succinic anhydride polymer (3,300) and polystyrene (2×10^5) showed the highest elongation of 8.4%, higher transmittance of 4.9%, comparatively lesser haze of 91.5% and good tensile strength

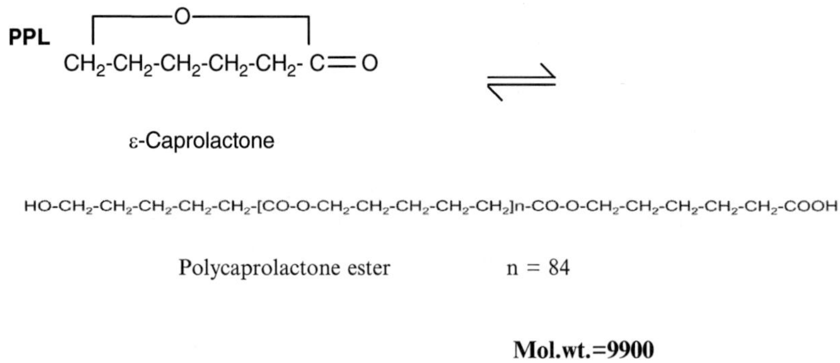

Scheme 5.2 Preparation of polycaprolactone ester

(160 psi) than the other enzymatically prepared polymer blends. However, for a good film, these characteristics had to be improved further.

Use of linker molecules led to various possibilities in the formation of the polymer chain. The acyl transfer reaction by the enzyme might aid in the monomer dicarboxylic acid being attached to one end of the polylactic acid chain. This resulted in a free carboxylic acid terminal on one polylactic acid chain, which facilitated reaction with the hydroxyl terminal of another polylactic acid chain. This reaction, although difficult, had occurred in case of malonic acid and succinic anhydride resulting in polymers of molecular weights of 3,795 and 3,300, respectively.

It is generally recognised that polymerisation of free hydroxy acids by enzymatic means is difficult. In case of enzymatic polymerisation of dicarboxylic acids like adipic acid or vinyl adipate with diols like 1,4-butanediol, molecular weights achieved were less than 2,000 (Seymour and Carrea 1984). In case of small- and medium-sized lactones and macrolides, polymerisation has been shown to be easier due to higher ring strain (Kobayashi et al. 1998). Molecular weights of the order of 25,000 were, in fact, achieved with higher macrolides (Chaudhary and Qadri 1990). Inoue (1996) has studied the chemical ring opening polymerisation of a large number of ring systems with and without initiators. They have generally observed that while six, seven or even higher-membered lactone rings can be easily polymerised, the chemical ring opening polymerisation of five membered lactones has never been successful even with initiators.

5.3 Poly-ε-caprolactone

A typical polymerisation with ε-caprolactone is depicted in Scheme 5.2 (Divakar 2004).

5.3.1 Shake-Flask- and Bench-Scale-Level Experiments

At shake-flask level, the yields were lesser, and the molecular weight build-up was also low which were overcome at bench-scale level. The experimental set-up employed has been described before (Divakar et al. 1999). Table 5.4 shows the results.

A typical reaction profile of ε-caprolactone polymerisation using PPL is shown in Fig. 5.2. Polycaprolactone esters prepared showed a maximum molecular weight of 11,004 by end-group analysis. This was much higher than what was observed with several monomers attempted so far. However, gel permeation chromatography showed a molecular weight of 3,448 for the same polymer.

A molecular weight of 10,000 probably might indicate the presence of 84 ε-caprolactone monomers formed on an average from two polycaprolactone ester chains of 42 residues each. This indicates limitation exhibited by lipase on the

Table 5.4 Data on polymerisation of ε-caprolactone by porcine pancreas lipase[a]

System (mole)	PPL g	Yield %	Period of incubation	Molecular weight[b]
ε - Caprolactone (0.26)	3	75.9	285	11,004 (3448)[f]
ε - Caprolactone + Succinic anh. (0.03) (0.006)	0.5	59	620	7,696
ε - Caprolactone + Malonic acid (0.33) (0.003)	0.5	59.2	370	8,700
ε - Caprolactone + Lactic acid[d] (0.03) (0.03)	0.5	31.3	518	6,197
ε - Caprolactone + Polylactic acid[e] (0.03) (0.06)	0.5	53.6	258	8,333
ε - Caprolactone + caprolactam (0.03) (0.03)	0.5	33.6	445	2,847
Polycaprolactone + Adipic acid Ester (0.04) (0.004)	0.5	22.2	291	9,769
Polycaprolactone + Erythritol Ester (0.04) (0.004)	0.5	63.5	744	7,162
Polycaprolactone + Erythritol Ester (0.04) (0.008)	0.5	76.2	216	8,222

[a]Solvent 150-ml benzene with 10-ml pyridine in some cases. Error in molecular weights will be ±5–10%
[b]Molecular weight determined by end-group analysis
[c]Polycaprolactone ester of molecular weight 11,004
[d]Crystalline lactic acid containing 4% water
[e]Polylactic acid of molecular weight 1,400
[f]Molecular weight determined by gel permeation chromatography

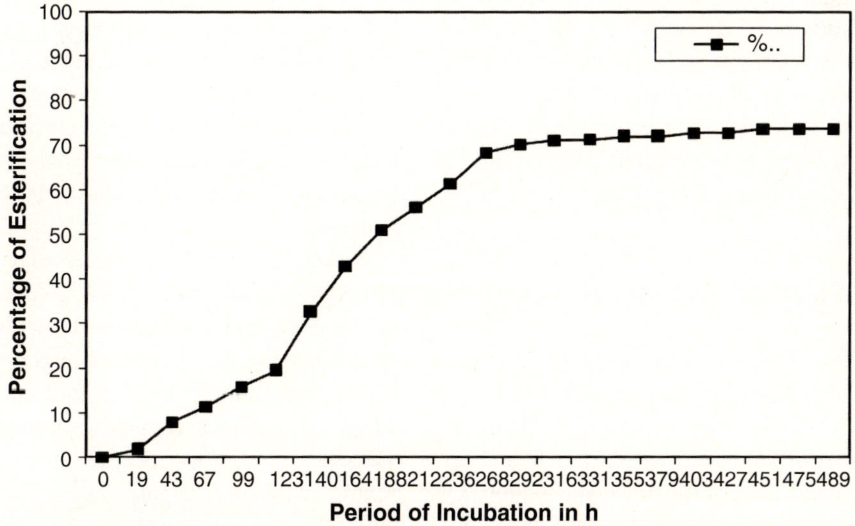

Fig. 5.2 Reaction profile of ε-caprolactone polymerisation by PPL. Reaction conditions: ε-caprolactone (0.26 mole); PPL: 3 g; solvent: benzene 150 mL

length of acyl groups transferred. In order to further enhance the molecular weight, polycaprolactone ester chains could be linked to other molecules like succinic anhydride, adipic acid, malonic acid, glycolic acid, lactic acid, caprolactam and erythritol and polylactides.

These molecules did not result increase in molecular weight. However, few monomers like caprolactam, glycolic acid polylactide and higher equivalents of erythritol showed decrease in molecular weights. Since most of the monomers employed were capable of efficient acyl transfer, they did not add to the terminal groups of the polymer chain to enhance further molecular weight build-up probably due to steric factors arising out of binding of long polymer chains as acyl acceptors in the active sites. Instead, the smaller molecules can effect transestrification reactions (acidolysis) leading to reduction in molecular weights.

5.3.2 Nuclear Magnetic Resonance Spectroscopy

Polycaprolactone esters prepared through lipase catalysis were characterised by ^{1}H and ^{13}C NMR. In the ^{1}H NMR spectrum, ε-caprolactone showed signals for 2-CH_2 at 2.58 ppm, 3,5-CH_2 at 1.86 ppm, 4-CH_2 at 1.60 ppm and 6-CH_2 at 4.32 ppm. On polymerisation, no monomer signals could be observed. Similarly, in ^{13}C NMR, monomer signals on polymerisation showed upfield shifts. C-1 signal showed shift due to conversion from lactone to ester. Similarly, other signals including C-6 showed upfield shifts due to polymerisation.

5.3.3 Polycaprolactone Ester Films

It was found that polycaprolactone esters of molecular weights in the range 4,000–11,000 could be cast into films only when blended with polystyrene of molecular weight 2×10^5 and cellulose acetate of molecular weight 39,000. On their own, they could not be cast into films. Various proportions of polystyrene and cellulose acetate were mixed with polycaprolactone ester in chloroform and cast into films and tested (Table 5.5).

Polycaprolactone ester copolymerised with polylactic acid (molecular weight 1,400) when blended with cellulose acetate (1:1) showed the highest tensile strength of 1,383 psi. This blend showed also better percentage elongation (2.59), percentage haze (5.1), percentage transmittance (83.9) and burst strength (0.527 kg cm^{-2}). Most of polycaprolactone ester blends were heat sealable and showed good heat seal strength

These studies clearly show that lipase-catalysed polymerisation of ε-caprolactone possesses good potential to be developed as a biodegradable polymer.

5.4 Poly-*p*-hydroxybenzoate

Liquid crystal polymers (LCP) are basically aromatic polyesters. They have crystalline structure both in molten and in the solid state. They find wide applications in the field of electronics. The most frequently used basic monomers in conventional LCP are hydroxybenzoic acid, aminobenzoic acid, hydroxynaphthoic acid, terephthalic acid, isophthalic acid and bisphenol A. Poly-*p*-hydroxybenzoate and its copolymers are used for printed circuits, wire coating, in dyes and plastics, as hot-melt adhesive to bond PVC sheets, to improve the adhesion to rubber, in transparent bottles, as an interior part in automobile or airplane and in films. Poly-*p*-benzamides and its copolymers are used as fibres, electrical insulation, in cosmetics and in films (Huang et al. 1992; Taesler et al. 1996; Kumar et al. 1996; Preston et al. 1992).

Porcine pancreas lipase-catalysed polymerisation of *p*-hydroxybenzoic (PHBA) acid monomer and its copolymers with hydroxy acids and lactones (Scheme 5.3) showed a highest yield of 5.2% with a number average molecular weight (M_n) of 184 was observed for the homo-oligomer (Divakar 2003). This showed that the reaction mixture contained largely unreacted monomer and its dimer, the former being the major component. With the increase in PPL amount, there was no molecular weight build-up.

Table 5.5 Properties of films formed with polycaprolactone ester blends

Samples[a]	Avg., max and min thickness (microns)	Tensile strength (psi)	Percentage elongation	Percentage haze	Percentage transmittance	Burst strength (kg cm^{-2})	Heat sealability	Heat seal strength (psi)
Polystyrene[b] (5%)	–	341	0.836	5.7	84.0	–	Yes	–
Poly (ε -CL–poly LA 1400), mol. wt. 8300+polystyrene[b] (1:1, 10%)	45.0, 80, 30	341	0.855	100	1.0	0.070	Yes	235.2
Poly (ε-CL–crystalline LA), mol. wt. 6200+polystyrene[b] (1:1, 10%)	52.0, 100, 25	190	1.038	100	0.7	1.054	Yes	88.4
Poly (ε-CL), mol. wt. 9900 +polystyrene[b] (1:1, 10%)	42.7, 60, 25	210	2.200	100	1.1	0.070	Yes	209.7
Cellulose acetate[c] (5%)	32.5, 95, 15	38	1.372	14.9	70.1	0.158	Yes	–
Poly (ε-CL–poly LA 1400), mol. wt. 8300+cellulose acetate[c] (1:1, 20%)	61.0, 90, 30	1,383	2.591	5.1	83.9	0.527	Yes	145.1
Poly (ε-CL–crystalline LA), mol. wt. 6200+cellulose acetate[c] (1:1, 10%)	126.0, 330, 30	80	1.931	3.1	85.5	0.141	Yes	45.6

LA Lactic acid, *ε-CL* ε-caprolactone; all molecular weight values determined by end-group analysis

[a]All samples were prepared enzymatically. Films were wet cast after dissolving in chloroform and were conditioned at 27°C and 65% R_H

[b]Molecular weight 2×10^5

[c]Molecular weight 39,000 (GPC), 33% acetylated

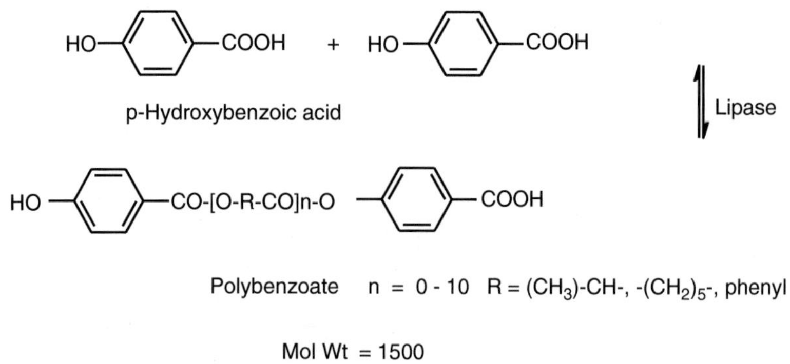

Scheme 5.3 Lipase-catalysed polymerisation of p-hydroxybenzoic acid

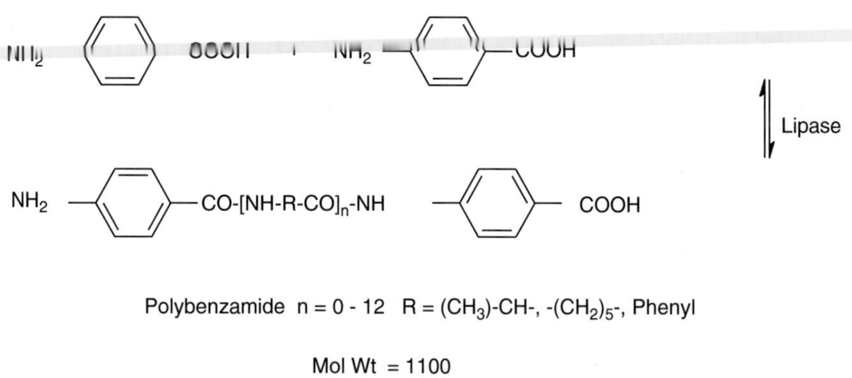

Scheme 5.4 Lipase-catalysed polymerisation of p-aminobenzoic acid

In order to improve the molecular weight, PHBA was reacted with different monomers like lactic acid, glycolic acid and ε-caprolactone in presence of PPL. Lactic acid exhibited a conversion of 65.1% with M_n 195 indicating again a mixture of unreacted monomers and probably their esters. Copolymerisation with glycolic acid was studied in different solvents at different ratios of the monomers. A maximum esterification of 60% with M_n 497 was obtained in 1:4 MIBK–hexane (v/v) system which was found to be the best solvent system for this reaction. However, the highest M_n of 525 was observed when 1:2 PHBA and glycolic acid were employed. Such low M_n values indicate only mixture of oligomers.

ε-Caprolactone, as a comonomer with PHBA gave higher M_n than the other monomers. Thus, PHBA and ε-caprolactone showed 60.4% esterification with M_n 1,486. Also, a 1:2 M equivalent of PHBA and ε-caprolactone showed 65.2% yield with M_n 1,398. Increase in ε-caprolactone content did not aid in further molecular weight build-up. Thus, ε-caprolactone along with PHBA showed greater propensity for copolymerisation.

5.5 Poly-*p*-benzamide

A homo-oligomer of *p*-aminobenzoic acid was prepared (Scheme 5.4) showing a conversion of 26.4% with M_n of 778 (Divakar 2003). With increase in incubation period, percentage conversion increased with molecular weight. In order to increase the molecular weight further, *p*-aminobenzoic acid was polymerised with lactic acid and ε-caprolactone. *p*-Aminobenzoic acid–lactic acid oligomer showed M_n 1,049 after 590 h at 52.6% conversion, and *p*-aminobenzoic acid–ε-caprolactone oligomer showed M_n 706 after 494 h at 36.1% conversion. Copolymer reaction with lactic acid showed the highest conversion yield and molecular weight probably due to smaller molecular size of lactic acid acting as both acyl acceptor and donor. The copolymerisation potentiality of *p*-aminobenzoic acid was found to be slightly better than that of PHBA although the observed low M_n values indicated only oligomer formation.

It was found that the oligomers of PHBA with ε-caprolactone (M_n) prepared enzymatically could not be cast into films as such even when a 10% solution in $CHCl_3$ was employed. But when blended with polystyrene (molecular weight – 2×10^5), soya protein (molecular weight – 30,000) and cellulose acetate (molecular weight – 39,000), films were obtained. Low molecular weight polymers of *p*-aminobenzoic acid (700–1,500) could not be cast into films even after blending in varying proportions with polystyrene and cellulose acetate. This may be probably because the hydrophobic interactions between the benzene ring of PHBA copolymer and polystyrene or cellulose acetate may be stronger than that of *p*-aminobenzoic acid.

Enzymatic methods generally have shown low molecular weight polymers even with monomers like lactones (Gross et al. 2001: Kumar and Gross 2000; Svirkin et al. 1996: Namekawa et al. 1996). The reason for obtaining low molecular weight copolymers by enzymatic methods in case of PHBA and ε-caprolactone and *p*-aminobenzoic acid and lactic acid (as in case of lactones also) is probably due to strong binding of especially the aromatic monomers to the active site of porcine lipase preventing facile acyl transfer to the hydroxyl or amino groups of the other monomers. Large amounts of enzymes than those employed in this work may probably be required for preparing higher molecular weight polymers. A better approach would be to derivatise the monomers to suitably activate the functional groups to attain higher molecular weight polymers.

5.6 Polyadipates

Adipic acid polymers (polyadipates) have recently been found to possess several desirable characteristics necessary for a wide variety of applications (Inoue 1996; Binns et al. 1994; Sandez-Adsuar and Martin-Martiz 2000; Bartz and Roehe 2003; Muhlfeld and Wagener 2000). Reports have shown that even low molecular weight polymers (molecular weight around 2,000) of polyadipates can be used along with polyurethane and other such polymers to make automobile interior parts, hot-melt adhesives for textiles, moulding compositions, packaging materials and medical goods to name a few (Bohl et al. 2000). A reaction between adipic acid and 1,6-hexanediol (Scheme 5.5) showed minimum yield of 33.7% with a number average molecular weight of 1,265. The highest number average molecular weight of 2,396 was obtained with 1,6-hexanediol. The reaction with 1,4-butanediol showed 55.5% conversion after 568 h, with a number average molecular weight of 1,645. Reaction between adipic acid and ethylene glycol showed 36.5% conversion with a number average molecular weight of 2,173 after 568 h. Although this reaction showed lower conversion, higher molecular weight polymers were obtained, when compared to reactions with the other diols (Table 5.6).

Other aromatic dicarboxylic acids like terephthalic acid and isophthalic acids were also reacted with ethylene glycol. In order to dissolve terephthalic acid and isophthalic acid, reactions were conducted with 10-mL pyridine and toluene–hexane (30:60) solvent mixture. Reactions were incubated for a period of 20–27 days. Reaction between ethylene glycol and terephthalic acid showed 36.6% conversion with a number average molecular weight of 927 after 666 h whereas reaction with isophthalic acid showed a conversion of 61.3% and a number average molecular weight of 1,488 after 735 h.

This study has shown that low molecular weight polyadipates of molecular weights 2,400 can be produced through lipase catalysis in

Table 5.6 Data on enzymatically prepared polyadipates and glycolates

Monomers (mole)		PPL (mg)	Solvent (ml)	(%) Yield (time – h)	Molecular weight
Adipic acid + 1,6-hexanediol		500	Benzene to pyridine	59 (568)	1,645
(0.01)	(0.01)		(9:1)		
Adipic acid + 1,6-hexanediol		500	EMK – 25	33.7 (473)	1,267
(0.01)	(0.01)		toluene to hexane (25:50)		
Adipic acid + 1,6-hexanediol		500	Benzene to pyridine	92 (500)	2,396
(0.025)	(0.025)		(9:1)		
Adipic acid + 1,6-hexanediol		500	Benzene to pyridine	65 (387)	570
(0.035)	(0.035)		(9:1)		
Adipic acid + 1,6-hexanediol		500	Benzene to pyridine	63 (360)	567
(0.05)	(0.05)		(9:1)		
Adipic acid + 1,4-butanediol		500	Benzene to pyridine	65 (744)	547
(0.025)	(0.025)		(9:1)		
Adipic acid + 1,4-butanediol		500	EMK – 25	55.5 (568)	1,645
(0.01)	(0.01)		toluene to hexane (25:50)		
Adipic acid + ethylene glycol		500	Benzene to pyridine	82 (620)	467
(0.025)	(0.025)		(9:1)		
Adipic acid + ethylene glycol		500	Benzene to pyridine	64 (616)	754
(0.025)	(0.025)		(9:1)		
Adipic acid + ethylene glycol		500	EMK – 25	36.5 (568)	2,173
(0.01)	(0.01)		toluene to hexane (25:50)		
Ethylene glycol + terephthalic acid		500	Pyridine – 10	36.6 (666)	927
(0.01)	(0.01)		toluene to hexane (30:60)		
Ethylene glycol + isophthalic acid		500	Pyridine – 10	61.3 (735)	1,488
(0.01)	(0.01)		toluene to hexane (30:60)		

EMK Ethyl methyl ketone. Error in molecular weights will be ±10%

Scheme 5.5 Lipase-catalysed polymerisation of adipic acid and 1,4-hexanediol

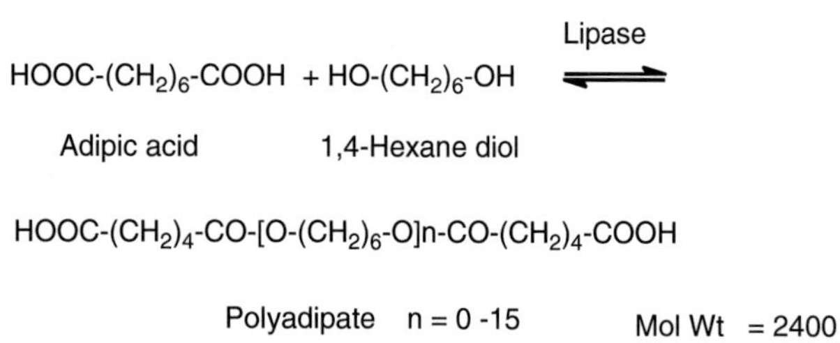

organic media. Even with purified monomers, higher molecular weights could not be obtained. Further increase in molecular weight may not be possible because of limitations in acyl transfer by the enzyme due to increase in chain length. However, activation of -COOH and OH groups can yield higher molecular weight polyadipate copolymers.

References

Bartz T, Roehe P (2003) Ger Offen DE 19854404. A1, 31 May 2000, p 6

Binns F, Roberts SM, Taylor A, Williams CF (1994) Studies leading to the large scale synthesis of polyesters using enzymes. J Chem Soc Perkin Trans 1:899–904

Bisht KS, Deng F, Gross RA, Kaplan DL, Swift G (1998) Ethyl glucoside as a multifunctional initiator for enzyme-catalyzed regio-selective lactone ring-opening polymerization. J Am Chem Soc 120:1363–1367

Bohl W, Partusch G, Fleishmann S (2000) Ger Offen DE 19928608 A1, 28 Dec

Chaudhary AN, Qadri RB (1990) Higher fatty acid esters of lactic acid. Pak J Ind Res 33:334–336

Divakar S (2003) Porcine pancreas lipase catalysed preparation of oligomers of p-hydroxybenzoic acid and p-aminobenzoic acid. Ind J Chem Sect B 42B:1467–1470

Divakar S (2004) Porcine pancreas lipase catalysed ring-opening polymerization of ε-caprolactone. J Macromol Sci Part A Pure Appl Chem A41(5):537–546

Divakar S, Kiran KR, Harikrishna S and Karanth NG (1999) An improved process for the preparation of esters of organic acids and alcohols. Indian Patent, 1243/DEL/99 No. 191078

Gowariker VR, Viswanathan NV, Sreedhar J (1992) Polymer science. Wiley Eastern Ltd., New Delhi

Gross RA, Kumar A, Kalra B (2001) Polymer synthesis by in vitro enzyme catalysis. Chem Rev 101(7):2097–2124

Huang K, Lin YG, Winter HH (1992) p-Hydroxy benzoate/ethylene terephthalate copolyester: structure of high-melting crystals formed during partially molten state annealing. Polymer 33:4533–4537

Inoue A (1996) JPN Kokai Tokkyo Koho Jp 08059847, A25, Mar 1996, Heirei 5, Japan (CA)

Kiran KR, Divakar S (2003) Lipase catalyzed polymerization of lactic acid and its film forming properties. World J Microbiol Biotechnol 19:859–865

Knani D, Gutman AL, Kohn DH (1993) Enzymatic polyesterification in organic media enzyme – catalyzed synthesis of linear polyesters. 1. Condensation polymerization of linear hydroxyesters. J Polym Sci Part A Polym Chem 31:1221–1232

Kobayashi S, Uyama H, Namckawa S, Hayakawa H (1998) Enzymatic ring-opening polymerization and copolymerization of 8-octanolide by lipase catalyst. Macromolecules 31:5655–5659

Kumar A, Gross RA (2000) Candida antarctica lipase B-catalyzed transesterification: new synthetic routes to copolyesters. J Am Chem Soc 122:11767–11770

Kumar A, Kumar N, Parmar V (1996) Preparative and mechanistic aspects of interesterification reactions on diols and peracetylated polyphenolic compounds catalysed by lipases. Pure Appl Chem 68:749–752

Lipinsky ES, Sinclair RG (1986) Is lactic acid a commodity chemical? Chem Eng Progr 82:26–32

Muhlfeld H, Wagener S (2000) Eur Pat Appl EP 1043349, A2, 11 Oct 2000, 5 pp

Namekawa S, Uyama H, Kobayashi S (1996) Lipase-catalyzed ring-opening polymerization and copolymerization of β-propiolactone. Polym J 28:730–731

Preston J, Kotek R, Krigbaum WR (1992) Macromol Synth 11:27

Rika M, Tadaki S, Kayoko Y (1996) Jpn Kokkai Tokkyo Koho JP 08:73,573

Sandez-Adsuar MS, Martin-Martiz JM (2000) J Adhes Sci Technol 14:8

Seiji O, Masahiro O (1992) Jpn Kokai Tokkyo Koho JP 06:172502

Seymour RB, Carrea CE Jr (1984) Structure relationship in polymers. Plannum Press, New York, pp 55–69

Svirkin YY, Xu J, Gross RA, Kaplan DL, Swift G (1996) Enzyme catalyzed stereo-selective ring-opening polymerization of α-methyl-β-propiolactone. Macromolecules 29:4591–4597

Taesler C, Wittich H, Jourgen C, Schulte K, Kricheldorf HR (1996) J Appl Polym Sci 61:783–792

Uyama H, Kobayashi S (1996) Enzymatic ring-opening polymerization of macrolides to polyesters. Front Biomed Biotechnol 3:5–15

Voss W, Spielan P (1958) German (East) Patent 14, 548

Zhang H, Deng X, Huang Z (1996) Ring-opening polymerization of ε-caprolactone by bacterial protease. Biotechnol Lett 18:1051–1054

Lipase-Catalysed Preparation of Aminoacyl Esters of Carbohydrates

6

Abstract

The esterification potentialities of lipases from *Rhizomucor miehei*, *Candida rugosa* and porcine pancreas (PPL) are discussed in detail with respect to the syntheses of L-alanyl, L-valyl, L-leucyl, L-isoleucyl, L-prolyl, L-phenylalanyl, L-tryptophanyl and L-histidyl esters of representative carbohydrates like hexo-pyranoses (D-glucose, D-galactose, D-mannose), ketose (D-fructose), pento-furanoses (D-arabinose, D-ribose), disaccharides (lactose, maltose, sucrose) and sugar alcohols (D-sorbitol, D-mannitol), using unprotected and unactivated amino acids and carbohydrates. Lipase-catalysed esterification reactions of L-alanyl-D-glucose, L-valyl-D-glucose, L-leucyl-D-glucose, L-phenylalanyl-D-glucose and L-phenylalanyl-lactose were optimised in terms of incubation period, solvents, lipase concentrations, substrate concentrations, buffer pH and its concentrations and lipase reusability. Spectroscopic investigations discuss the regioselectivity of the product esters formed. The relative merits of the three lipases under the reaction conditions employed, in terms of incubation period, substrate and enzyme concentrations, solvent, pH and buffer concentrations, reusability leading to the observed yield and nature and types of products formed, are discussed in detail.

6.1 Introduction

Carbohydrates are information-rich molecules, which are suited well for modification to new types of compounds with expected biological activity (Hurtley et al. 2001). Compared to fatty acid esters of carbohydrates, aminoacyl esters of carbohydrates can give added additional functionality to the side chain. Presence of hydroxyl as well as amine groups in the molecule helps in the polycondensation reactions (Park et al. 1999). Recently Shiraki et al. (2004) reported that aminoacyl esters prevent thermal inactivation and aggregation of lysozyme.

Aminoacyl esters of carbohydrates are used as sweetening agents, surfactants, microcapsules in pharmaceutical preparations, active nucleoside amino acid esters, antibiotics and in the delivery of biological active agents (Dordick 1989; Tamura et al. 1985; Kirk et al. 1992; Zaks and Dodds 1997). Chemical acylation of carbohydrates regioselectively is complex due to the presence of multiple hydroxyl groups, which require protection and deprotection (Tamura et al. 1985; Haines 1981). Use of lipases in the synthesis of sugar esters is industrially important due to regio- and stereoselectivity

S. Divakar, *Enzymatic Transformation*,
DOI 10.1007/978-81-322-0873-0_6, © Springer India 2013

Table 6.4 (continued)

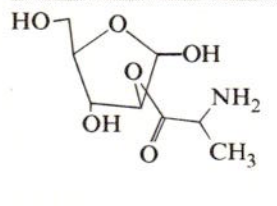 2-*O*-L-alanyl-D-arabinose (33)	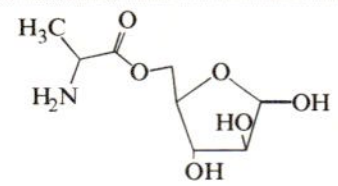 5-*O*-L-alanyl-D-arabinose(34)	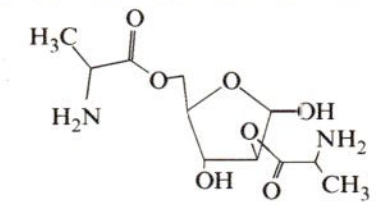 2,5-di-*O*-L-alanyl-D-arabinose (33)	9[c] (monoesters-6, diester-3)
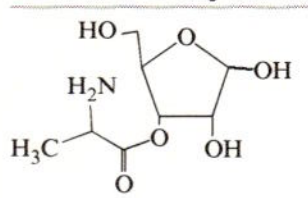 3-*O*-L-alanyl-D-ribose (16)	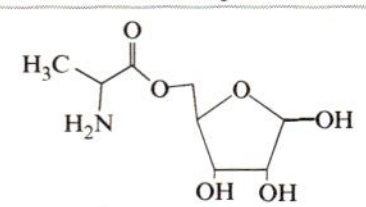 5-*O*-L-alanyl-D-ribose	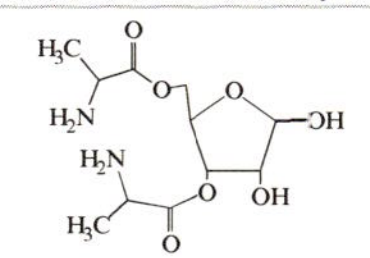 3,5-di-*O*-L-alanyl-D-ribose (52)	48 (monoesters-23, diester-25)
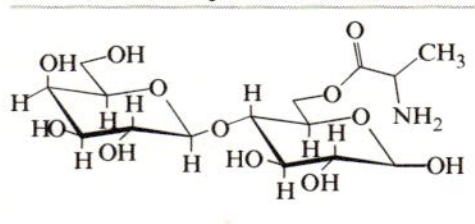 6-*O*-L-alanyl-lactose (34)	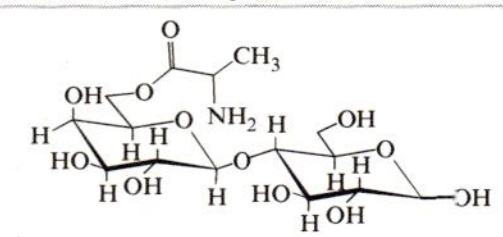 6′-*O*-L-alanyl-lactose (34)	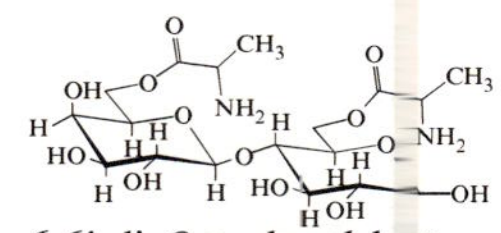 6,6′-di-*O*-L-alanyl-lactose (32)	20 (monoesters-14, diester-6)
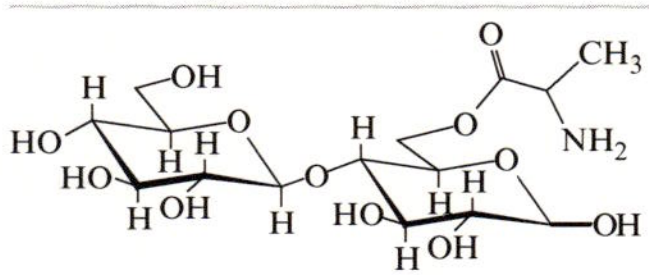 6-*O*-L-alanyl-maltose (34)	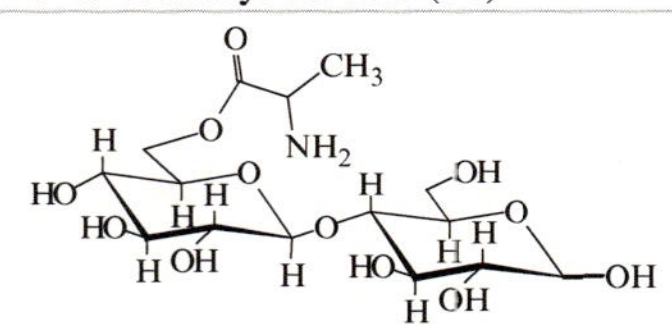 6′-*O*-L-alanyl-maltose (34)	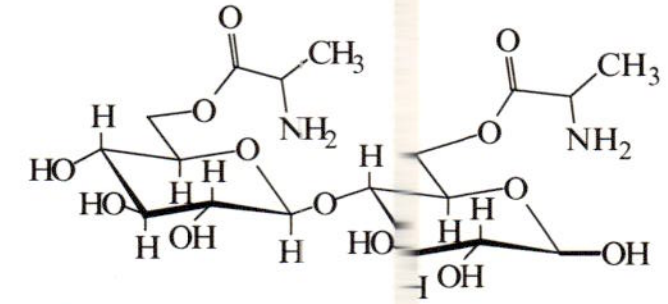 6,6′-di-*O*-L-alanyl-maltose (32)	56 (monoesters-38, diester-18)
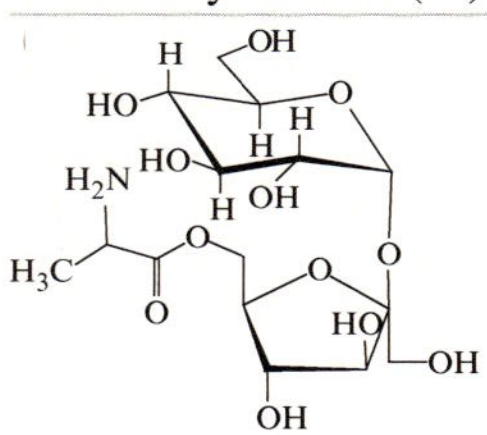 6-*O*-L-alanyl-sucrose			8 (only monoester)

[a]L-Alanine, 2 mmol; carbohydrates, 1 mmol; RML, 40% (w/w carbohydrate); buffer, 0.1-mM (0.1 mL of 0.1 M) acetate buffer (pH 4.0); CH_2Cl_2:DMF (v/v 90:10) at 40°C; incubation period, 72 h. Conversion yields were from HPLC with respect to L-alanine concentration

[b]Percentage proportions of individual esters were determined from the peak areas or from their cross peaks of the carbon-13 C6 and C5 (in case of pentoses) signals in the 2-D HSQCT spectrum

Table 6.5 Preparation of L-alanyl ester of carbohydrate using lipases from *Candida rugosa* and porcine pancreas

L-Alanyl ester of carbohydrate	CRL[a] % yield (mmol)	Crude PPL[b] % yield (mmol)
L-Alanyl-D-glucose	25 (0.50)	78 (0.78)
L-Alanyl-D-galactose	22 (0.44)	72 (0.72)
L-Alanyl-D-mannose	33 (0.66)	10 (0.10)
L-Alanyl-D-fructose	12 (0.24)	67 (0.67)
L-Alanyl-D-arabinose	3 (0.06)	27 (0.27)
L-Alanyl-D-ribose	12 (0.24)	38 (0.38)
L-Alanyl-lactose	27 (0.54)	58 (0.58)
L-Alanyl-maltose	17 (0.34)	28 (0.28)
L-Alanyl-sucrose	8 (0.16)	8 (0.83)

[a]L-Alanine, 2 mmol; carbohydrates, 1 mmol; CRL, 40% (w/w carbohydrate); buffer, 0.1-mM (0.1 mL of 0.1 M) phosphate buffer pH 7.0; CH_2Cl_2:DMF (v/v 90:10) at 40°C; incubation period, 72 h. Conversion yields were from HPLC with respect to L-alanine concentration

[b]L-Alanine, 1 mmol; carbohydrates, 1 mmol; crude PPL, 111% (w/w carbohydrate); buffer, 0.2-mM (0.2 mL of 0.1 M) acetate buffer pH 5.0; CH_2Cl_2: DMF (v/v 90:10) at 40°C; incubation period, 72 h

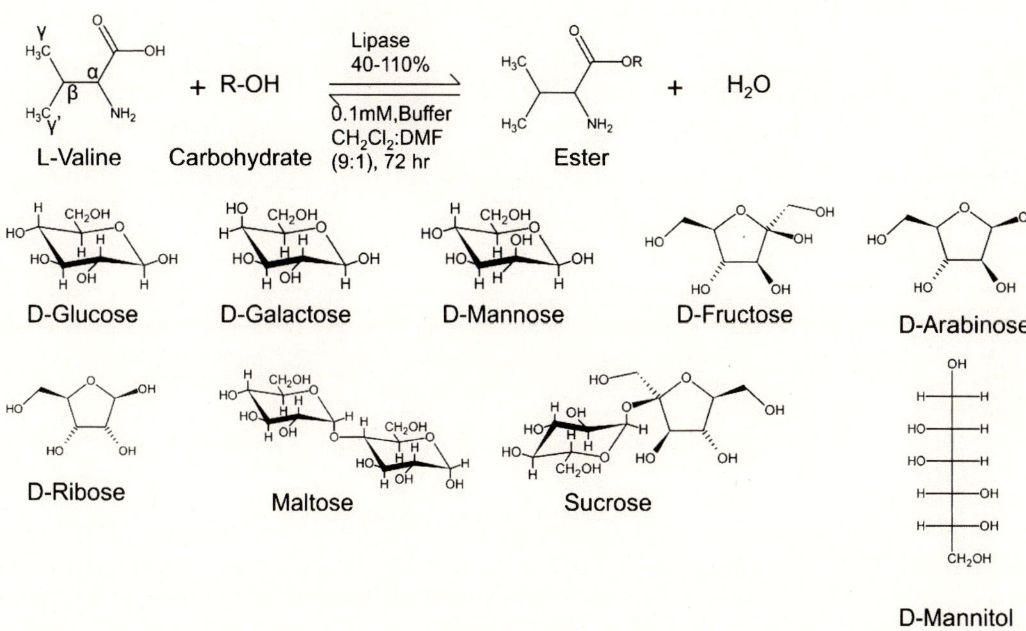

Scheme 6.2 Lipase-catalysed synthesis of L-valyl esters of carbohydrates

2007) and the esterification reaction was studied in detail using RML and PPL in terms of incubation period, lipase concentrations, buffer (pH and concentration) and substrate concentrations. Optimum conditions determined were:

A CRL concentration of 40% w/w D-glucose exhibited 68% conversion at 72 h at an initial rate of esterification of 0.0425 mmol h^{-1}; buffer – 0.1-mM (0.1 mL of 0.1 M) phosphate buffer (pH 7.0) – showed the highest yield of 68% (0.68 mmol); CRL steadily increased esterification up to 30% enzyme concentration (84%, 0.84 mmol) (Table 6.4); both 10% (w/w D-glucose) RML and PPL in the presence of 0.1-mM

Table 6.6 Effect of lipase concentration on the synthesis of L-valyl-D-glucose[a]

Lipase concentration (% w/w D-glucose)	*Candida rugosa* lipase (CRL) yield % (mmol)	*Candida rugosa* lipase (CRL)[b] yield % (mmol)	*Rhizomucor miehei* lipase (RML)[c] yield % (mmol)	Porcine pancreas lipase (PPL)[c] yield % (mmol)
10	2 (0.02)	73 (0.73)	59 (0.59)	62 (0.62)
20	6 (0.06)	77 (0.77)	41 (0.41)	45 (0.45)
30	14 (0.14)	84 (0.84)	31(0.31)	33 (0.33)
40	25 (0.25)	68 (0.68)	24 (0.24)	27 (0.27)
50	10 (0.10)	65 (0.65)	25 (0.25)	36 (0.36)

[a]D-Glucose, 1 mmol, and L-valine, 1 mmol; solvent, CH_2Cl_2:DMF (v/v 90:10) at 40°C
[b]Carried out in the presence of buffer with 100 mL of solvent containing 0.1-mM (0.1 mL of 0.1 M) phosphate buffer (pH 7.0)
[c]Carried out in the presence of buffer with 100 mL of solvent containing 0.1-mM (0.1 mL of 0.1 M) acetate buffer (pH 5.0)

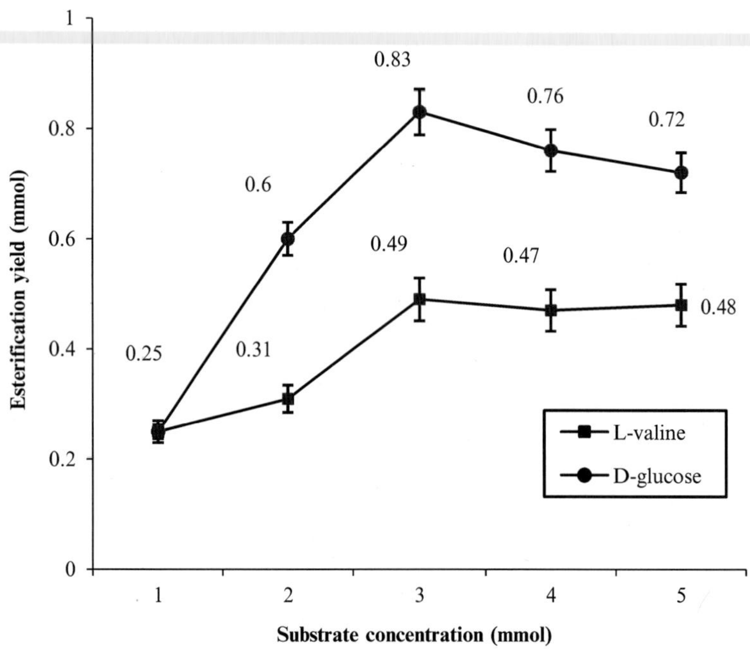

Fig. 6.4 Effect of substrate concentration on synthesis of L-valyl-D-glucose. CRL, 40% (w/w D-glucose); solvent, CH_2Cl_2:DMF (v/v 90:10), at 40°C. l-*Valine* (■), 1–5 mmol at 1 mmol D-glucose; d-*glucose* (•), 1–5 mmol at 1 mmol L-valine and a constant enzyme concentration of 72 mg

(0.1 mL of 0.1 M) acetate buffer (pH 5.0) showed a maximum yield of 59% (0.59 mmol) and 62% (0.62 mmol), respectively, which thereafter decreased up to 50% of lipase concentration (Table 6.6); 3–5 mmol of L-valine gave (Fig. 6.4) 0.49 mmol of the ester.

Similarly, 3 mmol D-glucose gave a yield of 0.83 mmol. With both L-valine and D-glucose variation, the esterification increased up to 3 equivalents, indicating that L-valine and D-glucose are not inhibitors to CRL.

6.3.1 Syntheses of L-Valyl Esters of Carbohydrates

L-Valine (L-2-amino-3-methylbutanoic acid) is a polar and an essential dietary amino acid containing isopropyl group as a side chain. Esterification of L-valine with carbohydrates was carried out using CRL and crude PPL under optimal conditions. Table 6.7 shows the HPLC ester yields, types of esters formed and percentage proportions of the individual esters from CRL-catalysed reaction (Somashekar and Divakar 2007).

6.4 L-Leucyl-D-Glucose

6.4.1 Optimum Conditions

In a *Rhizomucor miehei* (RML)- and porcine pancreas (PPL)-catalysed esterification (Scheme 6.3, Somashekar and Divakar 2007) optimisation of reaction conditions showed the highest yield of 10.8% (0.22 mmol) at 20% enzyme concentration. Better conversions (RML- 36%, PPL-24%) could be achieved with both the enzymes at 5 equivalents of L-leucine. Effect of buffer salts in the range 4.0–8.0 with 40% (w/w D-glucose) RML at 1:2 molar equivalent of D-glucose and L-leucine (Table 6.8) showed that 0.2-mM (0.1 mL of 0.1 M) acetate buffer (pH 5.0) gave the highest yield of 63% (0.63 mmol). The effect of various buffer volumes in the range 0.05 mM (0.05 mL) to 0.6 mM (0.6 mL) of phosphate buffer (pH 7.0) showed that the esterification yields increased with increase in the buffer concentration from 0.05-mM (0.05 mL) to 0.6-mM (0.6 mL) phosphate buffer (pH 7.0) with the highest yield of 85% (1.7 mmol).

6.4.2 Selectivity

An attempt to improve the selectivity of ester formation made by reducing the incubation period in case of L-alanyl-D-glucose synthesis using RML gave a conversion yield of 15%. From two-dimensional HSQCT NMR, the formation of three monoesters (6-*O*- 42%, 3-*O*- 33% and 2-*O*- 25%) with only β-anomer of D-glucose was found, the D-glucose employed being a 40:60 mixture of α- and β-anomers, respectively.

6.4.3 Determination of Critical Micellar Concentration (CMC)

Surfactant property of the aminoacyl esters of carbohydrate was evaluated by determining the critical micellar concentration (CMC) value for L-alanyl-β-D-glucose spectroscopically at 470 nm (Rosenthal and Loussale 1983). The CMC of L-alanyl-β-D-glucose was found to be 2.25 mM (0.056%).

6.4.4 Syntheses of L-Leucyl Esters of Carbohydrates

L-Leucine (L-2-amino-4-methylpentanoic acid) is a polar and an essential dietary amino acid containing 2-methylpropyl group as a side chain. Compared to L-alanine (solubility 127.3 g L^{-1} at 25°C) and L-valine (solubility 88.5 g L^{-1} at 25°C), the solubility of L-leucine in water is low (24.26 g L^{-1} at 25°C). However, the carbohydrate esters can be expected to exhibit higher solubility in water. Using optimum conditions, L-leucyl esters of different carbohydrates were prepared using CRL and crude PPL (Vijayakumar et al. 2004). Table 6.9 shows the HPLC ester yields, types of esters formed and percentage proportions of the individual esters from CRL catalysis (Somashekar and Divakar 2007).

6.5 Syntheses of L-Isoleucyl Esters of Carbohydrates

L-Isoleucine (L-2-amino-3-methylpentanoic acid) is a polar and an essential dietary amino acid containing 3-methylpropyl group as a side chain. Here also, the solubility of L-isoleucine is low (41.2 g L^{-1} at 25°C) like of that of L-leucine. Using optimum conditions, L-isoleucyl esters of different carbohydrates were prepared (Somashekar and Divakar 2007) by employing CRL and crude PPL (Scheme 6.4). Table 6.10 shows the HPLC ester yields, types of esters formed and percentage proportions of the individual esters from CRL-catalysed reaction.

Table 6.7 Syntheses of L-valyl esters of carbohydrates[a]

L-Valyl esters of carbohydrates (% proportions[b])			Yield (%)
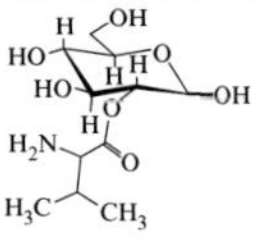 2-*O*-L-valyl-D-glucose (10)	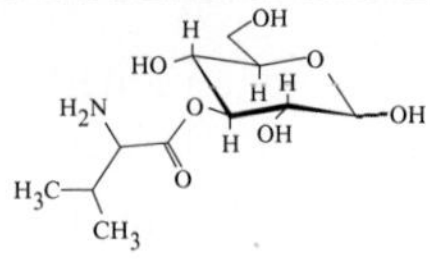 3-*O*-L-valyl-D-glucose (12)	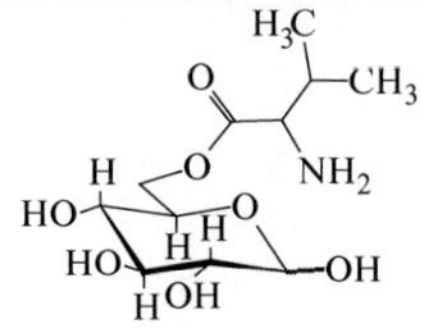 6-*O*-L-valyl-D-glucose (31)	68 (monoesters-36, diesters-32)
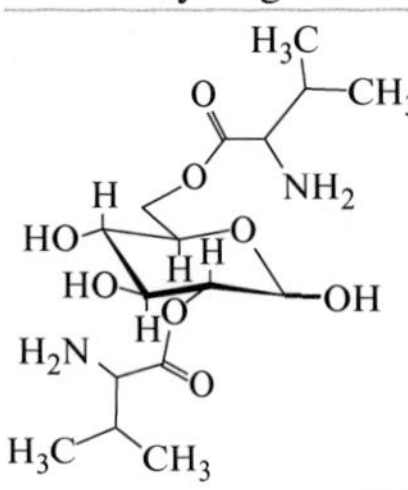 2,6-di-*O*-L-valyl-D-glucose (23)	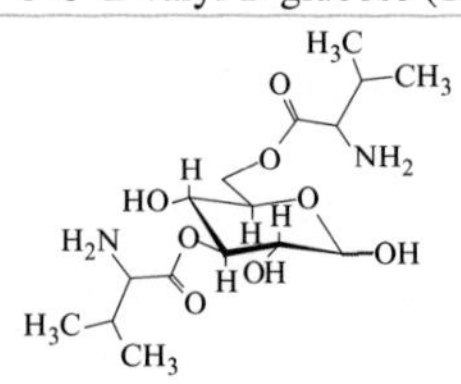 3,6-di-*O*-L-valyl-D-glucose (24)		
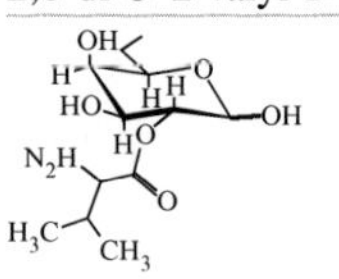 2-*O*-L-valyl –D-galactose (48)	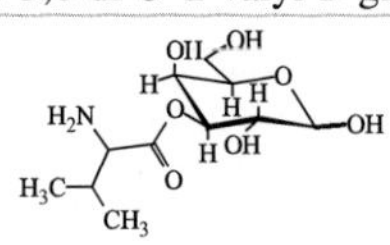 3-*O*-L-valyl-D-galactose (26)	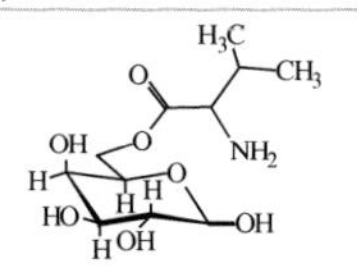 6-*O*-L-valyl-D-galactose (26)	30 (only monoesters)
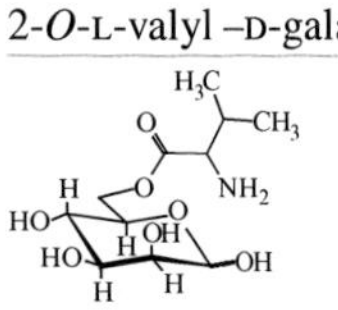 6-*O*-L-valyl-D-mannose			51 (only monoester)
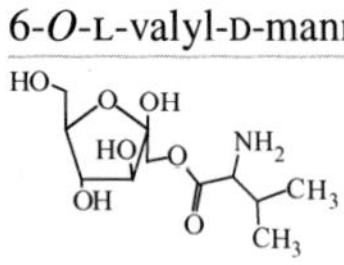 1-*O*-L-valyl-D-fructose (29)	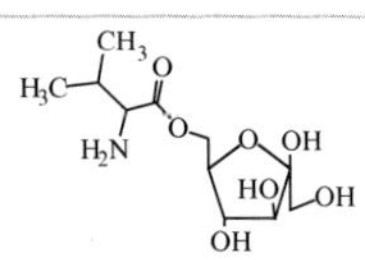 6-*O*-L-valyl-D-fructose (34)	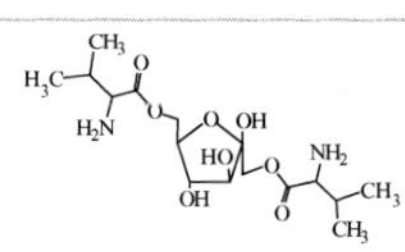 1,6-di-*O*-L-valyl-D-fructose (37)	34 (monoesters-21, diester-13)
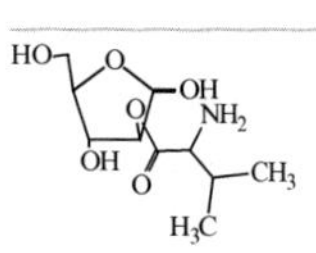 2-*O*-L-valyl-D-arabinose (32)	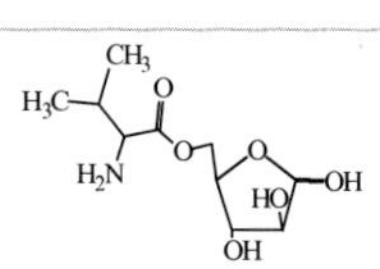 5-*O*-L-valyl-D-arabinose (25)	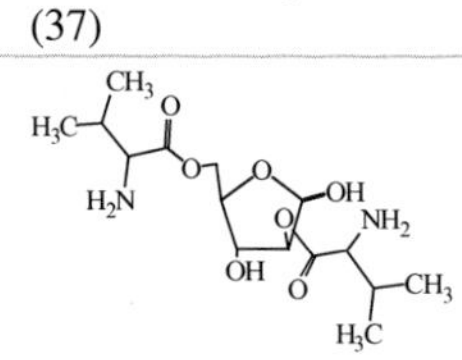 2,5-di-*O*-L-valyl-D-arabinose (43)	25[c] (monoesters-14, diester-11)
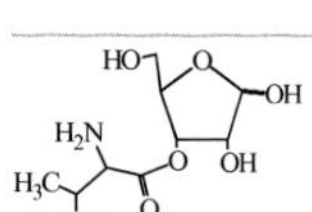 3-*O*-L-valyl-D-ribose (26)	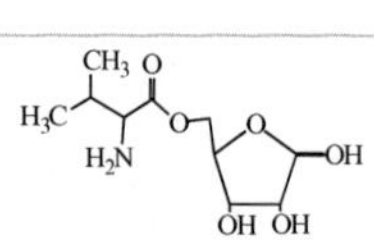 5-*O*-L-valyl-D-ribose (26)	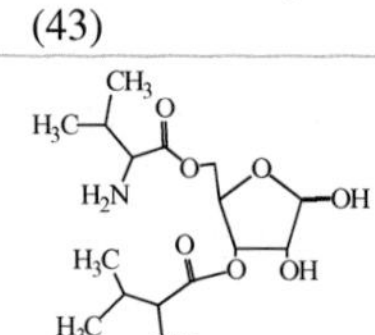 3,5-di-*O*-L-valyl-D-ribose (48)	33[c](monoesters-17, diester-16)

(continued)

Table 6.7 (continued)

L-Valyl esters of carbohydrates (% proportions[b])	Yield (%)
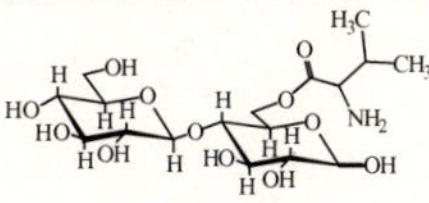 6-*O*-L-valyl-maltose (49) 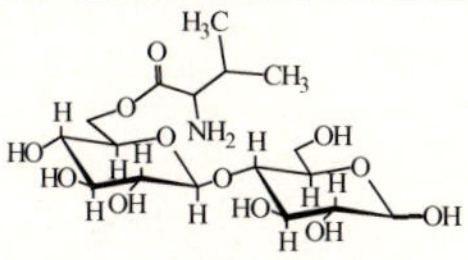 6′-*O*-L-valyl-maltose (51)	47 (only monoesters)
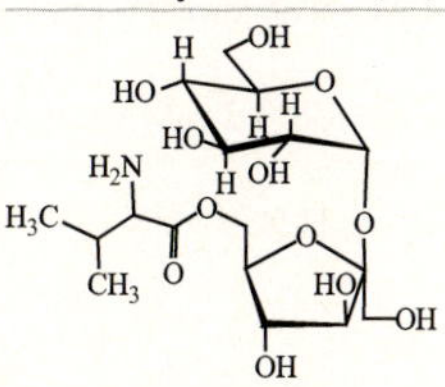 6-*O*-L-valyl-sucrose	60 (only monoester)
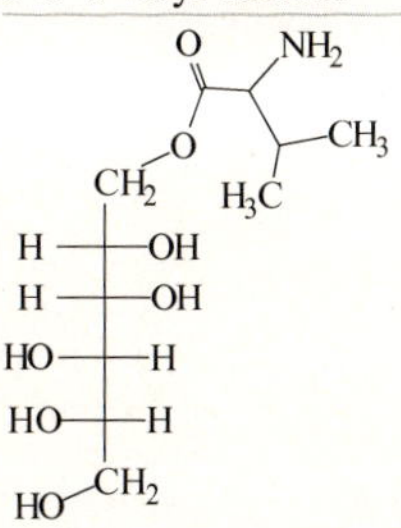 1-*O*-L-valyl-D-mannitol	52 (only monoester)

[a]L-Valine, 2 mmol; carbohydrates, 1 mmol; CRL, 40% (w/w carbohydrate); buffer, 0.1-mM (0.1 mL of 0.1 M) phosphate buffer pH 7.0; CH_2Cl_2:DMF (v/v, 90:10) at 40°C; incubation period, 72 h. Conversion yields were from HPLC with respect to L-valine concentration

[b]Percentage proportions of individual esters were determined from the peak areas or from their cross peaks of the carbon-13 C6 and C5 (in case of pentoses) signals in the 2-D HSQCT spectrum

[c]Several cross peaks, due to opening and/or degradation of the five-membered ring during esterification

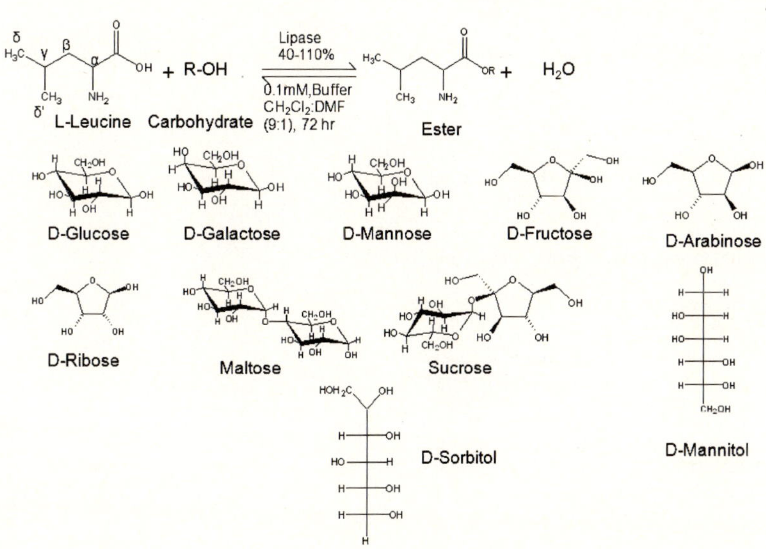

Scheme 6.3 Lipase-catalysed synthesis of L-leucyl-D-glucose esters in anhydrous organic media

Table 6.8 Effect of buffer salts (pH and buffer concentration) on the synthesis of L-leucyl-D-glucose[a]

pH[b]	Yield % (mmol)	pH 7.0[c] concn mM	Yield % (mmol)
4.0	16 (0.31)	0.05	5 (0.1)
5.0	63 (1.26)	0.1	21 (0.42)
6.0	16 (0.33)	0.2	14 (0.28)
7.0	14 (0.27)	0.4	32 (0.64)
8.0	7 (0.14)	0.6	85 (1.70)

[a]D-Glucose, 1 mmol, and L-leucine, 2 mmol; incubation period, 72 h; RML, 40% (w/w D-glucose) 100 mL of the solvent containing specified volumes, concentration and pH of the buffer

[b]Solvent, 100 mL C_6H_6:$CHCl_3$:DMF (v/v/v 45:45:10) at 60°C. Buffer, 0.2-mM (0.2 mL of 0.1 M) appropriate pH buffer

[c]Solvent, 100 mL. Buffer, 0.05 mL to 0.6 mL of 0.1 M phosphate (pH 7.0)

Table 6.9 Syntheses of L-leucyl esters of carbohydrates[a]

L-Leucyl esters of carbohydrates (% proportions[b])			Yield (%)
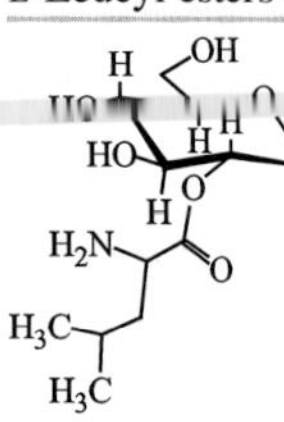 2-*O*-L-leucyl-D-glucose (17)	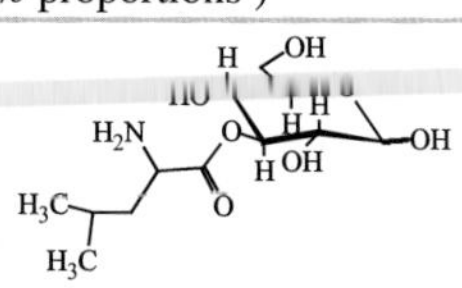 3-*O*-L-leucyl-D-glucose (20)	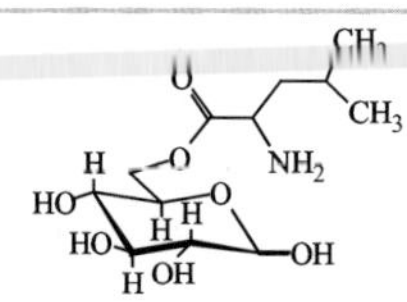 6-*O*-L-leucyl-D-glucose 42)	4[illegible] (monoesters 34, diesters-9)
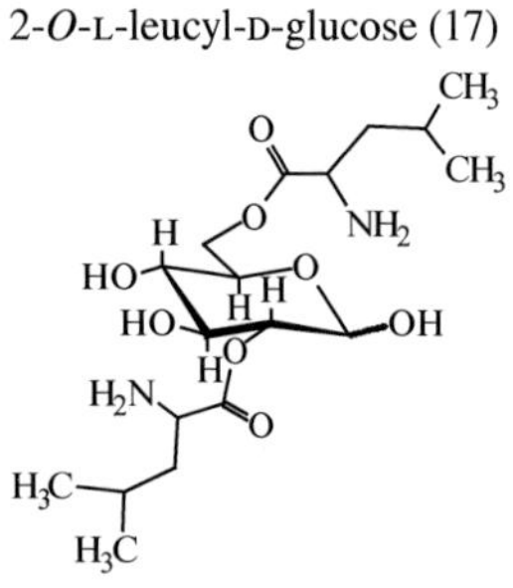 2,6-di-*O*-L-leucyl-D-glucose (10)	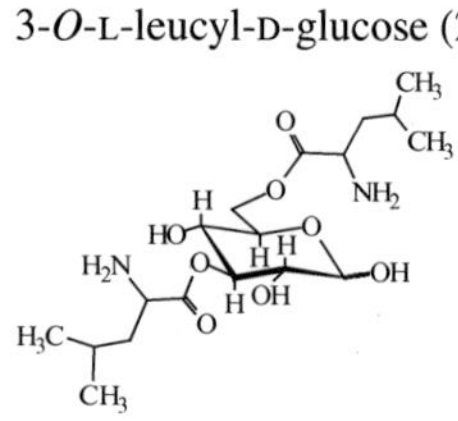 3,6-di-*O*-L-leucyl-D-glucose (11)		
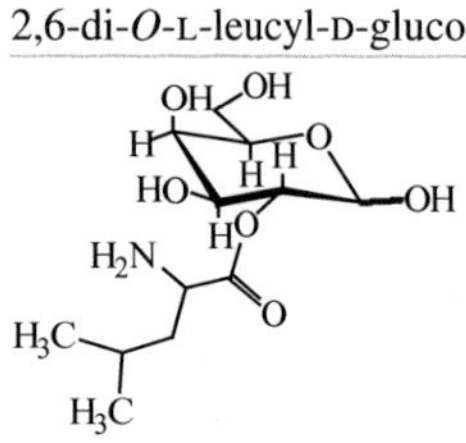 2-*O*-L-leucyl-D-galactose (48)	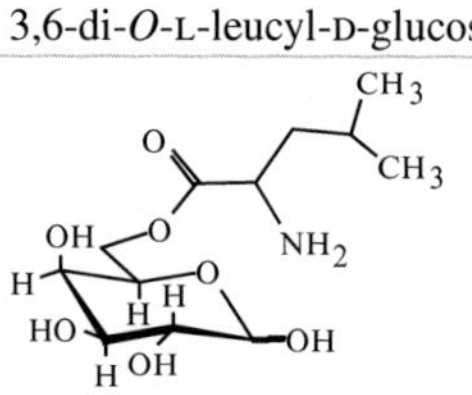 6-*O*-L-leucyl-D-galactose (52)		21 (only mono esters)
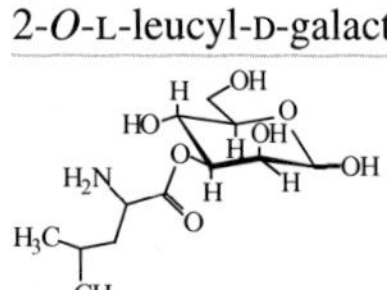 3-*O*-L-leucyl-D-mannose (28)	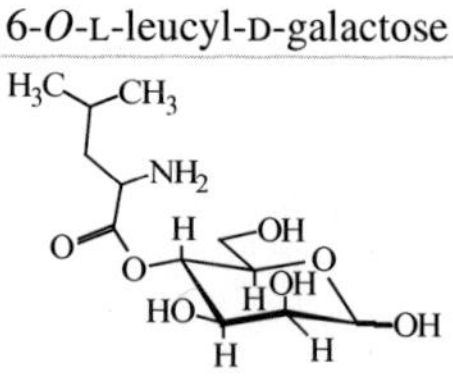 4-*O*-L-leucyl-D-mannose (30)	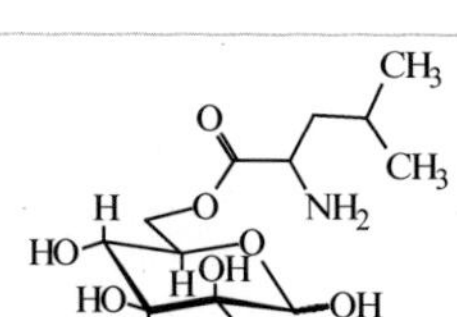 6-*O*-L-leucyl-D-mannose (42)	31 (only mono esters)
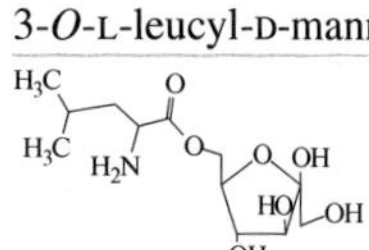 6-*O*-L-leucyl-D-fructose			48 (only monoester)

(continued)

Table 6.9 (continued)

L-Leucyl esters of carbohydrates (% proportions[b])			Yield (%)
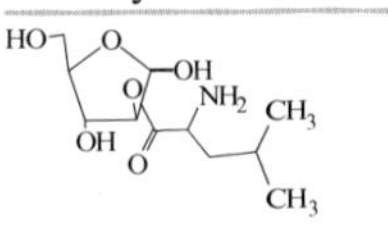 2-*O*-L-leucyl-D-arabinose (24)	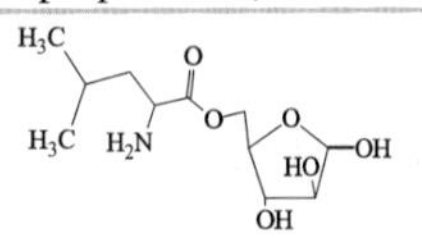 5-*O*-L-leucyl-D-arabinose (33)	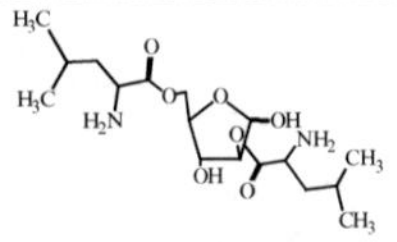 2,5-di-*O*-L-leucyl-D-arabinose (43)	42 (monoesters-24, diester-18)
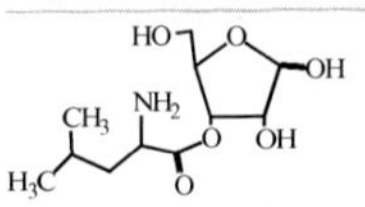 3-*O*-L-leucyl-D-ribose (16)	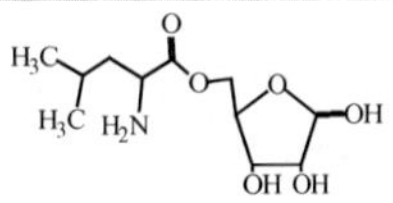 5-*O*-L-leucyl-D-ribose (32)	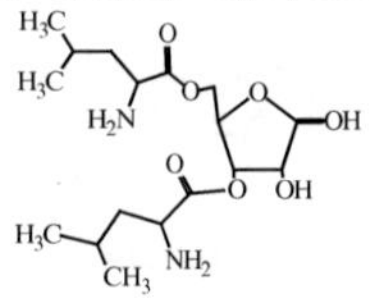 3,5-di-*O*-L-leucyl-D-ribose (52)	38 (mono esters-18, diester-20)
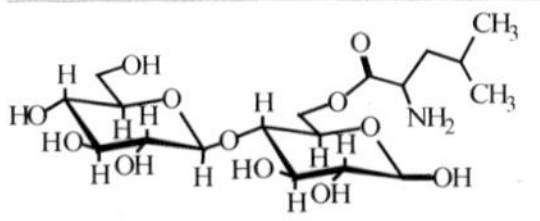 6-*O*-L-leucyl-maltose			44 (only monoester)
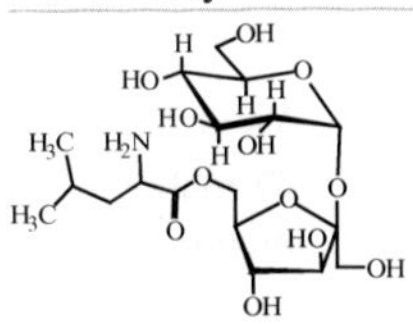 6-*O*-L-leucyl-sucrose			38 (only monoester)
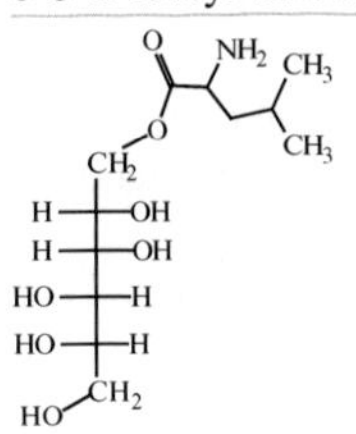 1-*O*-L-leucyl-D-mannitol (56)	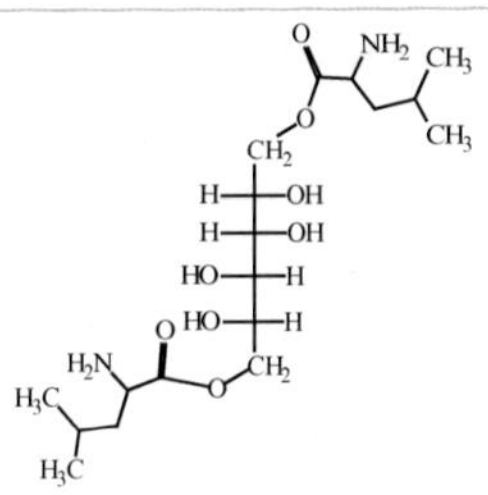 1,6-di-*O*-L-leucyl-D-mannitol (44)		45 (monoester-25, diester-20)
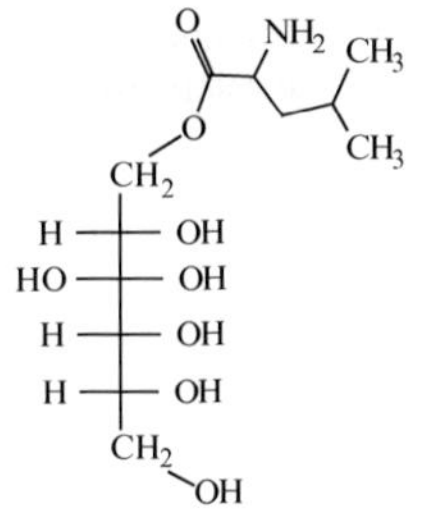 1-*O*-L-leucyl-D-sorbitol			25 (only monoester)

[a]L-Leucine, 2 mmol; carbohydrates, 1 mmol; CRL, 40% (w/w carbohydrate); buffer, 0.1-mM (0.1 mL of 0.1 M) phosphate buffer (pH 7.0); CH2Cl2:DMF (v/v 90:10) at 40°C; incubation period, 72 h. Conversion yields were from HPLC with respect to L-leucine concentration

[b]Percentage proportions of individual esters were determined from the peak areas or from their cross peaks of the carbon-13 C6 and C5 (in case of pentoses) signals in the 2-D HSQCT spectrum

[c]Several cross peaks, due to opening and/or degradation of the five-membered ring during esterification

Table 6.25 Percentage yields and proportions of L-alanyl esters of carbohydrates from RML and L-valyl, L-leucyl and L-isoleucyl esters of carbohydrates CRL catalysed reactions[a]

Carbohydrate	L-Alanyl		L-Valyl		L-Leucyl		L-Isoleucyl	
	Yield[b]	% Proportions	Yield[b]	% Proportions	Yield[b]	% Proportions	Yield[b]	% Proportions
D-glucose	30	2-*O*-Ester (20)	68	2-*O*-Ester (10)	43	2-*O*-Ester (17)	47	3-*O*-Ester (42)
		3-*O*-Ester (12)		3-*O*-Ester (12)		3-*O*-Ester (20)		6-*O*-Ester (58)
		6-*O*-Ester (47)		6-*O*-Ester (31)		6-*O*-Ester (42)		
		2,6-di-*O*-Ester (15)		2,6-di-*O*-Ester (23)		2,6-di-*O*-Ester (10)		
		3,6-*di*-*O*-Ester (6)		3,6-di-*O*-Ester (24)		di-*O*-Ester (12)		
D-galactose	21	2-*O*-Ester (33)	30	2-*O*-Ester (48)	21	2-*O*-Ester (48)	46	2-*O*-Ester (78)
		3-*O*-Ester (32)		3-*O*-Ester (26)		6-*O*-Ester (52)		3-*O*-Ester (10)
		6-*O*-Ester (35)		6-*O*-Ester (26)				6-*O*-Ester (12)
D-mannose	49	3-*O*-Ester (25)	51	6-*O*-Ester	31	3-*O*-Ester (28)	55	3-*O*-Ester (19)
		4-*O*-Ester (25)				4-*O*-Ester (30)		4-*O*-Ester (13)
		6-*O*-Ester (30)				6-*O*-Ester (42)		6-*O*-Ester (13)
		3,6-di-*O*-Ester (9)						3,6-di-*O*-Ester (27)
		4,6-di-*O*-Ester (11)						4,6-di-*O*-Ester (28)
D-fructose	52	1-O-Ester (34)	34	1-*O*-Ester (29)	48	6-*O*-Ester	43	1-*O*-Ester (36)
		6-O-Ester (34)		6-*O*-Ester (34)				6-*O*-Ester (30)
		1,6-di-O-Ester (32)		1,6-di-*O*-Ester (37)				1,6-di-*O*-Ester (34)
D-arabinose	9	2-*O*-Ester (33)	25	2-*O*-Ester (32)	42	2-*O*-Ester (24)	55	2-*O*-Ester (24)
		5-*O*-Ester (34)		5-*O*-Ester (25)		5-*O*-Ester (33)		5-*O*-Ester (33)
		2,5-di-*O*-Ester (33)		2,5-di-*O*-Ester (43)		2,5-di-*O*-Ester (43)		2,5-di-*O*-Ester (43)
D-ribose	48	3-*O*-Ester (16)	33	3-*O*-Ester (26)	38	3-*O*-Ester (16)	53	3-*O*-Ester (52)
		5-*O*-Ester (32)		5-*O*-Ester (26)		5-*O*-Ester (32)		5-*O*-Ester (20)
		3,5-di-*O*-Ester (52)		3,5-di-*O*-Ester (48)		3,5-di-*O*-Ester (52)		3,5-di-*O*-Ester (28)
lactose	20	6-*O*-Ester (34)					45	2-*O*-Ester (39)
		6′-*O*-Ester (34)	–	–	–	–		6-*O*-Ester (40)
		6,6′-di-*O*-Ester (32)						6′-*O*-Ester (21)
maltose	56	6-*O*-Ester (34)	47	6-*O*-Ester (49)	44	6-O-Ester	54	2-*O*-Ester (38)
		6′-*O*-Ester (34)		6′-*O*-Ester (51)				6-*O*-Ester (40)
		6,6′-di-*O*-Ester (32)						6′-*O*-Ester (22)
sucrose	8	6-*O*-Ester	60	6-*O*-Ester	38	6-*O*-Ester	22	6-*O*-Ester
D-mannitol	–		52	1-*O*-Ester	45	1-*O*-Ester (56)	52	1-*O*-Ester (62)
						1,6-di-*O*-Ester (44)		1,6-di-*O*-Ester (38)
D-sorbitol	–	–	–	–	25	1-*O*-Ester	–	–

Table 6.26 Percentage yields and proportions of L-prolyl, L-phenylalanyl, L-tryptophanyl and L-histidyl esters of carbohydrates from CRL catalysed reactions[a]

Carbohydrate	L-Prolyl		L-Phenylalanyl		L-Tryptophanyl		L-Histidyl	
	Yieldb	% Proportions	Yieldb	% Proportions	Yieldb	% Proportions	Yieldb	% Proportions
D-glucose	62	2-*O*-Ester (26)	79	2-*O*-Ester (19)	42	2-*O*-Ester (22)	32	2-*O*-Ester (25)
		3-*O*-Ester (26)		3-*O*-Ester (23)		3-*O*-Ester (21)		3-*O*-Ester (24)
		6-*O*-Ester (48)		6-*O*-Ester (25)		6-*O*-Ester (38)		6-*O*-Ester (28)
				2,6-di-*O*-Ester (17)		2,6-di-*O*-Ester (10)		2,6-di-*O*-Ester (12)
				3,6-di-*O*-Ester (16)		3,6-di-*O*-Ester (9)		3,6-di-*O*-Ester (11)
D-galactose	52	2-*O*-Ester (20)	45	2-*O*-Ester (19)	27	6-*O*-Ester	44	–
		3-*O*-Ester (12)		3-*O*-Ester (20)				
		6-*O*-Ester (68)		6-*O*-Ester (32)				
				2,6-di-*O*-Ester (16)				
				3,6-di-*O*-Ester (13)				
D-mannose	40	3-*O*-Ester (21)	62	3-*O*-Ester (18)	34	6-*O*-Ester	72	3-*O*-Ester (28)
		4-*O*-Ester (20)		4-*O*-Ester (18)				6-*O*-Ester (72)
		6-*O*-Ester (24)		6-*O*-Ester (39)				
		3,6-di-*O*-Ester (19)		3,6-di-*O*-Ester (13)				
		4,6-di-*O*-Ester (16)		4,6-di-*O*-Ester (12)				
D-fructose	61	1-*O*-Ester (31)	50	1-*O*-Ester (28)	18	1-*O*-Ester (45)	58	6-*O*-Ester
		6-*O*-Ester (42)		6-*O*-Ester (72)		6-*O*-Ester (55)		
		1,6-di-*O*-Ester (27)						
D-arabinose	–	–	64	2-*O*-Ester (35)	–	–	–	–
				5-*O*-Ester (44)				
				2,5-di-*O*-Ester (21)				
D-ribose	41	3-*O*-Ester (35)	–	–	–	–	–	–
		5-*O*-Ester (65)						
lactose	68	2-*O*-Ester (27)	61	6-*O*-Ester (42)	42	6-*O*-Ester (64)	–	–
		6-*O*-Ester (38)		6′-*O*-Ester (31)		6′-*O*-Ester (36)		
		6′-*O*-Ester (35)		6,6′-di-*O*-Ester (27)				

(continued)

Table 6.26 (continued)

Carbohydrate	L-Prolyl		L-Phenylalanyl		L-Tryptophanyl		L-Histidyl	
	Yield[b]	% Proportions	Yield[b]	% Proportions	Yield[b]	% Proportions	Yield[b]	% Proportions
maltose	66	2-*O*-Ester (29)	60	6-*O*-Ester (59)	70	2-*O*-Ester (13)	58	6-*O*-Ester (40)
		6-*O*-Ester (38)		6′-*O*-Ester (41)		6-*O*-Ester (38)		6′-*O*-Ester (32)
		6′-*O*-Ester (33)				6′-*O*-Ester (29)		6,6′-di-*O* Ester (28)
						6,6′-di-*O*-Ester (20)		
sucrose	–	–	–	–	7	6-*O*-Ester	–	–
D-sorbitol	20	1-*O*-Ester (73)	–	–	8	1-*O*-Ester (79)	8	–
		6-*O*-Ester (27)				6-*O*-Ester (21)		
D-mannitol	–	–	43	1-*O*-Ester (62)	–	–	62	1-*O*-Ester
				1,6-di-*O*-Ester (38)				

[a]Confirmation of esters and their percentage proportions were determined by 2-D HSQCT NMR. Conversion yields are an average from two experiments
[b]Yields from HPLC. Errors in yield measurements will be ±10%

substrate concentrations, pH and buffer concentration clearly explain the behaviour of the lipases.

Most of the parameters show that esterification increases up to a certain point, and thereafter they remain as such or decrease a little. This complex esterification reaction is not controlled by kinetic factors or thermodynamic factors or water activity alone. Use of lower enzyme concentrations did not result in thermodynamic yields. The thermodynamic binding equilibria regulate the concentrations of the unbound substrates at different enzyme and substrate concentrations and thereby conversion as the reaction proceeds with time. At lesser enzyme concentrations, for a given amount of substrates (enzyme/substrate ratio low), rapid exchange between bound and unbound forms of both the substrates with the enzyme (on a weighted average based on binding constant values of both the substrates) leaves substantial number of unbound substrate molecules at the start of the reaction which decrease progressively as conversion takes place (Romero et al. 2005; Marty et al. 1992). This becomes more so, if one of them binds more firmly to the enzyme than the other (higher binding constant value) as the respective enzyme/substrate ratios change (during the course of the reaction) unevenly till the conversion stops due to total predominant binding (inhibition). At intermediatory enzyme concentrations, such a competitive binding results in a favourable proportion of bound and unbound substrates to effect quite a good conversion. At higher enzyme concentrations, most of the substrates would be in the bound form leading to inhibition and lesser conversion (higher enzyme/substrate ratios). Also, the esterification reaction requires larger amount of enzyme compared to hydrolysis. While this leads to lesser selectivity, they also give rise to varying bound and unbound substrate concentrations till the conversion ends. For a given amount of enzyme and substrates, there is no increase in conversion beyond 72 h to 120 h. Longer incubation periods of especially lesser enzyme concentrations could also result in partial enzyme inactivation. However, not all the enzymes are inactivated before the end of the reaction.

The present study investigated the effect of buffer salts on this esterification reaction which rendered 'pH tuning' of the enzyme, besides providing optimum water activity necessary for the better performance of the enzyme. All the three lipases, RML, CRL and PPL, showed higher conversions when a small amount of buffer salt was employed. By imparting 'pH memory' or 'pH tuning', the catalytic activity of the lyophilised subtilisin Carlsberg in the pH range 5.0–11.0, in organic solvents like acetonitrile and 3-pentanone was reported to be enhanced (Xu and Klibanov 1996). The enzymatic activity of subtilisin cross-linked crystals in anhydrous 3-pentanone was accelerated by the addition of organic soluble (a mixture of a suitable acid and sodium salt) buffers (Xu and Klibanov 1996). The enantioselectivity of *Candida antarctica* lipase B in organic media was increased by 'pH tuning' of the enzyme by the addition of certain buffer salts which altered the protonation state of the enzyme and selectively tuned enantioselectivity and catalytic activity (Quiros et al. 2002). Similarly, the present work also showed enhanced activity of RML and CRL in the presence of buffer salts. However, buffer salts did not enhance esterification with PPL.

Besides imparting 'pH memory' or 'pH tuning', added water is essential for the integrity of the three-dimensional structure of the enzyme molecule and therefore its activity (Dordick 1989). Zaks and Klibanov (1986, 1988) reported that at low water activities, the lower the solvent polarity, the higher the enzyme activity. Beyond the critical water concentration, esterification decreases because the size of the water layer formed around the enzyme retards the transfer of acyl donor to the active site of the enzyme (Humeau et al. 1998; Camacho-Paez et al. 2003a, b) and also the water layer surrounding the enzymes makes enzyme to be more flexible by forming multiple hydrogen bonds and interacting with organic solvent causing denaturation (Valiveti et al. 1991). Increase in buffer volume affected this esterification reaction significantly. It could increase the water activity of the system in the initial stages by increasing the thickness of the micro-aqueous layer around the enzyme. Higher volumes of the buffer in the micro-aqueous layer could also cause slight inactivation of the enzyme

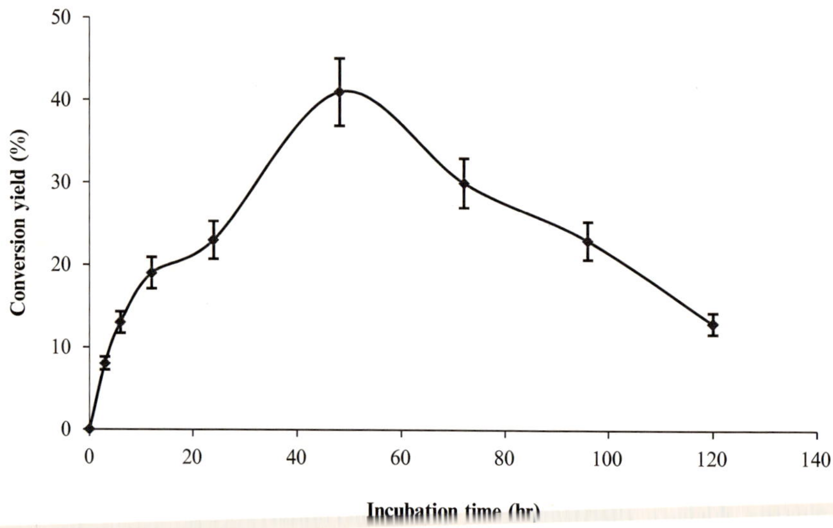

Fig. 7.2 A typical reaction profile for n-octyl-D-glucoside synthesis by the reflux method. Conversion yields were from HPLC with respect to 0.555 mmol of D-glucose. Reaction conditions: D-glucose –0.555 mmol; n-octanol – 0.027 mol; amyloglucosidase –50% (w/w D-glucose); 0.04 mM (0.4 mL of 10 mM buffer); pH 5.0 acetate buffer; solvent – di-isopropyl ether; and temperature –68°C

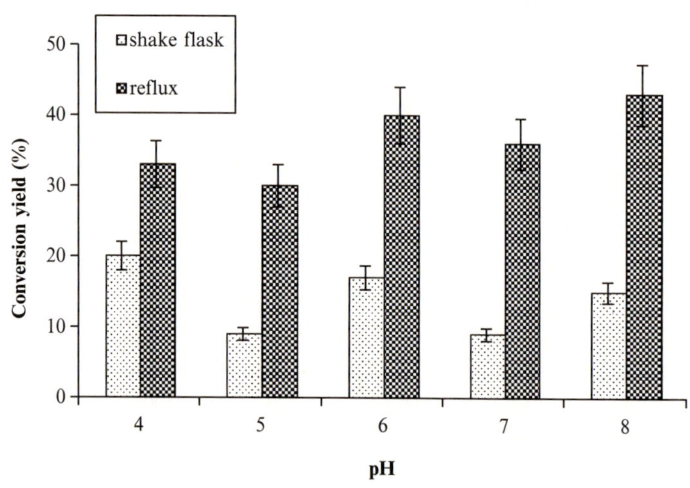

Fig. 7.3 Effect of pH on n-octyl-D-glucoside synthesis. Conversion yields were from HPLC with respect to 0.555 mmol of D-glucose. Reaction conditions: D-glucose –0.555 mmol: n-octanol –0.027 mol; buffer concentration –0.04 mM (0.4 mL of 10 mM buffer) for reflux method and 0.8 mM (0.4 mL of 10 mM buffer) for shake-flask method; and incubation –72 h. Reflux method: solvent – di-isopropyl ether and temperature –68°C. Shake-flask method: temperature –60°C

incubation period of 72 h. Similarly, the optimum conditions for the reflux method were found to be 30% (w/w D-glucose) amyloglucosidase (Tables 7.1 and 7.2, Fig 7.3) concentration and 0.04 mM (0.4 mL), pH 6.0 phosphate buffer at an incubation period of 72 h.

7.3 Synthesis of n-Octyl-D-Glucoside Using β-Glucosidase

n-Octyl-D-glucoside was synthesised by the reflux method using β-glucosidase isolated from sweet almond. The reaction mixture analysed by

HPLC showed 23% (230 μmol) conversion with respect to the D-glucose employed. Two-dimensional NMR (HSQCT) confirmed formation of n-octyl-β-D-glucoside.

7.4 Determination of Critical Micellar Concentration (CMC)

Since glycosides of long-chain alcohols serve as nonionic surfactants, critical micellar concentration (CMC) of n-octyl-D-glucoside determined (Rosenthal and Loussale 1983) was found to be 16.1 mM (0.47%).

7.5 Synthesis of n-Octyl Glycosides

The optimum conditions worked out for the synthesis of n-octyl-D-glucoside by the reflux method were employed for the synthesis of n-octanol, glycosides of various carbohydrates. The various carbohydrates employed were D-glucose, D-galactose, D-mannose, D-fructose, D-arabinose, D-ribose, maltose, sucrose, lactose, D-mannitol and D-sorbitol. n-Octyl maltoside (conversion yield 15%, 150 μmol) and n-octyl sucrose (conversion yield 13%, 130 μmol) were obtained by this procedure (Scheme 7.2).

The retention times for free carbohydrate and n-octyl glycosides in HPLC were D-glucose – 5.2 min; n-octyl-D-glucoside – 7.2 min; maltose – 7.4 min; n-octyl maltoside – 10.5 min; sucrose – 6.4 min; and n-octyl sucrose – 7.1 min. The structures of the glycosides formed, HPLC yield and product proportions are presented in Table 7.3. Other carbohydrates such as D-fructose, D-arabinose, D-ribose, lactose and D-sorbitol did not form any glycoside with n-octanol. However, HPLC indicated glycosylation of D-galactose, D-mannose and D-mannitol with conversion yields less than 5%.

7.6 Spectral Characterisation

The glycosides were characterised by UV, IR, 2-D NMR (HSQCT) and optical rotation. Two-dimensional HSQCT NMR gave information on the nature and proportions of the products formed (Table 7.3). Since n-octyl glycosides are surfactant molecules, the proton signals are broad due to aggregation at concentration above CMC employed, and hence, coupling constant values could not be determined satisfactorily for these compounds.

UV spectrum of n-octyl-α and β-D-glucosides, maltoside and sucroside synthesised by β-glucosidase showed σ→σ* band between 190 and 205 nm and n→π* band between 275 and 278.5 nm. The IR spectral band in the region 1,027–1,053 cm^{-1} corresponded to the glycosidic

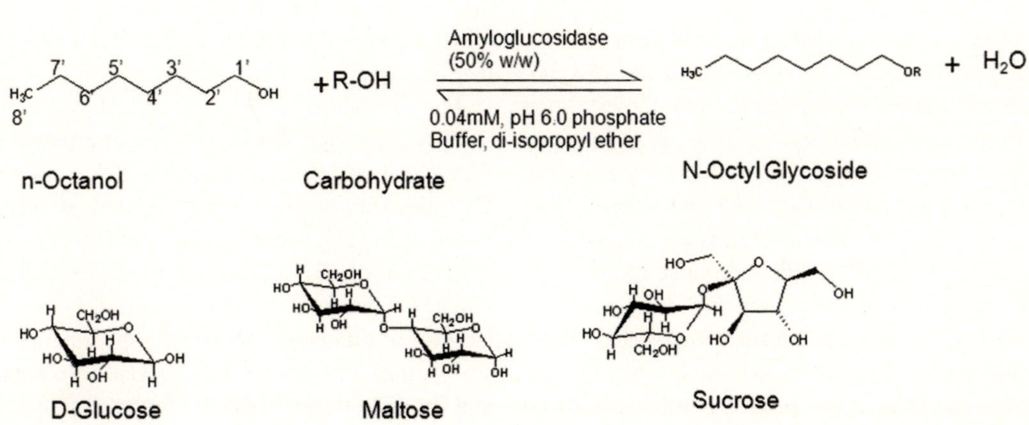

Scheme 7.2 Synthesis of n-octyl glycosides

Table 7.3 n-Octyl glycosides with conversion yields and product proportions[a]

Glycosides and product proportions[b]		Glycosylation yield (%)[c]
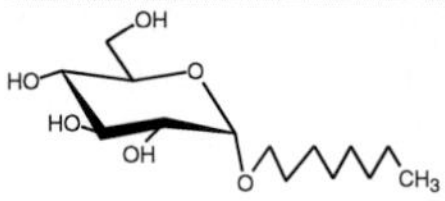 n-Octyl-α-D-glucoside (63)	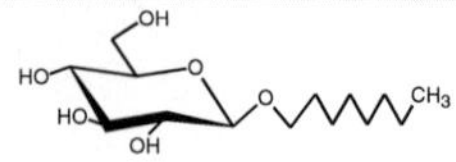 n-Octyl-β-D-glucoside (25)	46
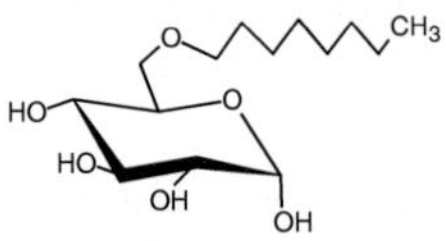 C6-O-octyl-D-glucose (12)		
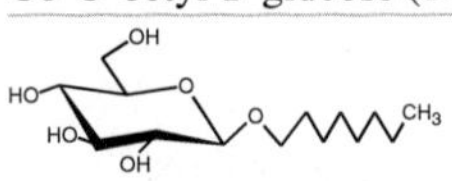 n-Octyl-β-D-glucoside[d]		23
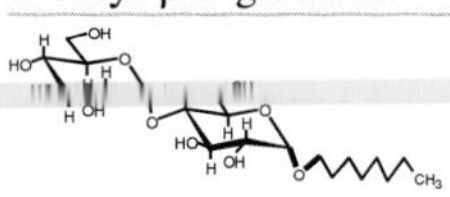 n-Octyl maltoside		15
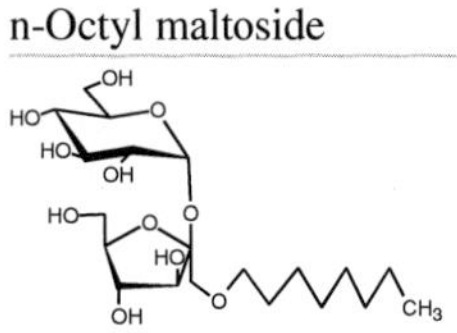 C1-O-octyl sucrose (44)	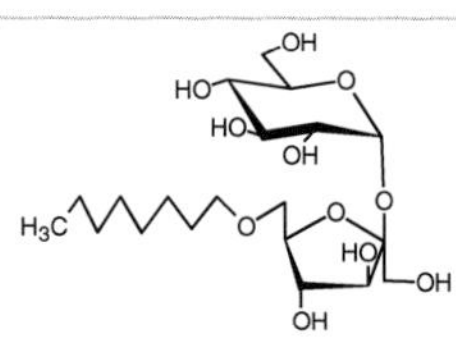 C6-O-octyl sucrose (56)	13

[a]Carbohydrate and n-octanol – 1:50 equivalents; amyloglucosidase – 30% (w/w carbohydrate); solvent – di-isopropyl ether; temperature – 68°C; and incubation period – 72 h

[b]The product proportions determined from 2D-HSQCT NMR C1/C6 cross peak areas are shown in brackets

[c]Conversion yields were from HPLC with respect to free carbohydrate

[d]The compound was synthesised by using β-glucosidase from sweet almonds

C–O–C symmetrical stretching and 1,225–1,259 cm^{-1} for the glycosidic C–O–C asymmetrical stretching frequencies.

In case of α-glucoside, the ^{1}H NMR area of the C1 anomeric cross peaks, confirmed that α:β anomeric composition obtained was 63:25. Apart from the free C1α and C1β signals, the cross peaks in the anomeric region with ^{13}C chemical shift values at 98.5 and 103.2 ppm with the corresponding ^{1}H values at 4.62 and 4.17 ppm clearly indicated C1α of the glucoside and C1β glucoside formation. The cross peak in the C6 region (^{13}C at 67.2 ppm and ^{1}H at 3.64 ppm) indicated C6-O-alkylated product. Mass value 294 [M+2] obtained from mass spectrum further confirmed the product formation

Two-dimensional HSQCT showed the downfield chemical shift for C1β at 103.2 ppm and the corresponding ^{1}H value at 4.18 ppm, indicating the formation of only C1β glucoside. No other product formation (C1α glucoside or C6-O-alkylated) was detected unlike the amyloglucosidase-catalysed reaction.

For the maltoside, the downfield chemical shift for C1α at 98.8 ppm and the corresponding ^{1}H value at 4.63 ppm indicated formation of only C1α maltoside and no C-6-O-alkylated products. A mass value 455 was obtained for [M+1] peak.

In case of the sucroside, two-dimensional HSQCT confirmed the formation of C1-O- and C6-O- of alkylated products. The chemical shift values for C1 at 62.8 ppm (^{1}H at 3.76 ppm) and C6 at 63.0 ppm (^{1}H at 3.25 ppm) indicated the formation C1-alkylated and C6-alkylated products, respectively. Also NMR data clearly showed that hydrolysis of sucrose has taken place and the hydrolyzed D-glucose is glycosylated at C1α (^{13}C at 98.7 ppm and ^{1}H at 4.64 ppm) and alkylated at the C6-O- position (^{13}C at 67.5 ppm and ^{1}H at 3.54 ppm). The mass value 455 for [M+1] peak confirmed the product formation.

7.7 Synthesis of n-Alkyl Glucosides Using Amyloglucosidase

n-Alkyl glucosides using alcohols of carbon chain length C1–C18 were synthesised with the following specific alcohols, namely, methyl alcohol, ethyl alcohol, n-propyl alcohol, n-butyl alcohol, n-amyl alcohol, n-hexyl alcohol, n-heptyl alcohol, n-octyl alcohol, n-nonyl alcohol, n-decyl alcohol, lauryl alcohol, cetyl alcohol and stearyl alcohol. The reaction mixtures were analysed by HPLC. Retention times are D-glucose – 5.2 min; n-methyl-D-glucoside – 7.0 min; n-ethyl-D-glucoside – 7.1 min; n-propyl-D-glucoside – 7.1 min; n-butyl-D-glucoside – 7.1 min; n-amyl-D-glucoside – 7.1 min; n-hexyl-D-glucoside – 7.2 min; n-heptyl-D-glucoside – 7.2 min; n-nonyl-D-glucoside – 7.5 min; n-decyl-D-glucoside – 7.6 min; lauryl-D-glucoside – 7.6 min; cetyl-D-glucoside – 7.7 min; and stearyl-D-glucoside – 7.7 min.

7.7.1 Shake-Flask Method

Alcohols of carbon chain lengths C2–C18 were employed for the synthesis of glucosides. The yields obtained (with respect to D-glucose) were found to be in the range 3–28% (16–156 μmol). The results showed that the yields are higher for ethanol (10%, 57 μmol), n-propyl alcohol (13%, 73 μmol) and n-butanol (9%, 52 μmol). For other medium-chain-length alcohols like n-amyl alcohol (3%, 16 μmol), n-hexyl alcohol (9%, 47 μmol) and n-heptyl alcohol (5%, 26 μmol), the yields were much lower. However, from n-octyl alcohol to n-decyl alcohol, the yields were the highest (20%, 109 μmol to 23%, 127 μmol) with n-nonyl alcohol giving the highest yield (28%, 155 μmol). The yields decreased slightly with further increase in alcohol chain lengths up to stearyl alcohol. The shake-flask method gave lesser yields at pH 4.0 in general for the carbon chain lengths up to C7. From n-octyl alcohol onwards, the yields increased with increase in chain lengths up to C10.

7.7.2 Reflux Method

Alkyl glucosides were also synthesised with various alcohols by the reflux method (Vijayakumar et al. 2007) in presence of 0.04 mM (0.4 mL), pH 4.0 and 5.0 acetate buffer (Figs. 7.4 and 7.5). At

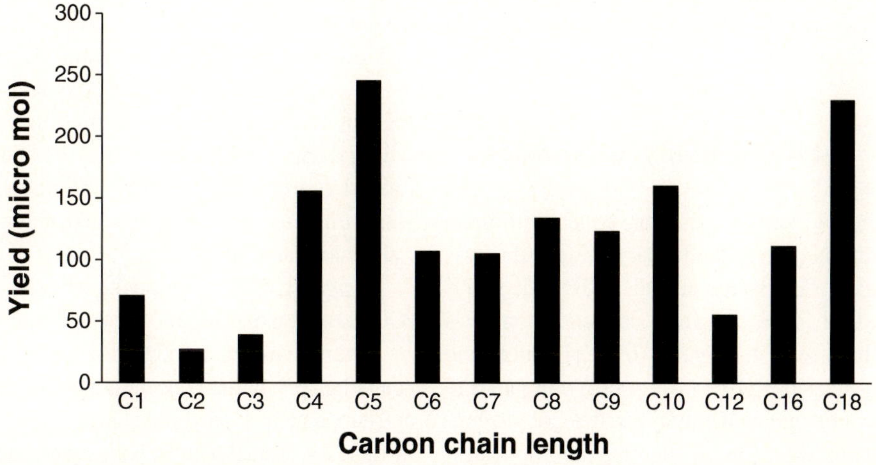

Fig. 7.4 Synthesis of n-alkyl glucosides by reflux method at pH 4.0. D-glucose (0.000555 mol) and alcohol in 1:50 M proportions along with 50% w/w of amyloglucosidase and 0.4 mL 0.01 M acetate buffer at pH 4.0 refluxed with 50 mL di-isopropyl ether (boiling point –68°C) for 72 h. Conversion yields by HPLC

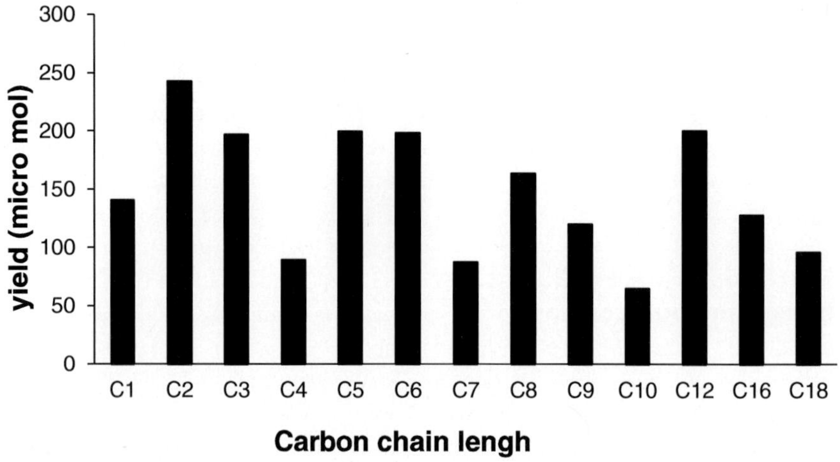

Fig. 7.5 Synthesis of n-alkyl glucosides by reflux method at pH 5.0, D-glucose (0.000555 mol) and alcohol in 1:50 M proportions along with 50% w/w of amyloglucosidase and 0.4 mL 0.01 M acetate buffer at pH 5.0 refluxed with 50 mL di isopropyl ether (boiling point 68°C) for 72 h. Conversion yields by HPLC

pH 4.0 the yields were lower for methyl (13%, 71 µmol), ethyl (5%, 27 µmol) and n-propyl alcohols (7%, 39 µmol). However, the yields were higher for the remaining alcohols. The highest yield was observed for n-amyl alcohol (44%, 245 µmol). In general, except lauryl alcohol (10%, 55 µmol), the yields were higher for n-octyl alcohol (24%, 134 µmol) onwards towards higher-chain-length alcohols.

At pH 5.0, the yields obtained were found to be in the range of 12% (65 µmol)–44% (242 µmol). Higher yields were observed for ethyl alcohol (44%, 242 µmol) and lauryl alcohol (36%, 200 µmol).

7.8 Cetyl and Stearyl Glucosides

The effect of increasing enzyme concentration on the synthesis of cetyl and stearyl glucosides was investigated (Vijayakumar et al. 2007). The yields obtained at 40% enzyme concentration was higher in case of both cetyl (6%, 34 µmol) and stearyl glucosides (19%, 105 µmol) compared to the other enzyme concentrations. The yields generally were higher at all the enzyme concentrations for stearyl alcohol compared to cetyl alcohol. This could be because the longer-chain-length alcohol functioned as a better nucleophile for accepting a D-glucose molecule than the shorter-chain-length alcohol.

7.9 Optimisation of n-Octyl-D-Glucoside Synthesis Using Response Surface Methodology

Ismail et al. (1998) have reported the synthesis of butyl glucoside by RSM. Using β-galactosidase, Chahid et al. (1994) have reported the synthesis of a mixture of octylglucoside and octylgalactoside through a transglycosylation reaction involving lactose and n-octanol. The present work deals with a detailed RSM analysis of the synthesis of n-octyl-D-glucoside using amyloglucosidase from *Rhizopus* sp. by the shake-flask method (Vijayakumar et al. 2005).

Design: Central composite rotatable design (CCRD), 32 experiments, 5 variables at 5 levels

Variables: n-Octanol concentration, enzyme concentration, pH, buffer concentration (buffer volume) and temperature. Coded and actual values are shown in Table 7.4. Actual set of experiments undertaken as per the CCRD with coded values and the glucosylation yields obtained is given in Table 7.5.

Table 7.4 Coded values of the variables and their corresponding actual values used in the design of experiments

Variables	−2	−1	0	1	2
n-Octanol (eq)	15	30	45	60	75
Amyloglucosidase (mg)	20	40	60	80	100
pH	4.0	5.0	6.0	7.0	8.0
Buffer volume (mL)	0.2	0.4	0.6	0.8	1.0
Temperature (°C)	30	40	50	60	70

Table 7.5 Experimental design with experimental and predicted yields of n-octyl-D-glucoside

Expt no	n-Octanol	Amylo glucosidase	pH	Buffer volume	Temperature	Yield [a]experimental	Yield predicted
1	−1	−1	−1	−1	1	20.0	22.4
2	−1	−1	−1	1	−1	10.7	12.3
3	−1	−1	1	−1	−1	27.2	30.3
4	−1	−1	1	1	1	0.9	5.8
5	−1	1	−1	−1	−1	37.5	35.6
6	−1	1	−1	1	1	1.8	1.8
7	−1	1	1	−1	1	14.4	13.8
8	−1	1	1	1	−1	13.1	11.6
9	1	−1	−1	−1	−1	31.6	30.8
10	1	−1	−1	1	1	1.2	4.2
11	1	−1	1	−1	1	26.6	27.1
12	1	−1	1	1	−1	17.8	17.4
13	1	1	−1	−1	1	18.1	15.6
14	1	1	−1	1	−1	17.9	14.5
15	1	1	1	−1	−1	37.2	33.3
16	1	1	1	1	1	2.5	2.4
17	0	0	0	−2	0	40.9	43.8
18	0	0	0	2	0	12.1	7.1
19	0	0	−2	0	0	14.1	15.9
20	0	0	2	0	0	20.0	18.1
21	0	−2	0	0	0	27.6	21.5
22	0	2	0	0	0	7.1	15.1
23	−2	0	0	0	0	26.3	24.3
24	2	0	0	0	0	26.2	28.1
25	0	0	0	0	−2	13.8	18.4
26	0	0	0	0	2	0	0
27	0	0	0	0	0	27.6	24.1
28	0	0	0	0	0	21.4	24.1
29	0	0	0	0	0	21.1	24.1
30	0	0	0	0	0	30.1	24.1
31	0	0	0	0	0	23.4	24.1
32	0	0	0	0	0	17.1	24.1

[a]Conversion yields were from HPLC with respect to 0.555 mmol of D-glucose. The experimental yields were an average from two experiments

Table 7.6 Validation of experimental data

n-Octanol (eq)	Amylo glucosidase (mg)	pH	Buffer volume (mL)	Temperature (°C)	Yield predicted	Yield[a] experimental
50	50	4.5	0.3	55	27.3	23.8
40	30	6.5	0.5	65	13.6	6.09
65	50	7.5	0.3	65	25.2	25.3
55	45	4.5	0.3	45	34.2	30.1
55	45	4.5	0.3	35	31.0	20.9
10	65	5.5	0.7	65	2.8	4.2
50	55	4.5	0.9	65	0	5.0
75	20	7.8	0.2	50	53.5	53.8

[a]Conversion yields were from HPLC with respect to 0.555 mmol of D-glucose. The experimental yields were an average from two experiments

Equation: A second-order polynomial equation was developed to study the effects of the variables on the esterification yields in terms of linear, quadratic and cross product terms.

$$Y=-98.557-0.780X_1 + 1.280X_2 + 17.752X_3 - 82.346X_4 + 4.005X_5 + 0.002X_1X_1 - 0.004\,X_2X_2 - 1.783\,X_3X_3 + 14.526X_4X_4 - 0.043X_5X_5 - 0.002X_1X_2 + 0.124X_1X_3 + 0.051X_1X_4 + 0.0003\,X_1X_5 - 0.066X_2X_3 + 0.049\,X_2X_4 - 0.009\,X_2X_5 + 1.341\,X_3X_4 + 0.035\,X_3X_5 + 0.166\,X_4X_5$$

where X_1 – n-octanol concentration; X_2 – amyloglucosidase concentration; X_3 – pH; X_4 – buffer volume; and X_5 – temperature.

Coefficients: Microsoft Excel software, Version 5.0

Analysis of Variance (ANOVA): Microsoft Excel software ANOVA showed the model is significant at $P<0.01$.

Optimisation: Microsoft Excel Solver function

The experimental data fitted the second-order polynomial equation well as indicated by a R^2 value of 0.895 (Table 7.5).

The maximum yield predicted based on the response model 53.5% is obtained at an n-octanol concentration of 75 eq 20 mg amyloglucosidase concentration, 0.30 mM (0.2 mL of 10 mM buffer) pH 7.8 buffer and 50°C. The experiments conducted at the above optimum conditions resulted in 53.8% yield. Apart from the above experiment, validation of the response model was carried out at selected random process conditions. These validation experiments closely agreed with the predicted yields (Table 7.6).

The effect of n-octanol and enzyme concentration on the glucosylation of n-octanol to n-octyl-D-glucoside at 0.6 mL of 10 mM, pH 6.0 at 50°C is shown in Fig. 7.6. The extent of conversion increased with the increase in n-octanol equivalents at lower enzyme concentrations. With increase in enzyme concentration from 20 to 60 mg (20–60% with respect to w/w D-glucose), the extent of glucosylation reached a maximum of 27.0% at 60 mg (60% w/w D-glucose) enzyme concentration. With further increase in enzyme concentration to 100 mg (100% w/w D-glucose), the yield decreased. This was the behaviour at all equivalents of n-octanol. Around an enzyme concentration of 60 mg (60% w/w D-glucose), increase in n-octanol equivalents exhibited a trough with a slight dip around an n-octanol concentration of 45 eq. The surface plot illustrates that at lower enzyme concentrations, sufficient amount of free n-octanol is available for transfer of D-glucose to n-octanol resulting in increase in glucosylation. At higher enzyme concentrations, a complete binding of n-octanol to the enzyme would effectively reduce the concentration of free n-octanol, thereby causing a reduction in the transfer of D-glucose molecule to n-octanol leading to reduced glucosylation.

The catalytic efficiency of the enzyme was found to be the highest at 50°C, and it decreased to very low values at 70°C at all enzyme concentrations (Fig. 7.7). Other conditions kept constant were n-octanol – 45 eq, pH – 6.0 and

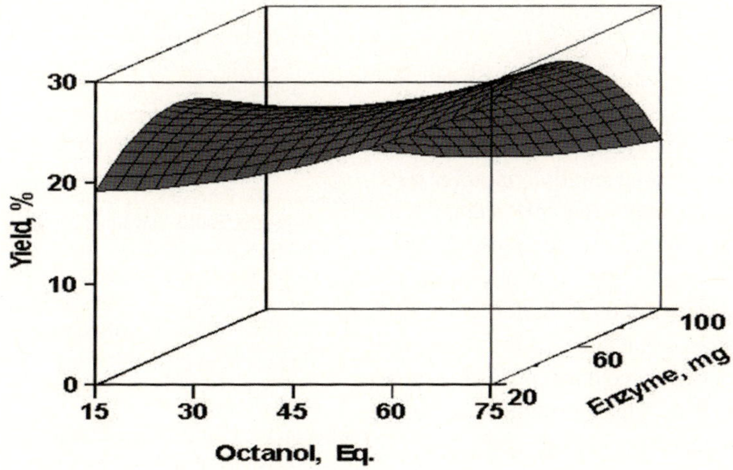

Fig. 7.6 Three-dimensional surface plot showing the effect of n-octanol concentration and enzyme concentration on the extent of glucosylation (pH –6.0, buffer volume –0.6 mL, temperature –50°C)

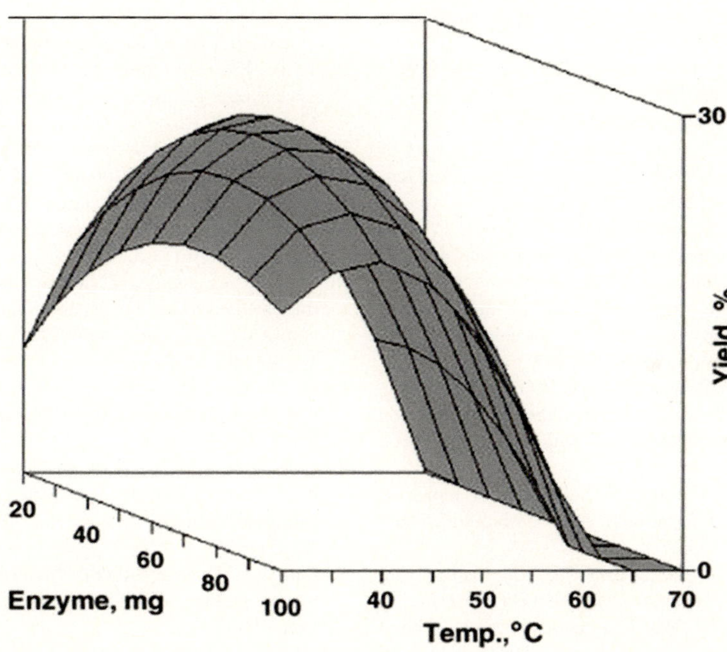

Fig. 7.7 Three-dimensional surface plot showing the effect of enzyme concentration and temperature on the extent of glucosylation (n-octanol – 45 eq, pH – 6.0, 0.6 mL of 10 mM buffer)

buffer concentration – 1.3 mM (0.6 mL of 10 mM buffer). At 30°C, glucosylation occurred at all enzyme concentrations. The highest conversion was with 60 mg (60% w/w D-glucose) of enzyme. At lower enzyme concentrations, the conversions were lower. However, they increased with increase in temperature up to 50°C indicating that the optimum temperature required for glucosylation is 50°C. At a higher temperature of 70°C, the enzyme most probably underwent denaturation leading to a loss in catalytic activity.

The effect of buffer volume and pH on the extent of glucosylation at an n-octanol concentration of 45 eq, 60 mg enzyme and 50°C effectively indicated the role of concentration of buffer salts and their effect on importing 'pH memory' to amyloglucosidase. With increase in pH, glucosylation decreased at all buffer volumes. An optimum buffer volume of 0.6 mL buffer was found to be the best for this reaction.

Thus, this study showed that this model is very good in predicting the glucosylation of n-octanol by the amyloglucosidase enzyme.

References

Balzar D (1991) Alkylglucosides, their physico-chemical properties and their uses. Tenside Surfactant Deterg 28:419–427

Busch P, Hensen H, Khare J, Tesmann H (1994) Alkylpolyglycosides-a new cosmetic concept for milderness. Agro-Food-Ind Hi-Tech 5:20–28

Chahid Z, Montet D, Pina M, Graille J (1992) Effect of water activity on enzymatic synthesis of alkylglycosides. Biotechnol Lett 14(4):281–284

Chahid Z, Montet D, Pina M, Bonnot F, Graille J (1994) Biocatalyzed octylglycoside synthesis from a disaccharide. Biotechnol Lett 16:795–800

Crout DHG, Vic G (1998) Glycosidases and glycosyl transferases in glycoside and oligosaccharides synthesis. Biocatal Biotransform 2:98–111

Gygax D, Spies P, Winkler T, Pfaar U (1991) Enzymatic synthesis of β-D-glucuronides with *in situ* regeneration of uridine 5′-diphosphoglucuronic acid. Tetrahedron 47:5119–5122

Haines AH (1976) Relative reactivities of hydroxyl groups in carbohydrates. Adv Carbohydr Chem Biochem 33:11–109

Ichikawa Y, Look GC, Wong CH (1992) Enzyme-catalysed oligosaccharide synthesis. Anal Biochem 202:215–238

Ismail A, Ghoul M (1996) Enzymatic synthesis of butyl glycosides by glycosidases. Biotechnol Lett 18:1199–1204

Ismail A, Soultani S, Ghoul M (1998) Optimization of the enzymatic synthesis of butyl glucoside using response surface methodology. Biotechnol Prog 14:874–878

Ismail A, Linder M, Ghoul M (1999) Optimization of butylgalactoside synthesis by β-galactosidase from *Aspergillus oryzae*. Enzyme Microb Technol 25:208–213

Konstantinovic S, Predojevic J, Gojkovic S, Ratkovic Z, Mojsilovic B, Pavlovic V (2001) Synthesis of C7-C16 alkyl 2,3 dideoxy glucosides from glucose and fatty acids. Indian J Chem 40B:1242–1244

Kosary J, Stefanovits-Banyai E, Boross L (1998) Reverse hydrolytic process for O-alkylation of glucose catalyzed by immobilized α- and β-glucosidases. J Biotechnol 66:83–86

Laroute V, Willemot RM (1992) Glucoside synthesis by glucoamylase or β-glucosidase in organic solvents. Biotechnol Lett 14:169–174

Madsen T, Petersen G, Seiero C, Torslov J (1996) Biodegradability and aquatic toxicity of glycoside surfactants and a nonionic alcohol etherate. J Am Oil Chem Soc 73:929–933

Matsumura S, Imai K, Yoshikawa S, Kawada K, Uchibori T (1990) Surface activities, biodegradability and antimicrobial properties of n-alkyl glucosides, mannosides and galactosides. J Am Oil Chem Soc 67:996–1001

Monsan PF, Paul F, Pelenc P, Bouler E (1996) Enzymatic production of α-butyl glucoside and its fatty acid esters. Ann N Y Acad Sci 799:633–641

Mutua LN, Akoh CC (1993) Synthesis of alkyl glucoside fatty acid esters in non aqueous media by *Candida* sp. lipase. J Am Oil Chem Soc 70:43–46

Nilsson KGI (1987) A simple strategy for changing the regioselectivity of glycosidase catalyzed formation of disaccharides. Carbohydr Res 167:95–103

Rantwijk FV, Oosterom MW, Sheldon RA (1999) Glycosidase-catalyzed synthesis of alkyl glycosides. J Mol Catal B: Enzym 6:511–532

Roitsch T, Lehle L (1989) Structural requirements for protein N-glycosylation. Influence of acceptor peptides on cotranslational glycosylation of yeast influence and site-directed mutagenesis around a sequon sequence. Eur J Biochem 181:525–529

Rosenthal KS, Loussale F (1983) Critical micelle concentration determination of non-ionic detergent with comassive brilliant blue-G250. Anal Chem 55: 1115–1117

Rubio E, Fernandez MA, Klibanov AM (1991) Effect of the solvent on enzyme regio-selectivity. J Am Chem Soc 113:695–696

Shin HK, Kong JY, Lee JD, Lee TH (2000) Synthesis of hydroxy benzyl-α-glucosides by amyloglucosidase-catalysed transglycosylation. Biotechnol Lett 22:321–325

Sivakumar R, Divakar S (2006) Glycosylation of vanillin by amyloglucosidase in organic media. Tetrahedron Lett 47:695–699

Stevenson DE, Stanley RA, Furneaux RH (1993) Optimization of alkyl β-D-galactopyranoside synthesis from lactose using commercially available β-galactosidase. Biotechnol Bioeng 42:657–666

Suzuki Y, Ki YH, Uchida K, Takami M (1996) Enzymatic synthesis of glucosylated and phosphatidylated biologically active compounds. J Appl Glycosci 43:273–282

Thiem J (1995) Applications of enzymes in synthetic carbohydrate chemistry. FEMS Microbiol Rev 16:193–211

Vic G, Crout DHG (1995) Synthesis of allyl and benzyl β-D-glucopyranosides and allyl β-D-galactopyranoside from D-glucose or D-galactose and the corresponding alcohol using almond β-D-glucosidase. Carbohydr Res 279:315–319

Vic G, Thomas D (1992) Enzyme-catalyzed synthesis of alkyl-β-D-glucosides in organic media. Tetrahedron Lett 33:4567–4570

Vic G, Hastings JJ, Crout DHG (1996) Glycosidase catalyzed synthesis of glycosides by an improved procedure for reverse hydrolysis. Application to the chemoenzymatic synthesis of galactopyranosyl-(1→4)-O-α-galactopyranoside derivatives. Tetrahedron Asymmetry 7:1973–1984

Vic G, Thomas D, Crout DHG (1997) Solvent effect on enzyme-catalyzed synthesis of β-D-glucosides using the reverse hydrolysis method application to the preparative-scale synthesis of 2-hydroxybenzyl and octyl β-D-glucopyranosides. Enzyme Microb Technol 20:597–603

Vijayakumar GR, Divakar S (2005) Synthesis of guaiacol-α-D-glucoside and curcumin- bis-α-D-glucoside by an amyloglucosidase from *Rhizopus*. Biotechnol Lett 27:1411–1415

Vijayakumar GR, Manohar B, Divakar S (2005) Amyloglucosidase catalyzed synthesis of n-octyl-D-glucoside-analysis using response surface methodology. Eur Food Res Technol 220:272–277

Vijayakumar GR, George C, Divakar S (2007) Synthesis of n-alkyl glucosides by amyloglucosidase. Indian J Chem B 46B:314–319

activity and modification is needed in its application. Guaiacol, eugenol, curcumin and α-tocopherol are lipophilic and insoluble in water, which limits their further pharmacological exploitation and practical applications. Currently, very few reports are available on the synthesis of curcuminyl glycosides (Kaminaga et al. 2003; Mohri et al. 2003) and eugenol glycosides (Orihara et al. 1992; Sato et al. 2003) and that too by cell suspension and chemical methods.

Chemical preparation of glycosides cannot meet EC food regulations, and therefore chemical preparation of glycosides is not applicable in the food industry. Further, regio- and stereoselectivity are two main criteria which are difficult to achieve by chemical methods due to multiple hydroxyl groups in sugar molecule which require protection and deprotection (Mohri et al. 2003). Hence, enzyme-catalysed reactions appear to be good alternatives.

This chapter describes synthesis of glycosides of phenols such as guaiacol, eugenol, curcumin, N-vanillylnonanamide, capsaicin, vanillin, DL-dopa, L-dopa, dopamine, serotonin and epinephrine by reflux method using amyloglucosidase from Rhizopus mould, β-glucosidase and immobilised β-glucosidase with carbohydrates: D-glucose, D-galactose, D-mannose, D-fructose, D-arabinose, D-ribose, maltose, sucrose, lactose, D-mannitol and D-sorbitol (Vijayakumar et al. 2006; Vijayakumar and Divakar 2005, 2007, 2011a, b; Sivakumar et al. 2006a, b, 2009; Sivakumar and Divakar 2007, 2009a, b). The reactions were investigated in terms of incubation period, pH, buffer, enzyme and substrate concentrations, regio- and stereoselectivity.

8.1.1 Guaiacyl Glycosides

8.1.1.1 Guaiacyl-D-Glucoside

Synthesis of guaiacyl-α-D-glucoside was studied in detail to understand phenolic glycosylation of such molecules (Vijayakumar and Divakar 2005). Guaiacyl-α-D-glucoside synthesis in di-isopropyl ether using amyloglucosidase was optimised in terms of incubation period, pH, buffer concentration, enzyme concentration and D-glucose concentration (Scheme 8.1). Optimum conditions determined were 1:50 equivalents of D-glucose and guaiacol, 50% (w/w D-glucose) of amyloglucosidase, 0.06-mM (0.6 mL) pH 7.0 phosphate buffer and 72-h incubation period (Table 8.1, Fig. 8.1).

8.1.1.2 Synthesis of Guaiacyl Glucoside Using β-Glucosidase

Guaiacyl-D-glucoside synthesised by the reflux method using β-glucosidase isolated from sweet almonds involved considering D-glucose and guaiacol in 1:50 M ratio along with 50% (w/w D-glucose) enzyme and 0.06-mM (0.6-mL) pH 7.0 buffer in di-isopropyl ether solvent and refluxed for 72 h. The reaction mixture analysed by HPLC showed 22% conversion with respect to the D-glucose concentration employed.

8.1.1.3 Synthesis of Guaiacyl Glycosides

Aldohexoses (D-galactose and D-mannose), ketohexose (D-fructose), aldopentoses (D-arabinose and D-ribose), disaccharides (maltose, sucrose and lactose) and carbohydrate alcohols (D-mannitol and D-sorbitol) were employed for the synthesis of guaiacyl glycosides under the optimum conditions obtained for guaiacyl-α-D-glucoside synthesis (Scheme 8.1). D-Galactose reacted with guaiacol and gave the conversion yield of 17%. The retention times for D-galactose and guaiacyl-D-galactoside were 5.2 and 8.1 min, respectively. Other carbohydrates did not undergo any glycosylation with guaiacol (Table 8.2).

8.1.1.4 Spectral Characterisation

UV spectrum of guaiacyl-α-D-glucoside showed $\sigma \rightarrow \sigma^*$ band in the region 201–210 nm and $n \rightarrow \pi^*$ band between 270 and 282 nm (free guaiacol 279.5 nm). The hypsochromic shift observed for the glycoside indicated that guaiacol had undergone glycosylation. IR spectral band in the region 1,030–1,041 cm^{-1} corresponded to glycosidic aryl–alkyl C–O–C symmetrical stretching and 1,236–1,250 cm^{-1} to glycosidic aryl–alkyl C–O–C asymmetrical stretching frequencies. NMR spectral data confirmed that guaiacol formed C1α-glucoside, C1β-D-glucoside and C6-O-arylated product with D-glucose. Mass

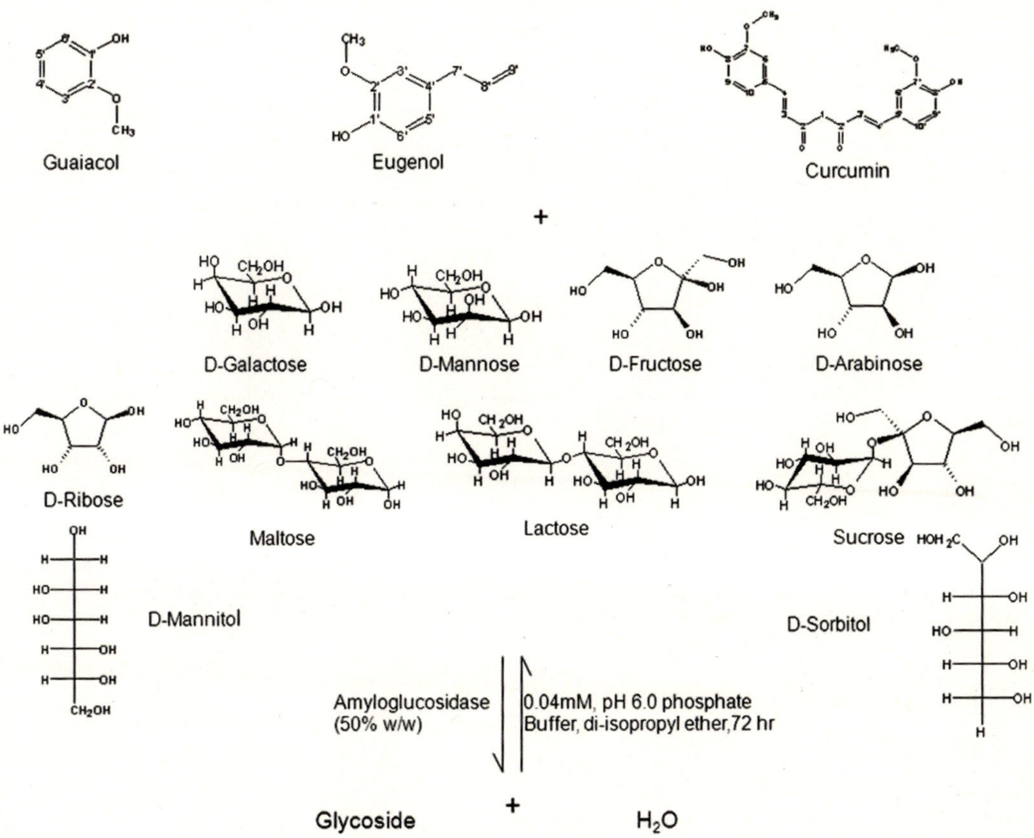

Scheme 8.1 Synthesis of guaiacol, eugenol and curcumin glycosides

Table 8.1 Effect of buffer pH and buffer concentration on guaiacyl-α-D-glucoside synthesis[a]

pH[b]	Yield, %	Buffer concentration[c] (mM) at pH 7.0	Yield, %
4.0	30	0.01	3
5.0	43	0.02	11
6.0	31	0.04	51
7.0	51	0.06	52
8.0	32	0.08	39
		0.1	11

[a]Conversion yields from HPLC with respect to 0.55 mmol of D-glucose. Error in yield measurements ± 5–10%. Guaiacol, 0.027 mol; enzyme, 50% (w/w D-glucose); incubation, 72 h; temperature, 68 °C

[b]0.04 mM (0.4 mL of 10-mM buffers added in 100-mL reaction mixture)

[c]0.1–1.0 mL of pH 7.0 phosphate buffer

spectrum for guaiacyl-D-glucoside showed *m/z* peak at 286 $[M]^+$ supporting the NMR data for the monoglycoside formation.

Downfield chemical shift of C1α signal at 95.2 ppm (^{1}H at 4.97 ppm) indicated that a C1 glycosylated product was formed with galactose and C6-O- signal at 67.0 ppm (^{1}H at 3.70 ppm) indicated C6-O-arylated product. Mass spectrum showed *m/z* peak at 286 $[M]^+$ indicating monogalactoside formation.

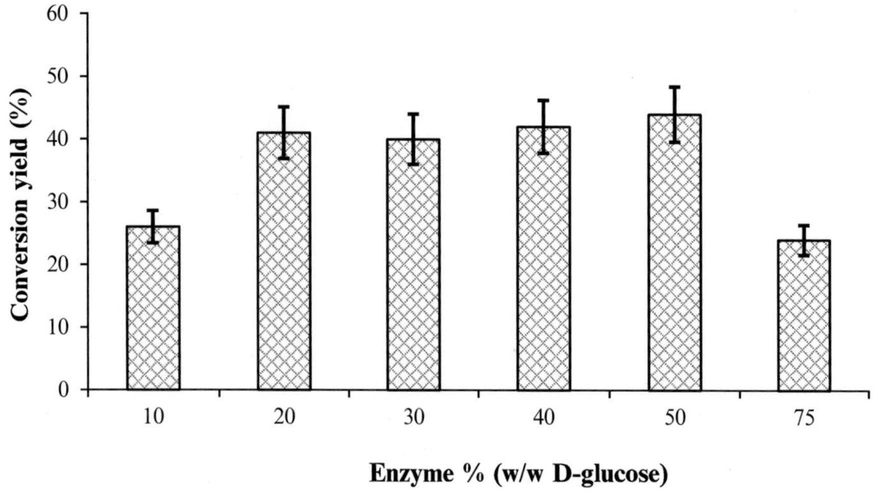

Fig. 8.1 Effect of amyloglucosidase concentration on guaiacyl-α-D-glucoside synthesis. Conversion yields were from HPLC with respect to 0.555 mmol of D-glucose. Reaction conditions: D-glucose, 0.555 mmol; guaiacol, 0.027 mol; 0.05-mM (0.5 mL of 10-mM buffer in 100-mL reaction mixture) pH 7.0 phosphate buffer; solvent, di-isopropyl ether; incubation, 72 h; temperature, 68°C

Table 8.2 Guaiacyl glycosides with conversion yields and product proportions

Glycosides and product proportions (%)[a]		Glycosylation yield[b] (%)
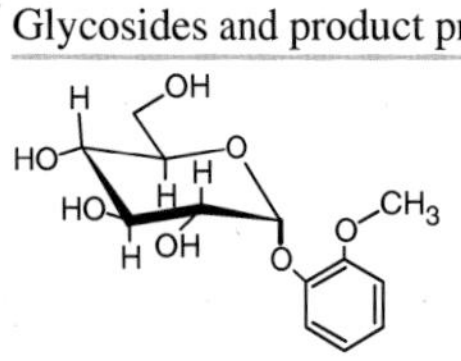 Guaiacyl-α-D-glucoside (52)	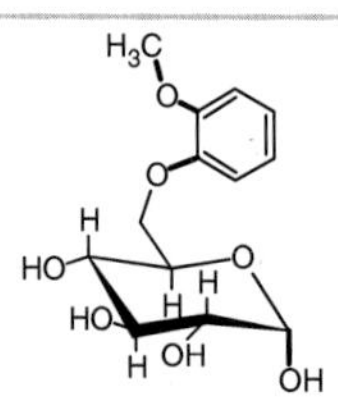 C6-O-guaiacyl-D-glucose (48)	52
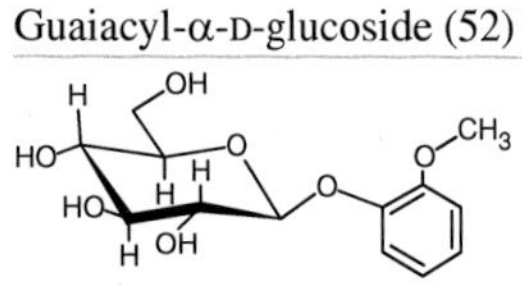 Guaiacyl-β-D-glucoside[c]		22
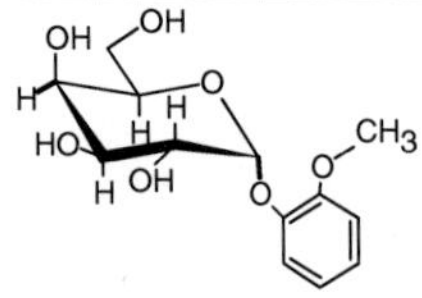 Guaiacyl-α-D-galactoside (95)	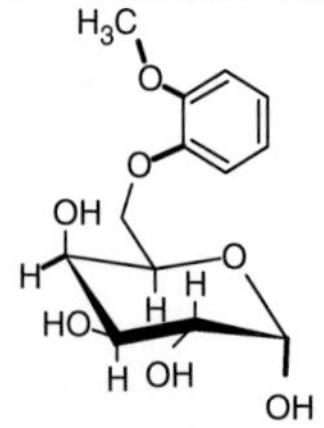 C6-O-guaiacyl-D-galactose (5)	17

[a]Product proportions shown in brackets were determined from 2-D HSQCT NMR C1/C6 cross peak areas
[b]Conversion yields were from HPLC with errors in yield measurements ±5–10%
[c]The compound was synthesised by using β-glucosidase from sweet almonds

8.1.2 Eugenyl Glycosides

Eugenyl glycosides were synthesised using amyloglucosidase (Vijayakumar and Divakar 2007).

8.1.2.1 Eugenyl Maltoside

Amyloglucosidase-catalysed synthesis of eugenyl maltoside was optimised in terms of incubation period, pH, buffer concentration and enzyme concentration as a prototype reaction for detailed investigation. The optimum conditions determined were as follows: glycosylation yield increased with the increase in incubation period from 3 h (7% yield) to 72 h (39% yield) and was the highest at 72 h at an initial rate of 19.2 μmol/h; conversion yield increased from 14% for 0.04 mM (0.4 mL) to 39% for 0.1 mM (1.0 mL); maximum glycosylation (39% yield) occurred at pH 5.0 acetate buffer; between 10 and 80% (w/w maltose) of enzyme concentration, 40% enzyme was found to be the best (conversion yield of 39%); and all other enzyme concentrations gave very less conversion yields (less than 19%).

8.1.2.2 Synthesis of Eugenyl Glucoside Using β-Glucosidase

β-Glucosidase isolated from sweet almond employed for the synthesis of eugenol glucoside in di-isopropyl ether solvent gave 19% conversion (from HPLC).

8.1.2.3 Synthesis of Eugenyl Glycosides

Eugenyl glycosides of carbohydrates were synthesised (Scheme 8.1). Retention times for the carbohydrates and glycosides are D-glucose, 5.20 min; eugenyl-D-glucoside, 7.47 min; D-mannose, 4.9 min; eugenyl-D-mannoside, 5.87 min; sucrose, 6.4 min; eugenyl sucrose, 8.02 min; D-mannitol, 5.3 min; and eugenyl-D-mannitol, 7.12 min. From HPLC results, it was confirmed that D-galactose, D-fructose, D-arabinose, D-ribose, lactose and D-sorbitol did not undergo glycosylation with eugenol (Table 8.3).

8.1.2.4 Spectral Characterisation

UV spectra of the eugenyl glycosides showed shifts in the wavelength for $\pi \rightarrow \pi^*$ (extended conjugation) band in the range 265–279 nm (275.5 nm for free eugenol) confirming product formation. IR C–O–C symmetrical stretching frequencies in the range 1,020–1,066 cm^{-1} and asymmetrical stretching frequencies in the range 1,262–1,268 cm^{-1} indicated that eugenol had undergone glycosylation. Allylic C=C stretching frequency bands for the glycosides were also detected in the range 1,631–1,676 cm^{-1}. 2-D HSQCT spectral data showed formation of C1 glycoside and C6-O-arylated products. The mass *m/z* peaks (experimental section) further confirmed the formation of above-mentioned monoglycosides.

In the presence of β-glucosidase, D-glucose reacted with eugenol, and the only product, C1β-glucoside, was found to be formed. In the anomeric region, the only cross peak (^{13}C at 103.3 ppm and ^{1}H at 4.19 ppm) apart from the free C1α and C1β cross peaks indicated the formation of C1β-glucoside.

8.1.3 Curcuminyl Glycosides

8.1.3.1 Synthesis of 1,7-*O*-(bis-β-D-Glucopyranosyl)Curcumin Using β-Glucosidase

Synthesis of 1,7-*O*-(bis-β-D-glucopyranosyl)curcumin using β-glucosidase from sweet almond was optimised in terms of incubation period, pH, buffer, enzyme and curcumin concentration in di-isopropyl ether solvent (Vijayakumar and Divakar 2005, 2007). In summary the essential findings are as follows: incubation period showed increase in conversion yields – 4% (12 h), 8% (24 h), 12% (48 h) and 25% (72 h) with increase in incubation period (Fig.8.2a) at an initial rate of 3.24 μmol/h; variation in buffer pH from pH 4.0 to 8.0 at 0.04 mM (0.4 mL) showed that the highest conversion yield of 29% was obtained at pH 4.0; effect of buffer concentration at pH 4.0 showed that conversion yield increased with increase in buffer concentration from 0.01 to 0.1 mM (0.1–1.0 mL) with the highest yield of 48% at 0.1 mM (1.0 mL); increasing amyloglucosidase concentration showed that (Fig. 8.2b) 50% (w/w D-glucose) enzyme was required to achieve a maximum conversion of 48%; higher curcumin concentrations could be inhibitory to the enzyme as the conversion yield decreased from 37% for 0.2 and

Table 8.3 Eugenyl glycosides with conversion yields and product proportions

Glycosides and product proportions (%)[a]			Glycosylation yield[b] (%)
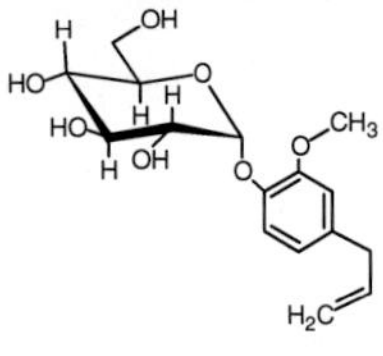 Eugenyl-α-D-glucoside (53)	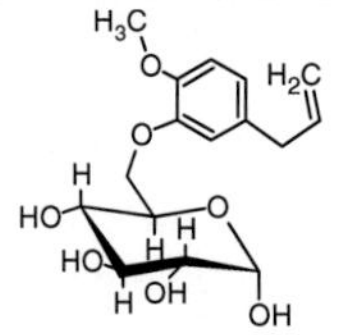 C6-O-eugenyl-D-glucose (47)		32
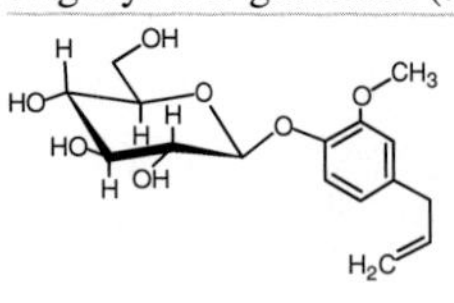 Eugenyl-β-D-glucoside[c]			19
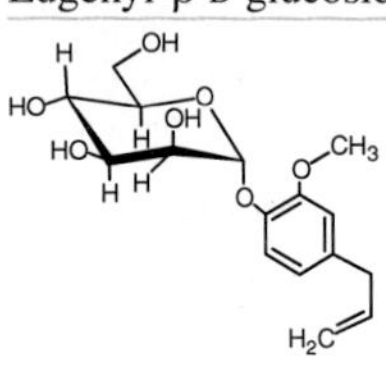 Eugenyl-α-D-mannoside			8
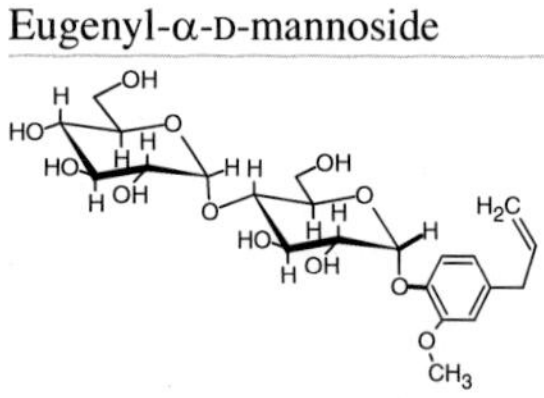 Eugenyl maltoside (52)	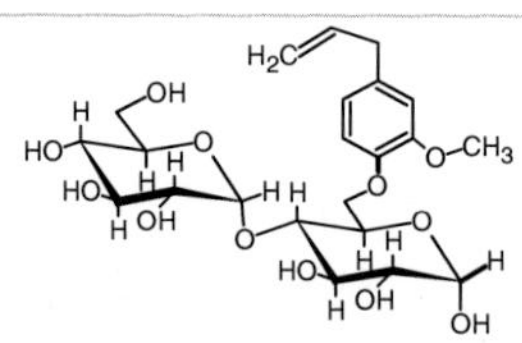 C6-O-eugenyl maltose (28)	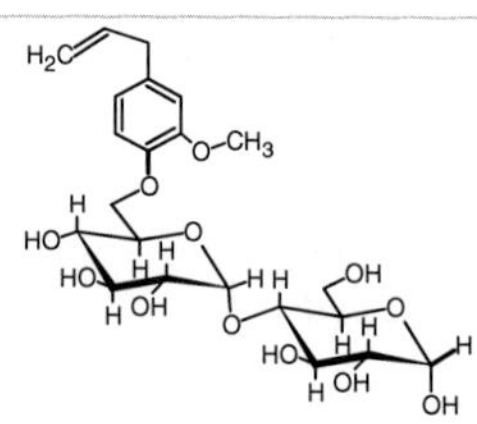 C6″-O-eugenyl maltose (20)	17
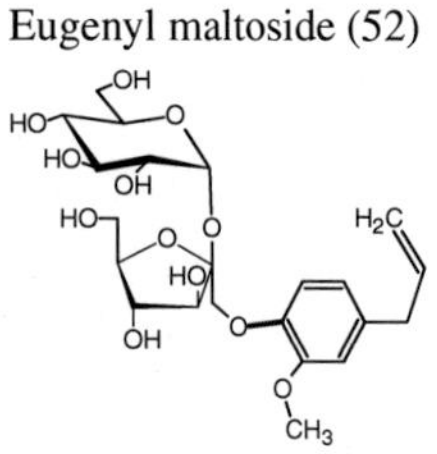 C1-O-eugenyl sucrose (35)	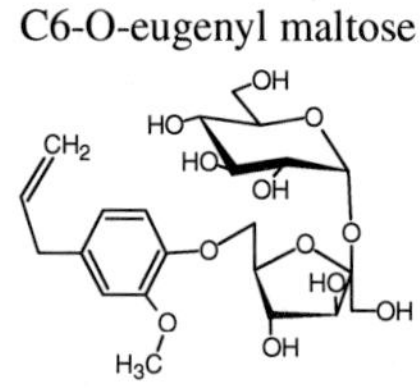 C6-O-eugenyl sucrose (45)	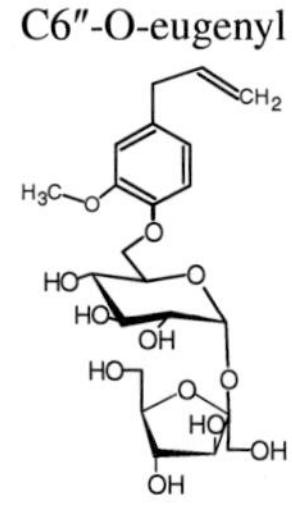 C6″-O-eugenyl sucrose (20)	7
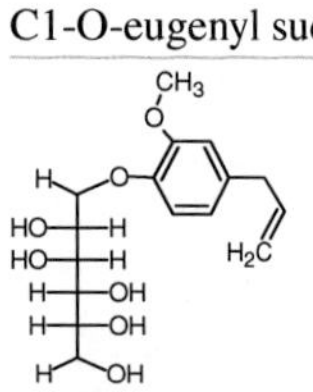 C1-O-eugenyl mannitol			7

[a]Product proportions determined from 2-D HSQCT NMR C1/C6 cross peak areas are shown in brackets
[b]Conversion yields were from HPLC with errors in yield measurements ±5–10%
[c]The compound was synthesised by using β-glucosidase from sweet almonds

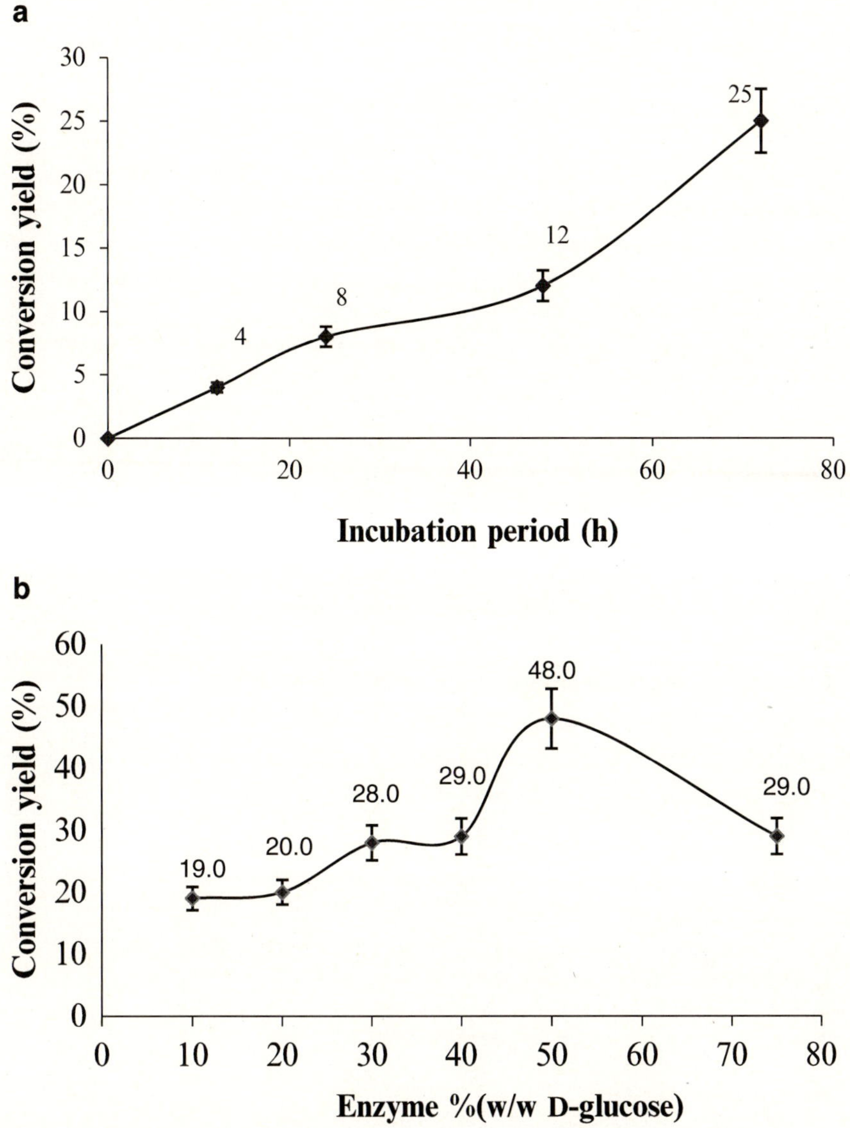

Fig. 8.2 Curcuminyl-bis-α-D-glucoside synthesis. Conversion yields were from HPLC with respect to 1-mmol D-glucose. (**a**) Effect of incubation period. Reaction conditions: D-glucose, 1 mmol; curcumin, 0.5 mmol; 0.06-mM (0.6-mL), pH 6.0 phosphate buffer. (**b**) Effect of enzyme concentration. Reaction conditions: D-glucose, 1 mmol; curcumin, 0.5 mmol; incubation period, 72 h; 0.1 mM (1.0 mL) of pH 4.0 acetate buffer

0.4 mmol curcumin to 34% (0.8 mmol), 29% (1.0 mmol), 22% (1.2 mmol) and 9% at 1.5 mmol curcumin; and maximum glycoside yield of 42% was obtained between 1.5 and 2.0 mmol of D-glucose concentrations.

The optimum conditions for this reaction were found to be curcumin and D-glucose in the 1:2 M ratio, 50% (w/w D-glucose) amyloglucosidase, 0.1-mM (1.0-mL), pH 4.0 acetate buffer and 72-h incubation period.

8.1.3.2 Antioxidant Activity

Antioxidant activity of curcumin and curcuminyl-bis-α-D-glucoside was evaluated by DPPH (2,2-diphenyl-1-picrylhydrazyl) radical scavenging method (Moon and Terao 1998). An antioxidant activity of 80% for curcuminyl-bis-α-D-glucoside was obtained. The antioxidant activity of the glycoside was found to be comparable to that of free curcumin (79%). Standard BHA activity was found to be 82%.

8.1.3.3 Solubility Test

Determination of the water solubility of curcuminyl-bis-α-D-glucoside showed that it is soluble to the extent of 14 g L^{-1}. Hence, curcuminyl-bis-α-D-glucoside was found to exhibit higher solubility than curcumin in water which exhibited negligible solubility.

8.1.3.4 Total Colour Test

Curcuminyl-bis-α-D-glucoside exhibited a total colour of 8.8 in DMSO with curcumin as the standard. The colour test was carried out according to the AOAC Official Methods of Analysis (1995).

8.1.3.5 Curcuminyl Glycosides of Other Carbohydrates

Amyloglucosidase-catalysed synthesis of curcuminyl glycosides was synthesised (Scheme 8.1). The extent of glycosylation was analysed by HPLC. The retention times are D-glucose, 5.20 min; curcuminyl-bis-D-glucoside, 7.52 min; D-mannose, 4.9 min; curcuminyl-bis-D-mannoside, 6.78 min; maltose, 7.4 min; curcuminyl-bis-maltoside, 8.3 min; sucrose, 6.4 min; curcuminyl-bis-sucrose, 8.44 min; D-mannitol, 5.3 min; and curcuminyl-bis-D-mannitol, 8.35 min. From HPLC results, it was confirmed that D-galactose, D-fructose, D-arabinose, D-ribose, lactose and D-sorbitol did not undergo glycosylation with curcumin (Table 8.4). The carbohydrates, which underwent glycosylation, are D-glucose, D-mannose, maltose, sucrose and D-mannitol.

Table 8.4 Curcuminyl glycosides with conversion yields and product proportions

Glycosides and product proportions (%)[a]	Glycosylation yield[b] (%)
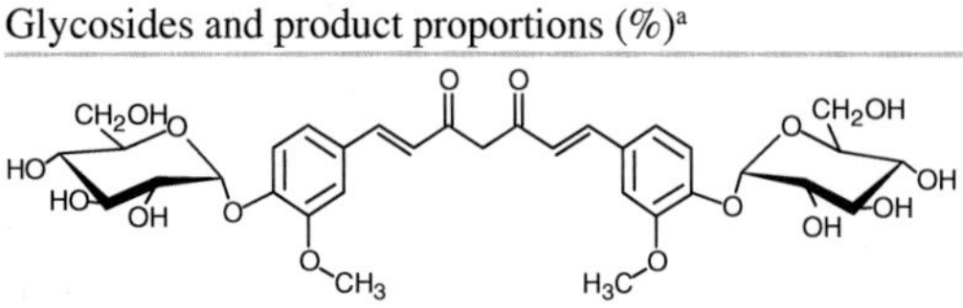 Curcuminyl-bis-α-D-glucoside (62) 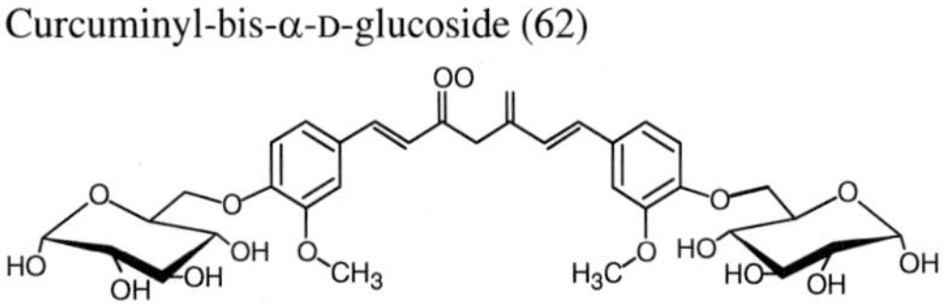 C6-O-curcuminyl-bis-D-glucose (38)	48
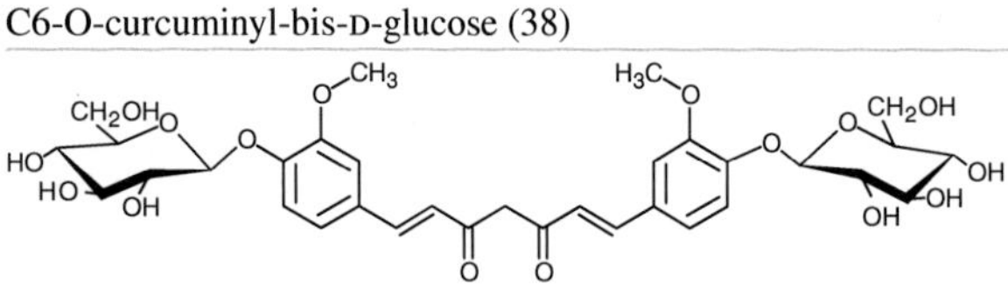 Curcuminyl-bis-β-D-glucoside[c]	11
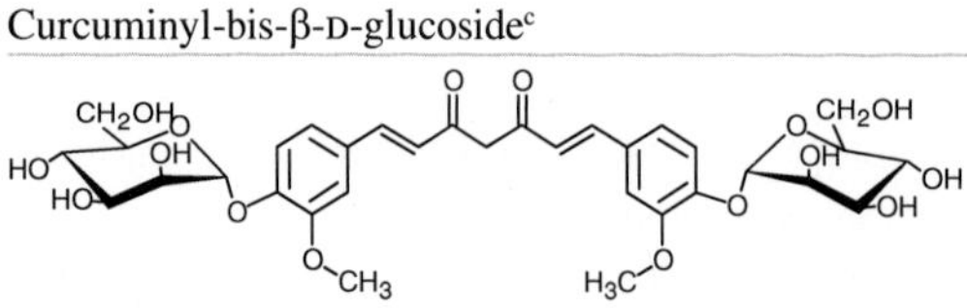 Curcuminyl-bis-α-D-mannoside	9

(continued)

Table 8.4 (continued)

Glycosides and product proportions (%)[a]	Glycosylation yield[b] (%)
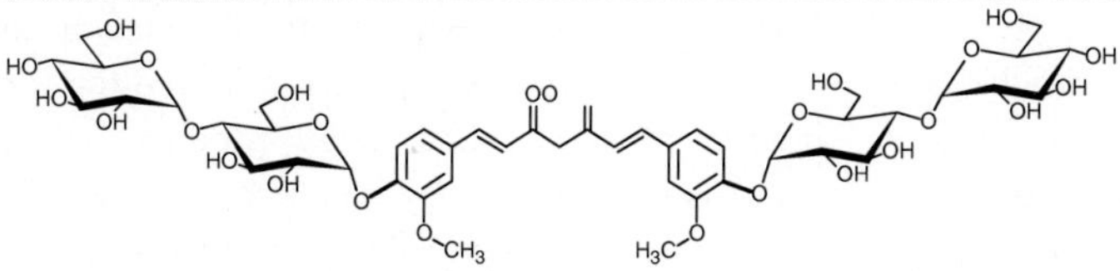 Curcuminyl-bis-maltoside (37) 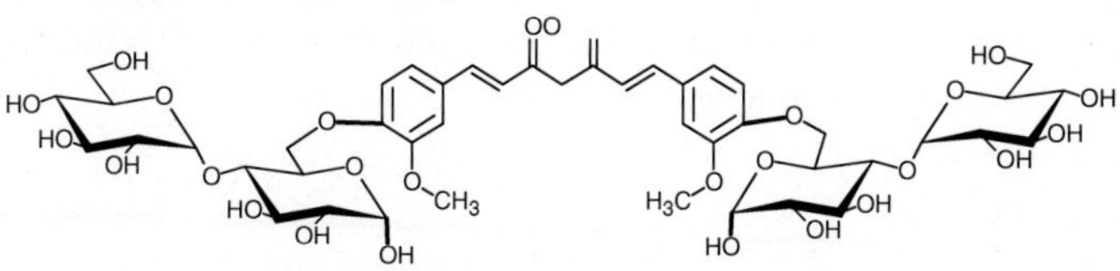 C6-O-curcuminyl-bis-maltose (36) 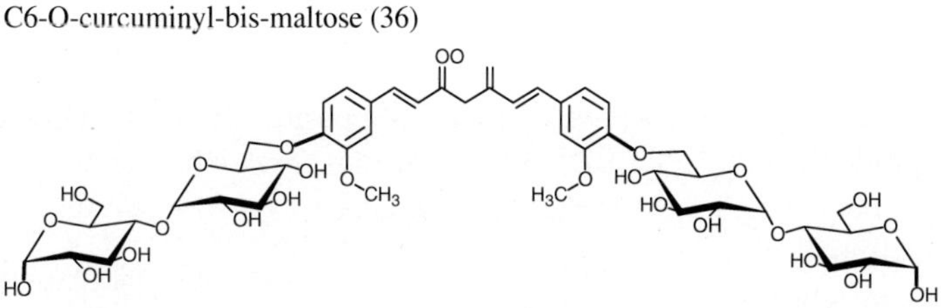 C6″-O-curcuminyl-bis-maltose (27)	19
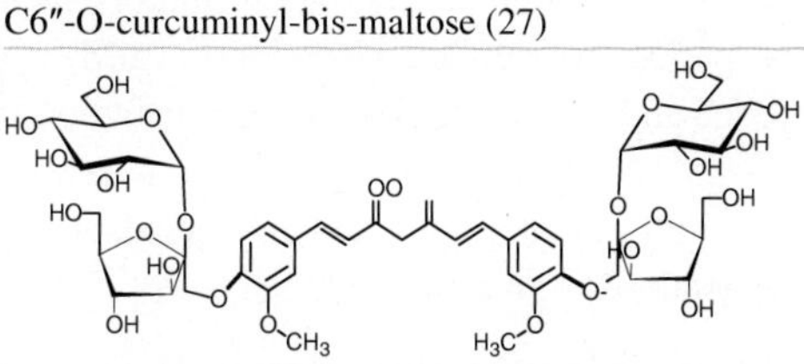 C1-O-curcuminyl-bis-sucrose (12) 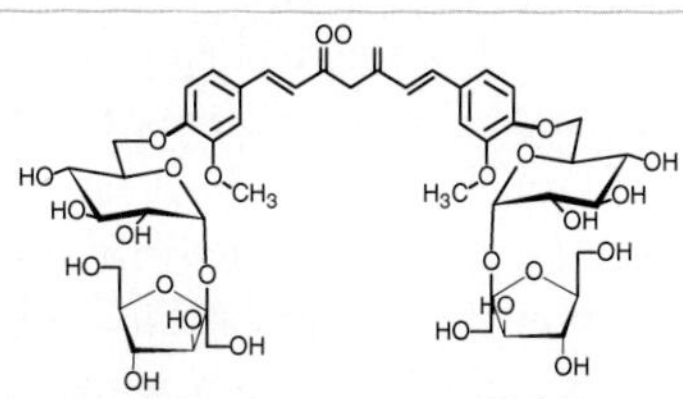 C6″-O-curcuminyl-bis-sucrose (18) 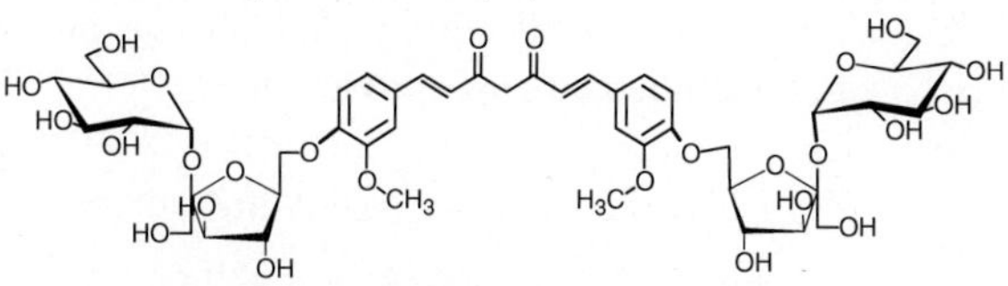 Curcuminyl-bis-sucrose (70)	19
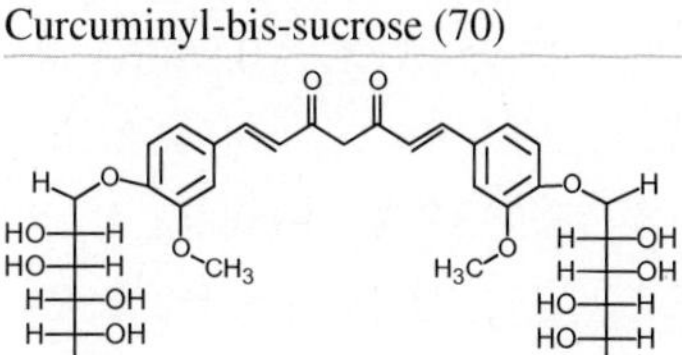 C1-O-curcuminyl-bis-D-mannitol	14

[a]Product proportions determined from ^{13}C 2-D HSQCT NMR C1/C6 peak areas or their cross peaks are shown in brackets

[b]Conversion yields were from HPLC with errors in yield measurements ±5–10%

8.1.3.6 Spectral Characterisation

The structures of the glycosides formed, HPLC yield and product proportions are presented in Table 8.4. Curcumin, a higher phenyl propanoid homologue of eugenol, also reacted with the same carbohydrate molecules D-glucose, D-mannose, maltose, sucrose and D-mannitol as eugenol.

From the UV spectra, shift in the wavelength for the $\pi \rightarrow \pi^*$ extended conjugation band in the 410–430 nm (433 nm for free curcumin) confirmed glycoside formation. IR C–O–C symmetrical stretching frequencies in the range 1,020–1,071 cm^{-1} and asymmetrical stretching frequencies in the range 1,254–1,262 cm^{-1} also indicated that curcumin had undergone glycosylation. The symmetrical chemical shift values in the curcumin region of the product glycosides in the 2-D HSQCT spectra indicated that bis-glycosylated and bis-C6-O-arylated products were formed. Curcumin glycosides did not change colour on treatment with dilute alkali indicating that both the phenolic OH groups are glycosylated. Two-dimensional HSQCT NMR spectrum for the glucoside showed downfield chemical shift values for C8 and C8′ at 150 ppm (149.4 ppm for free curcumin) and C1α at 99.0 ppm (^{1}H at 4.66 ppm), respectively, confirming that curcumin was glucosylated symmetrically at the C8 and C8′ position with the C1 carbon of the α-D-glucose anomer. Chemical shift values of the C6 cross peak (^{13}C at 66.5 ppm and ^{1}H at 3.52 and 3.70 ppm) confirmed C6-O-arylated product also being formed. Mass spectrum showed *m/z* value 691.5 $[M-1]^+$ for the bis-glycosylated products.

D-Mannose formed only a bis-C1α product (^{13}C at 101 ppm and ^{1}H at 4.99 ppm). Mass spectrum showed *m/z* value at 692.4 $[M]^+$ for the bis-D-mannoside. Maltose formed C1α-maltoside, C6-O-arylated and C6″-O-arylated products. In the mass spectrum, the mass *m/z* peak at 1,016 $[M]^+$ corresponded to the bis product. A nonreducing sugar, sucrose, reacted with curcumin, which resulted in the formation of three arylated products (Table 8.4) as evidenced from the shift in the peaks for C1-O- (^{13}C at 65.1 ppm and ^{1}H at 3.54 ppm), C6-O- (^{13}C at 63.2 ppm and ^{1}H at 3.25 ppm) and C6″-O- (^{13}C at 65.0 ppm and ^{1}H at 3.41 ppm). Mass spectrum showed the *m/z* peak at 1,039 $[M+Na]^+$ confirming the bis-arylated product. In D-mannitol the C1 cross peak at 65.0 ppm (^{1}H at 3.45 ppm) indicated the formation of bis-C1-O-arylated product, which was also confirmed by the *m/z* peak at 719 $[M+Na]^+$. Curcuminyl-bis-β-D-glucoside synthesised by using β-glucosidase from sweet almonds was confirmed from the C1 ^{13}C value at 103.2 ppm and the corresponding ^{1}H value at 4.16 ppm.

8.1.3.7 Optimisation of Curcuminyl-bis-α-D-Glucoside Synthesis Using Response Surface Methodology

Design: Central Composite Rotatable Design (CCRD), 32 experiments, 5 variables at 5 levels (Vijayakumar et al. 2006)

Variables: Amyloglucosidase amount (% w/w of D-glucose), curcumin concentration (mmol), incubation period (h), buffer concentration (mM) and pH coded and actual values are shown in Table 8.5. Actual set of experiments undertaken as per the CCRD with coded values and the glucosylation yields obtained is given in Table 8.6.

Equation: A second-order polynomial equation was developed to study the effects of the variables on the esterification yields in terms of linear, quadratic and cross product terms.

$$\begin{aligned} Y = &-61.7385 + 2.064X_1 + 28.920X_2 \\ &+0.264X_3 + 195.663X_4 + 4.439X_5 \\ &-0.014X_1X_1 - 22.954X_2X_2 - 0.0006X_3X_3 \\ &+168.79X_4X_4 - 0.330X_5X_5 + 0.446X_1X_2 \\ &-0.003X_1X_3 - 8.333X_1X_4 - 0.039X_1X_5 \\ &-0.213X_2X_3 - 381.2\,X_2X_4 + 5.062\,X_2X_5 \\ &+4.661X_3X_4 - 0.022X_3X_5 + 22.5X_4X_5 \end{aligned}$$

where X_1, amyloglucosidase concentration; X_2, curcumin concentration; X_3, incubation period; X_4, buffer concentration; and X_5, pH.

Coefficients: Microsoft Excel software, Version 5.0

Table 8.5 Coded values of the variables and their corresponding actual values used in the design of experiments

Variables	−2	−1	0	1	2
Amyloglucosidase (%)	10	25	40	55	70
Curcumin (mmol)	0.2	0.4	0.6	0.8	1.0
Incubation period (h)	24	48	72	96	120
Buffer concentration (mM)	0.02	0.04	0.06	0.08	0.1
pH	4.0	5.0	6.0	7.0	8.0

Table 8.6 Experimental design with experimental and predicted yields of glucosylation based on the response surface equation

Expt. no.	Enzyme %	Curcumin (mmol)	Incubation (h)	Buffer concn (mM)	pH	Yield[a] experimental (%)	Yield predicted (%)
1	25	0.4	48	0.08	5.0	28.6	29.9
2	25	0.4	48	0.04	7.0	17.4	19.9
3	25	0.4	96	0.04	5.0	20.9	20.7
4	25	0.4	96	0.08	7.0	42.9	44.5
5	25	0.8	48	0.04	5.0	22.1	22.2
6	25	0.8	48	0.08	7.0	35.3	37.1
7	25	0.8	96	0.08	5.0	37.8	37.0
8	25	0.8	96	0.04	7.0	25.6	26.0
9	55	0.4	48	0.04	5.0	28.1	27.2
10	55	0.4	48	0.08	7.0	28.6	29.4
11	55	0.4	96	0.08	5.0	36.8	34.9
12	55	0.4	96	0.04	7.0	24.5	23.9
13	55	0.8	48	0.08	5.0	33.1	31.5
14	55	0.8	48	0.04	7.0	42.1	41.8
15	55	0.8	96	0.04	5.0	35.0	32.0
16	55	0.8	96	0.08	7.0	40.2	38.9
17	10	0.6	72	0.06	6.0	23.6	19.6
18	70	0.6	72	0.06	6.0	21.3	25.2
19	40	0.2	72	0.06	6.0	28.9	27.0
20	40	1.0	72	0.06	6.0	34.4	36.1
21	40	0.6	24	0.06	6.0	33.8	31.4
22	40	0.6	120	0.06	6.0	33.8	36.1
23	40	0.6	72	0.06	4.0	27.7	30.7
24	40	0.6	72	0.06	8.0	40.3	37.2
25	40	0.6	72	0.02	6.0	28.7	29.2
26	40	0.6	72	0.10	6.0	47.1	46.5
27	40	0.6	72	0.06	6.0	38.2	35.3
28	40	0.6	72	0.06	6.0	40.6	35.3
29	40	0.6	72	0.06	6.0	33.3	35.3
30	40	0.6	72	0.06	6.0	35.0	35.3
31	40	0.6	72	0.06	6.0	30.3	35.3
32	40	0.6	72	0.06	6.0	34.9	35.3

[a]Conversion yields were from HPLC with respect to 1 mmol of D-glucose. The experimental yields are an average from two experiments. Error in yield measurements ±5–10%

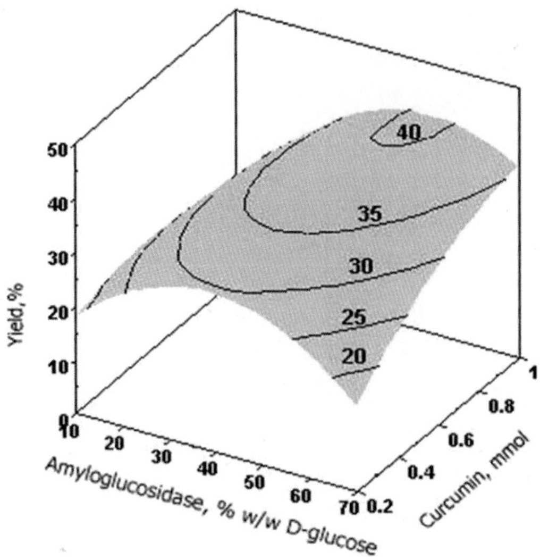

Fig. 8.3 Three-dimensional surface and contour plot showing the effect of amyloglucosidase concentration and curcumin concentration on the extent of glucosylation at incubation period, 72 h; buffer concentration, 0.06 mM (0.6 mL of 0.01-M buffer), pH 6.0

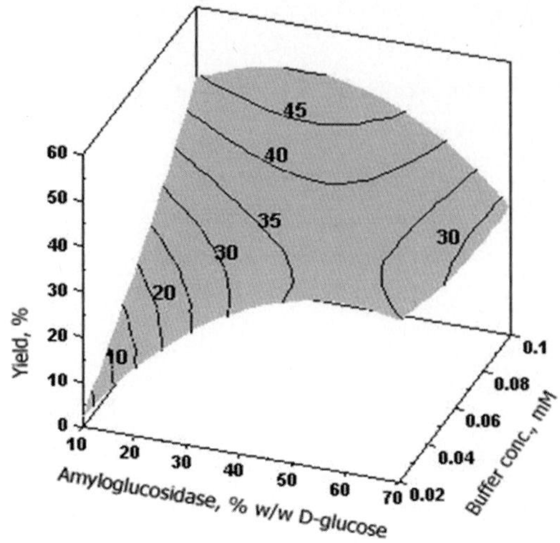

Fig. 8.4 Three-dimensional surface and contour plot showing the effect of amyloglucosidase concentration and buffer concentration on the extent of glucosylation at curcumin, 0.6 mmol; incubation period, 72 h; and pH, 6.0

Analysis of Variance (*ANOVA*): Microsoft Excel software. ANOVA shows the model is significant at $P<0.01$.

Optimisation: Microsoft Excel Solver function

The experimental data fitted the second-order polynomial equation well as indicated by an R^2 value of 0.9.

Average absolute deviation between predicted and experimental yields was found to be 6.0. Amyloglucosidase requires some amount of water to be present for its optimum activity, and this is achieved by adding buffer of certain volume, salt concentration and pH (Vic et al. 1997; Chahid et al. 1992). Salient features of this amyloglucosidase-catalysed reaction are described in the three-dimensional surface and contour plots.

Effect of various amyloglucosidase and curcumin concentrations on the extent of glucosylation of curcumin shows a maximum conversion of 35% for curcumin concentrations above 0.55 mmol at 35–60% (w/w D-glucose) amyloglucosidase concentrations (Fig. 8.3). Isoglucosylation regions of 15–25% yield could be predicted for amyloglucosidase concentrations below 30% and above 60% at all curcumin concentrations in the range 0.2–1.0 mmol. While lower amyloglucosidase converted less, higher amyloglucosidase could be inhibitory to curcumin at a constant D-glucose concentration.

Effect of amyloglucosidase concentration and pH on the extent of glycosylation also exhibited a similar pattern. A narrow range of amyloglucosidase concentration of 35–55% showed maximum glucosylation of 35% at above pH 5.5. Above and below this enzyme concentration range, the conversion yield was lesser in the pH range 4.0–8.0. For the hydrolytic activity, the pH optimum for amyloglucosidase is 5.0, and the isoelectric point for the enzyme is 4.2. Above pH 5.5 and towards higher pH 8.0, the ionisable groups especially from acidic amino acid residues (Glu314 and Glu544) in the active site could exist in anionic forms, enabling abstraction of the anomeric hydroxyl proton of glucose to aid in the facile transfer of glucose molecule to curcumin (Frandsen et al. 1994; Sierks et al. 1990).

Figure 8.4 shows the effect of amyloglucosidase concentration and buffer concentration on the conversion yield of curcuminyl-bis-α-D-glucoside. Buffer concentration included both the effects of the concentrations of the buffer salts and volume of the buffer. While buffer volume determined the effect of water activity on amyloglucosidase, pH controlled the extent of ionisation of the charged amino acid residues of the amyloglucosidase in the active site and on the surface, the latter being also affected by the buffer concentration. Together, both these quantities explain the role of the active conformation of the enzyme, on catalysis at the refluxing temperature of the solvent (at 68°C). After 120 h of incubation period, amyloglucosidase was found to lose only 20% of its activity. Between an amyloglucosidase concentration range of 15–45%, maximum conversion of 45% could be observed in the buffer concentration range 0.095–0.1 mM (0.95–1.0 mL of 0.01-M pH 6.0 buffer). The same amyloglucosidase concentration range could not give good conversion yields at buffer concentrations less than 0.08 mM (0.8 mL of 0.01-M pH 6.0 buffer). Besides, amyloglucosidase concentrations above 60% also showed lesser yields in the buffer concentration range 0.03–0.1 mM (0.3–1.0 mL of 0.01-M pH 6.0). This feature clearly shows that the extent of glucosylation could be dictated by a critical buffer (0.95–1.0 mL of 0.01-M pH 6.0) to enzyme ratio (15–45% w/w D-glucose).

A maximum yield of 65.6% was predicted based on the response model for an amyloglucosidase concentration of 16.9% (w/w D-glucose), 0.33-mmol curcumin concentration, 120-h incubation period and 0.1-mM (1.0 mL of 0.01-M pH 6.0 buffer) buffer concentration at pH 7.5. Experiment conducted at the above optimum conditions resulted in a yield of 56.3%. Validation of the response model was also tested by carrying out experiments at selected random conditions. The yields obtained from validation experiments also agreed with the predicted yields with an average absolute deviation of 8.5% (Table 8.7).

Thus, this study showed that this model is very good in predicting the glucosylation of curcumin by the amyloglucosidase enzyme.

8.1.4 Syntheses of N-Vanillyl-Nonanamide Glycosides

Capsaicin {(*E*)-N-[(4-hydroxy-3-methoxyphenyl) methyl]-8-methyl-6-nonemide} (Ohnuki et al. 2001), a fat-soluble phenolic compound present as the major pungent principle in fruits of *Capsicum* species, can also be used as a hypoglycemic

Table 8.7 Validation of experimental data

Expt. no.	Enzyme (%)	Curcumin (mmol)	Incubation period (h)	Buffer conc. (mM)	pH	% yield predicted	% yield[a] experimental
1	40	0.6	72	0.06	6.0	35.4	39.2
2	40	0.3	72	0.06	5.0	29.5	31.0
3	25	0.6	72	0.07	6.0	34.6	37.1
4	40	0.9	110	0.06	6.0	35.2	28.3
5	40	0.5	20	0.06	6.0	28.5	31.4
6	40	0.6	70	0.06	6.0	35.3	35.7
7	60	0.9	72	0.06	6.0	35.5	36.7
8	20	0.5	72	0.06	6.0	27.3	31.3
9	35	0.6	72	0.06	6.5	35.4	28.0
10	20	0.6	72	0.06	5.5	26.5	32.5
11	40	0.6	65	0.06	6.0	35.0	41.5
12	15	0.6	50	0.06	6.0	20.9	27.5
13	40	0.6	72	0.09	7.0	45.3	46.5
14	40	0.6	72	0.04	5.0	30.1	34.2
15	16.9	0.33	120	0.1	7.5	65.6	56.3

[a]Conversion yields were from HPLC with respect to 1 mmol of D-glucose. The experimental yields are an average from two experiments. Error in yield measurements ±5–10%

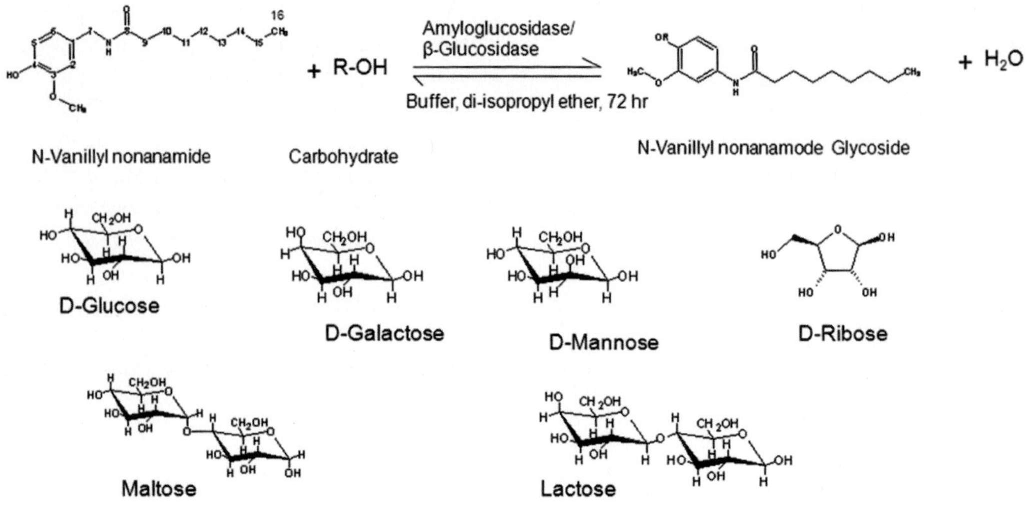

Scheme 8.2 Syntheses of N-vanillyl-nonanamide glycosides

(Lan et al. 2004), mutagenic and carcinogenic agent (Gannett et al. 1988); as a topical analgesic in pharmaceutical preparations (Rashid et al. 2003); and as antioxidant (Lee et al. 1995), anti-inflammatory (Heyes et al. 2004) and antifungal agent (Bhabadesh et al. 1996). It also can increase catecholamine secretion and suppress body fat accumulation (Watanabae et al. 1994), inhibit NADH oxidase in plasma membranes (Morre et al. 1995) and reduce the peri-renal adipose tissue weight and serum triacylglycerol by enhancing energy metabolism through a β-adrenergic action (Kawada et al. 1986). N-Vanillyl-nonanamide {N-[(4-hydroxy-3-methoxyphenyl)methyl] nonanamide}, a synthetic substitute of capsaicin, exhibits hypotensive and anti-nociceptive effects as that of natural capsaicin (Chen et al. 1992). Capsaicin (Iorizzi et al. 2001) has also been converted into its corresponding glucoside by cell suspension cultures (Kometani et al. 1993a; Hamada et al. 2003) and chemical methods (Hamada et al. 2001). Capsaicin and usage of its derivatives have been limited due to its low water solubility and high pungency.

Hence, a study on the synthesis of 4-*O*-(D-glucopyranosyl)N-vanillyl-nonanamide was undertaken (Scheme 8.2, Sivakumar and Divakar 2007).

HPLC analysis (Fig. 8.5) showed the following retention times: D-glucose, 6.9 min; 4-*O*-(D-glucopyranosyl)N-vanillyl-nonanamide, 8.2 min; D-galactose, 7.1 min; 4-*O*-(D-galactopyranosyl) N-vanillyl-nonanamide, 8.3 min; D-mannose, 6.7 min; 4-*O*-(β-D-mannopyranosyl)N-vanillyl-nonanamide, 8.9 min; D-ribose, 7.4 min; 4-*O*-(D-ribofuranosyl) N-vanillyl-nonanamide, 8.9 min; maltose, 8.5 min; 4-*O*-(α-D-glucopyranosyl-(1′→4)D-glucopyranosyl) N-vanillyl-nonanamide, 15.2 min; lactose, 9.3 min; and 4-*O*-(β-D-galactopyranosyl-(1′→4)β-D-glucopyranosyl)N-vanillyl-nonanamide, 17.9 min.

8.1.4.1 Synthesis of 4-*O*-(D-Glucopyranosyl)N-Vanillyl-Nonanamide Using Amyloglucosidase

Glucosylation of N-vanillyl-nonanamide with D-glucose using amyloglucosidase (Table 8.8) showed the following optimum conditions: incubation period, 72 h, and buffer, 0.2-mM, pH 7 phosphate buffer (2 mL of 0.01-M buffer in 100-mL di-isopropyl ether solvent); at pH 7, increase in buffer concentration from 0.03 to 0.25 mM (0.3–2.5 mL) showed increase in conversion with increase in buffer concentration to a maximum of 56% at 0.2-mM (2-mL) buffer; at 10% (w/w D-glucose) amyloglucosidase concentration conversion yield was the lowest of 16% which then increased to a maximum yield of 56% at 40% (w/w D-glucose) enzyme concentration (Table 8.8); increase

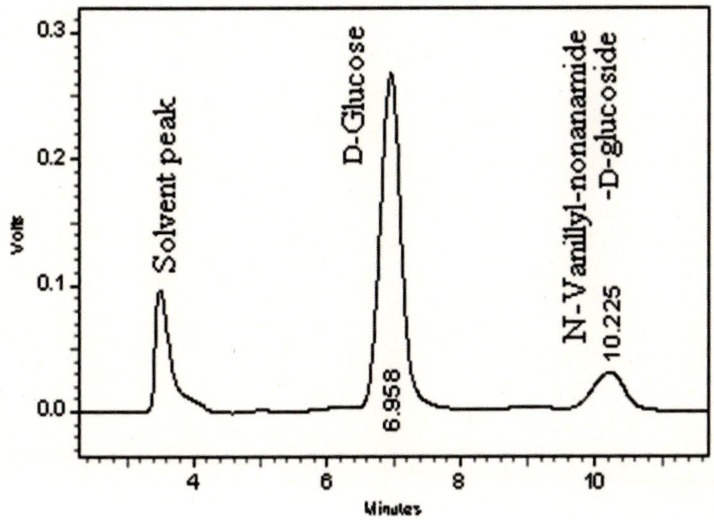

Fig. 8.5 HPLC chromatogram for the reaction mixture of D-glucose and 4-*O*-(D-glucopyranosyl)N-vanillyl-nonanamide. HPLC conditions: aminopropyl column (10 μm, 250 mm×4.6 mm); solvent, $CH_3CN:H_2O$ (70:30 v/v); flow rate, 1 mL min^{-1}; RI detector. Retention times: D-glucose, 7 min, and 4-*O*-(D-glucopyranosyl) N-vanillyl-nonanamide, 8.2 min

in N-vanillyl-nonanamide concentration (Table 8.8) exhibited increase in glucosylation from 0.2 mmol (12% yield) to 0.6 mmol (23%) and thereafter remained constant up to 1 mmol (23%).

8.1.4.2 Solubility of 4-*O*-(D-Glucopyranosyl)N-Vanillyl-Nonanamide

4-*O*-(D-Glucopyranosyl)N-vanillyl-nonanamide was found to be soluble in water to the extent of 7.7 g L^{-1}. Thus the water-insoluble and predominantly fat-soluble N-vanillyl-nonanamide has been rendered water-soluble through this glycosylation reaction.

8.1.4.3 Syntheses of N-Vanillyl-Nonanamide Glycosides of Other Carbohydrates Using Amyloglucosidase and β-Glucosidase and Their Spectral Characterisation

Other N-vanillyl-nonanamide glycosides were prepared using amyloglucosidase and β-glucosidase.

UV spectra of N-vanillyl-nonanamide glycosides showed shifts for σ→σ* band in the 192.5–203.5-nm (201.5 nm for free N-vanillyl-nonanamide) region, σ→π* band in the 225–227-nm (225 nm for free N-vanillyl-nonanamide) region and π→π* band in the 263–283.5-nm (279 nm for free N-vanillyl-nonanamide) region and IR C–O–C asymmetrical stretching frequencies in the 1,254–1,279-cm^{-1} region and symmetrical stretching frequencies in the 1,027–1,038-cm^{-1} region indicating that N-vanillyl-nonanamide had undergone glycosylation. From the 2-D HSQCT spectra of the N-vanillyl-nonanamide glycosides, C1α- and C1β-glycosides besides C6-*O*-arylated products were detected. Mass spectra also confirmed the formation of the above-mentioned glycosides (Table 8.9).

8.1.5 Capsaicin Glycosides

Capsaicin glycosides were synthesised enzymatically using β-glucosidase from sweet almond to obtain capsaicin derivatives whose water solubility is greater than that of capsaicin (Scheme 8.3). Optimum reaction conditions used for the synthesis of capsaicin glycosides are capsaicin 0.25 mmol; carbohydrate 0.25 mmol; β-glucosidase 40% (w/w carbohydrate); 1-ml (0.01-M) pH

Table 8.8 Optimisation of reaction conditions for the synthesis of 4-*O*-(D-glucopyranosyl)N-vanillyl-nonanamide

Reaction conditions	Variable parameter[b]	Yield (%)[c]
	Incubation period (h)	
N-Vanillyl-nonanamide – 0.5 mmol	3	25
D-Glucose – 0.5 mmol	6	29
pH – 7	12	31
Buffer concentration – 0.2 mM (2 mL)	24	33
Enzyme – 40% w/w D-glucose	48	45
	72	56
	96	34
	120	28
	pH (0.01 M)	
N-Vanillyl-nonanamide – 0.5 mmol[a]	4	18
D-Glucose – 0.5 mmol	5	18
Enzyme – 40% w/w D-glucose	6	25
Buffer – 0.1 mM (1 mL)	7	39
Incubation period – 72 h	8	19
	Buffer concentration (mM)	
N-Vanillyl-nonanamide – 0.5 mmol	0.03	26
D-Glucose – 0.5 mmol	0.06	37
Enzyme – 40% w/w D-glucose	0.1	42
pH – 7	0.15	49
Incubation period – 72 h	0.2	56
	0.25	50
	Enzyme concentration (% w/w D-glucose)	
N-Vanillyl-nonanamide – 0.5 mmol	10	16
D-Glucose – 0.5 mmol	20	24
pH – 7	30	39
Buffer concentration – 0.2 mM (2 mL)	40	56
Incubation period – 72 h	50	21
	75	10
	N-Vanillyl-nonanamide (mmol)	
pH – 7	0.2	12
Buffer concentration – 0.2 mM (2 mL)	0.4	14
D-Glucose – 0.2 mmol	0.6	23
Enzyme – 40% w/w D-glucose	0.8	24
Incubation period – 72 h	1	23

[a]Initial reaction conditions
[b]Other variables are the same as under reaction conditions, except the specified ones
[c]HPLC yields expressed with respect to 0.5-mmol D-glucose employed except for the last experiment where N-vanillyl-nonanamide concentrations were varied with respect to only 0.2-mmol D-glucose

7.0 phosphate buffer; 50-ml di-isopropyl ether; and 72-h incubation period. Glycosylation of capsaicin was carried out with D-glucose, D-galactose and lactose using β-glucosidase. Capsaicin glycosides yields were in the 15–29% range. Excellent regioselectivity was observed with D-glucose and lactose (Table 8.10).

8.1.6 Syntheses of Vanillyl Glycosides

Vanillin(4-hydroxy-3-methoxybenzaldehyde) is used as an additive in food and beverages (60%), considerable amounts as flavour and fragrances (20–25%) and 5–10% as an intermediate for pharmaceuticals. It possesses a wide range of

Table 8.9 Syntheses of N-vanillyl-nonanamide glycosides[a]

Glycosides	Amyloglucosidase catalysis		β-Glucosidase catalysis	
	Product (% proportion)[b]	Yields (%)[c]	Product (% proportion)[b]	Yields (%)[c]
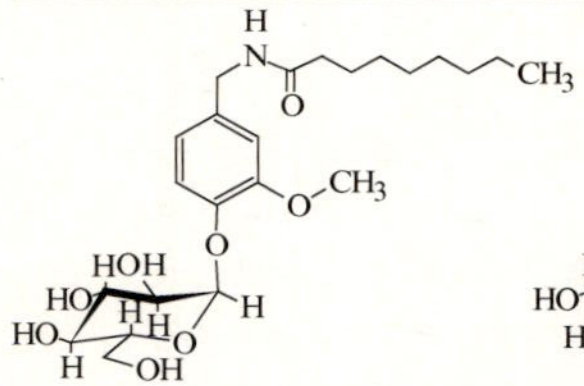 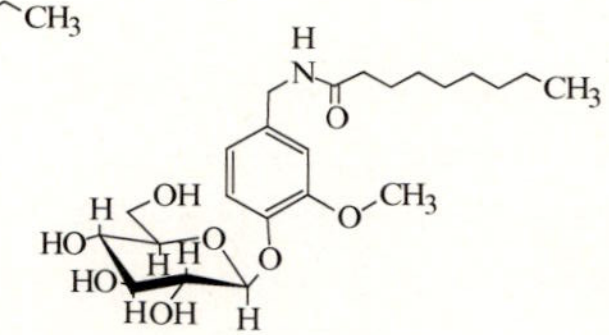 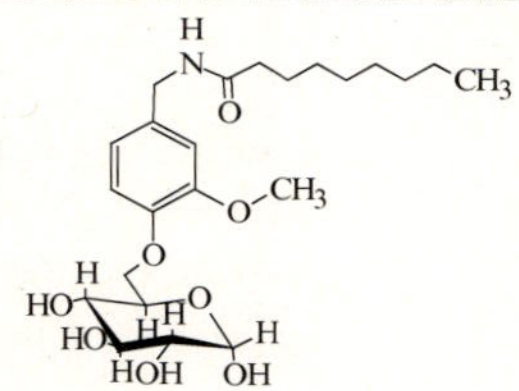 4-*O*-(α-D-Glucopyranosyl)N-vanillyl-nonanamide 4-*O*-(6-D-Glucopyranosyl)N-vanillyl-nonanamide 4-*O*-(β-D-Glucopyranosyl)N-vanillyl-nonanamide	C1α-glucoside (70), C6-*O*-arylated (30)	56	C1β-glucoside	35
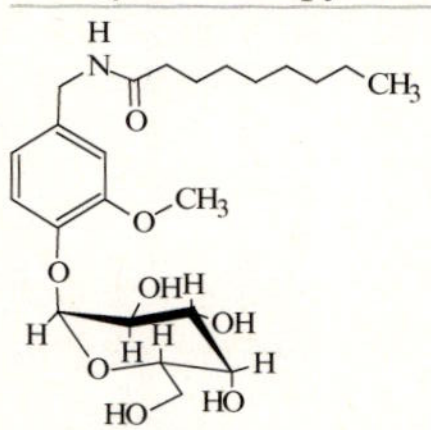 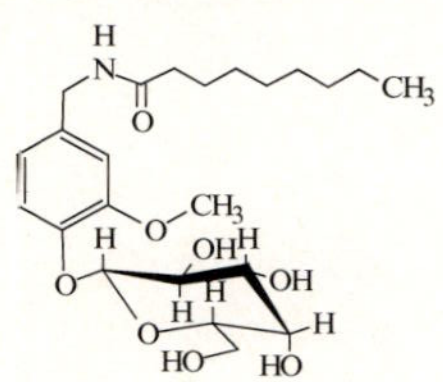 4-*O*-(α-D-Galactopyranosyl)N-vanillyl-nonanamide 4-*O*-(β-D-Galactopyranosyl)N-vanillyl-nonanamide	C1α-galactoside (42), C1β-galactosides (58)	14	C1β-galactoside	26
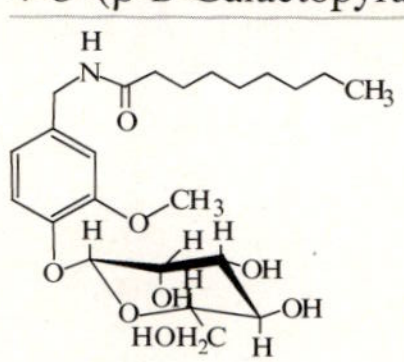 4-*O*-(β-D-Mannopyranosyl)N-vanillyl-nonanamide	–	–	C1β-mannoside	24

(continued)

Table 8.9 (continued)

Glycosides	Amyloglucosidase catalysis		β-Glucosidase catalysis	
	Product (% proportion)[b]	Yields (%)[c]	Product (% proportion)[b]	Yields (%)[c]
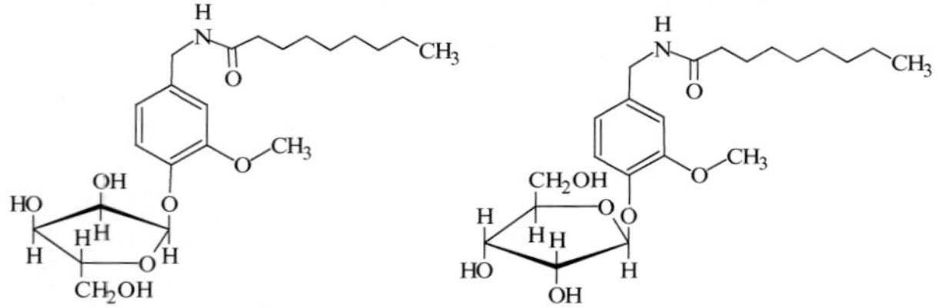 4-*O*-(α-D-Ribofuranosyl)N-vanillyl-nonanamide 4-*O*-(β-D-Ribofuranosyl)N-vanillyl-nonanamide	C1α riboside (33), C1β riboside (67)	9	C1α riboside (30), C1β riboside (70)	10
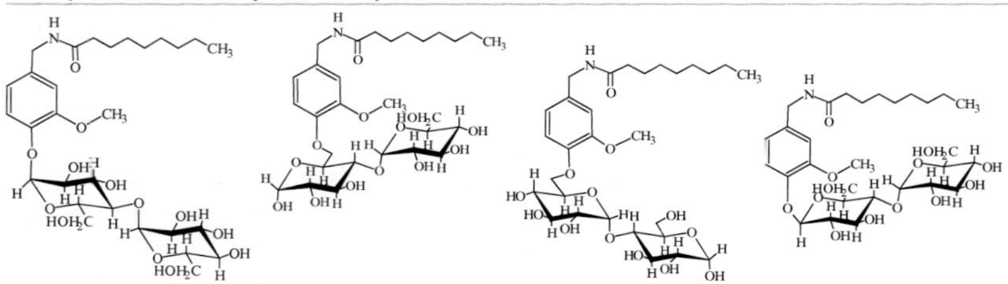 4-*O*-(α-D-Glucopyranosyl-(1′→4)α-D-glucopyranosyl)N-vanillyl-nonanamide 4-*O*-(α-D-Glucopyranosyl-(1′→4)6-D-glucopyranosyl)N-vanillyl-nonanamide 4-*O*-(α-D-Glucopyranosyl-(1′→4)6′-D-glucopyranosyl)N-vanillyl-nonanamide 4-*O*-(α-D-Glucopyranosyl-(1′→4)β-D-glucopyranosyl)N-vanillyl-nonanamide	C1α-maltoside (11), C6-*O*-arylated (40), C6′-*O*-arylated (49)	15	C1β-maltoside	9
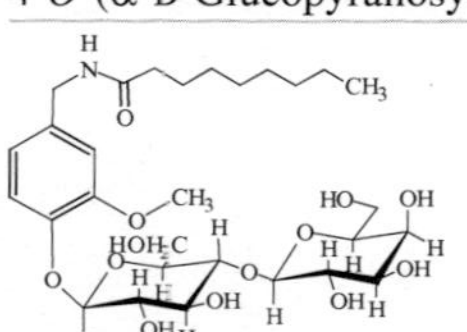 4-*O*-(β-D-Galactopyranosyl-(1′→4)β-D-glucopyranosyl)N-vanillyl-nonanamide	–	–	C1β-lactoside	28

[a]N-Vanillyl-nonanamide and carbohydrate, 0.5 mmol each; enzyme concentration, 40% w/w of carbohydrate; solvent, di-isopropyl ether; buffer, 0.2-mM (2-mL) pH 7 phosphate buffer; incubation period, 72 h

[b]The product proportions were determined from the area of respective $^{1}H/^{13}C$ signals

[c]Conversion yields were from HPLC with respect to free carbohydrate. Error in yield measurements is ±10%

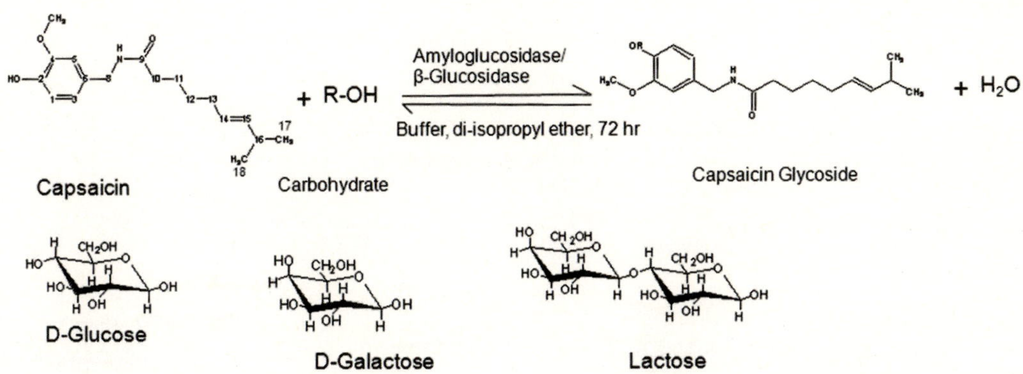

Scheme 8.3 Syntheses of capsaicin glycosides

Table 8.10 β-Glucosidase-catalysed syntheses of capsaicin glycosides[a]

Glycosides	Product (% proportion)[b]	Yields (%)[c]
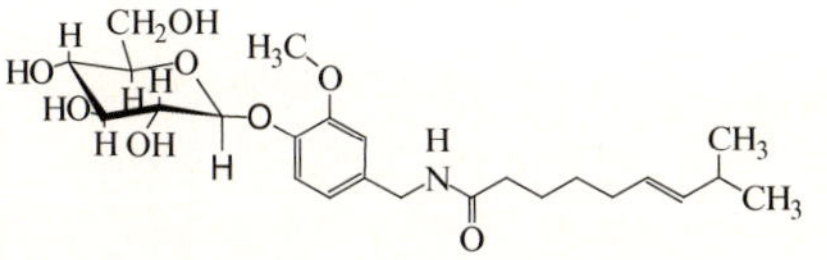 3-*O*-(β-D-Glucopyranosyl)capsaicin	C1β	15
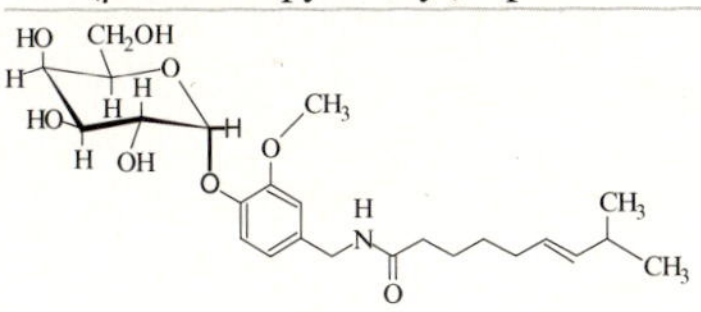 3-*O*-(α-D-Galactopyranosyl)capsaicin	C1α (42) and C1β (58)	13
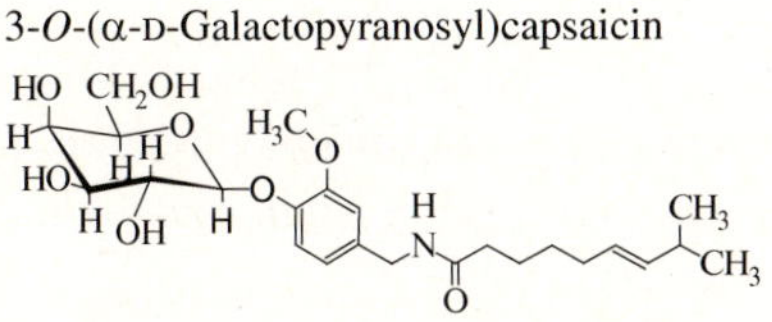 3-*O*-(β-D-Galactopyranosyl)capsaicin		
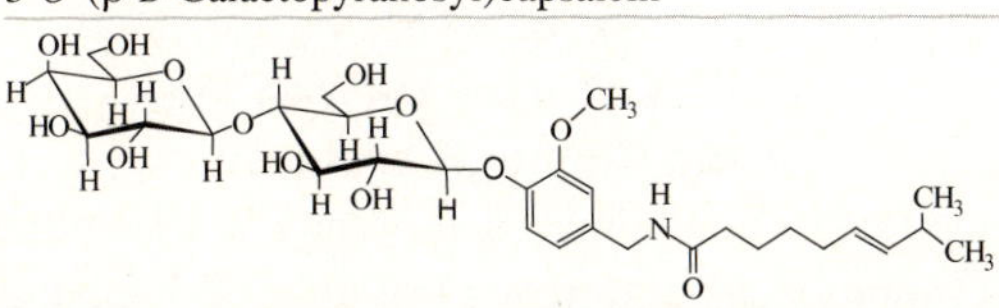 3-*O*-(β-D-Galactopyranosyl-(1′→4)β-D-glucopyranosyl)capsaicin	C1β	29

Capsaicin, 0.25 mmol; carbohydrate, 0.25 mmol; solvent, di-isopropyl ether
[a]β-glucosidase concentration, 40% w/w carbohydrates; 0.10-mM (1.0-ml) pH 7.0 phosphate buffer; incubation period, 72 h
[b]product proportions were calculated from the area of respective $^{1}H/^{13}C$ signals. Conversion yields were from HPLC with respect to the carbohydrate. Error in yield measurements is ±10%

pharmacological activities such as antimicrobial (Lopez-Malo et al. 1998), anticarcinogenic (Stephan and Peter 2003), antioxidant (Burri et al. 1999), antifungal (Fitzgerald et al. 2005) and antimutagenic (Kometani et al. 1993b). The solubility of vanillin in water varies from 3 g L^{-1} at 4.4°C to 62.5 g L^{-1} at 80°C (The Merck Index 2006). Thus the solubility and bioavailability of vanillin limit its pharmacological applications. Glycosylation is a useful tool to improve the

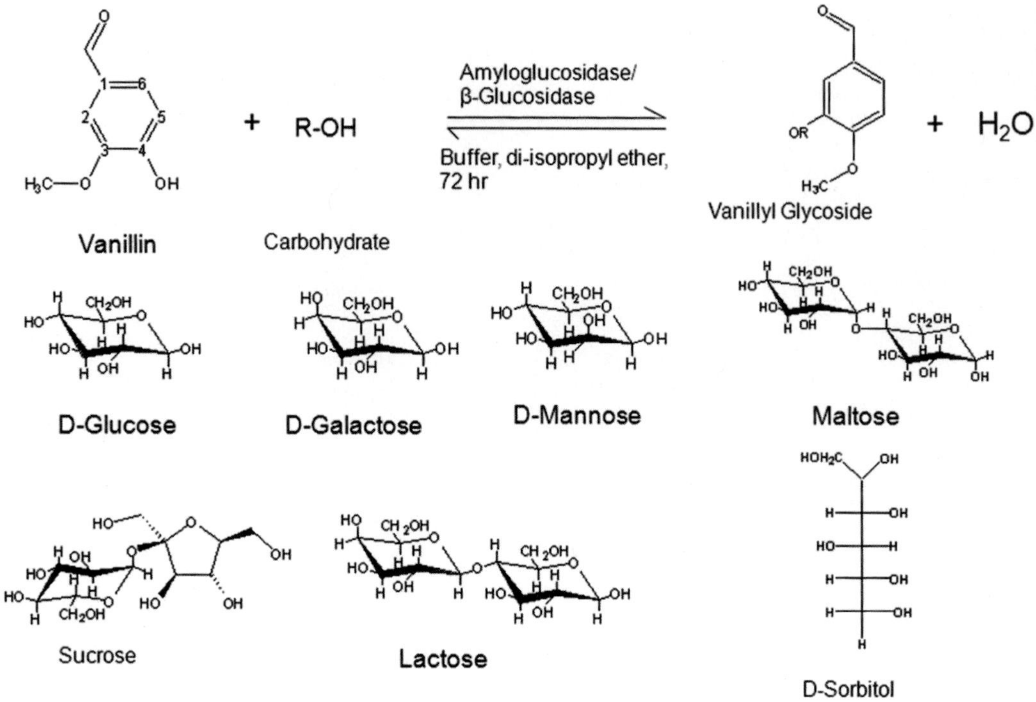

Scheme 8.4 Synthesis of vanillyl glycosides

water solubility and bioavailability (Kometani et al. 1993a; Tietze et al. 2003) of vanillin.

Preparation of vanillyl glycosides has been reported by cell suspension cultures (Tietze et al. 2003), plant cell tissue cultures, organ cultures (Sommer et al. 1997) and chemical methods (Reichel and Sckickle 1943). However, preparation by enzymatic methods has not been previously reported. This section deals with the enzymatic method using amyloglucosidase from a *Rhizopus* mould and β-glucosidase from sweet almond for the preparation of vanillyl glycosides in a nonpolar solvent (Scheme 8.4, Sivakumar and Divakar 2006).

HPLC analysis of the glycosides (Fig. 8.6) showed the following retention times: D-glucose, 5.4 min; 4-*O*-(D-glucopyranosyl)vanillin, 7.8 min; D-galactose, 5.3 min; 4-*O*-(D-galactopyranosyl) vanillin, 7.5 min; D-mannose, 4.9 min; 4-*O*-(D-mannopyranosyl)vanillin, 7.8 min; maltose, 8.5 min; 4-*O*-(α-D-glucopyranosyl-(1′→4)D-glucopyranosyl) vanillin, 17.1 min; sucrose, 9.8 min; 4-*O*-(D-fructofuranosyl-(2→1′)α-D-glucopyranosyl) vanillin, 14.3 min; lactose, 9.9 min; 4-*O*-(β-D-galactopyranosyl-(1′→4)β-D-glucopyranosyl) vanillin, 8.2 min; D-sorbitol, 6.7 min; and 4-*O*-(D-sorbitol)vanillin, 10.1 min.

8.1.6.1 Synthesis of 4-*O*-(D-Glucopyranosyl)Vanillin Using Amyloglucosidase

Optimum conditions determined are vanillin (1 mmol) and D-glucose (1 mmol); 72-h incubation period (Fig. 8.7a, Table 8.11); reflux temperature of di-isopropyl ether at 68°C; pH 4.0; buffer concentration, 0.1 mM (1 mL); and 53% at 40% (w/w D-glucose) enzyme concentration (Fig. 8.7b, Table 8.11).

8.1.6.2 Solubility of 4-*O*-(D-Glucopyranosyl)Vanillin

Determination of water solubility of 4-*O*-(D-glucopyranosyl)vanillin showed that it is soluble to the extent of 35.2 g L^{-1}. Hence,

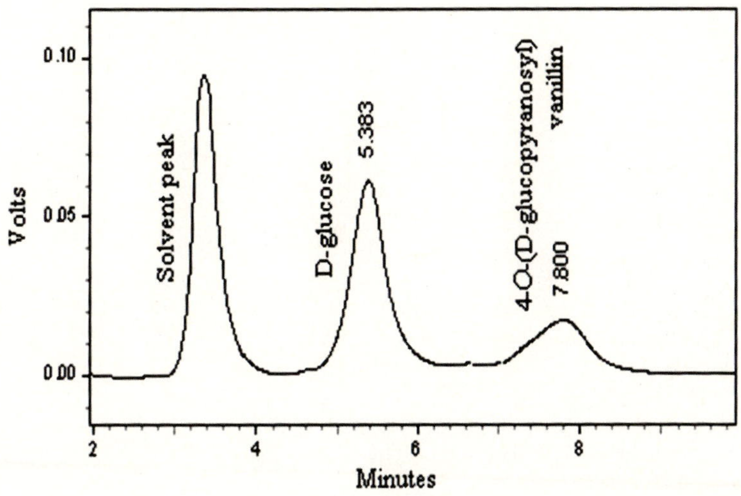

Fig. 8.6 HPLC chromatogram for the reaction mixture of D-glucose and 4-*O*-(D-glucopyranosyl)vanillin. HPLC conditions: aminopropyl column (10 μm, 300 mm × 3.9 mm); solvent, $CH_3CN{:}H_2O$ (80:20 v/v); flow rate, 1 mL min^{-1}; RI detector. Retention times: D-glucose, 5.4 min, and 4-*O*-(D-glucopyranosyl)vanillin, 7.8 min

4-*O*-(D-glucopyranosyl)vanillin was found to be more soluble than vanillin (2 g L^{-1}) at 25°C in water.

8.1.6.3 Syntheses of Vanillyl Glycosides of Other Carbohydrates Using Amyloglucosidase and β-Glucosidase and Their Spectral Characterisation

Vanillyl glycosides of other carbohydrates (Scheme 8.4) were synthesised using amyloglucosidase and β-glucosidase (Table 8.12).

UV spectra of vanillyl glycosides showed shifts in σ→σ* band in the 193.5–198.5-nm (195 nm for free vanillin) range, σ→π* band in the 222–224.5-nm (228.5 nm for free vanillin) range, π→π* band in the 273–283.5-nm (272 nm for free vanillin) range, IR C–O–C asymmetrical stretching frequencies in the 1,249–1,265-cm^{-1} range and symmetrical stretching frequencies in the 1,024–1,038-cm^{-1} range indicating that vanillin had undergone glycosylation. In 2-D Heteronuclear Single Quantum Coherence Transfer (HSQCT) spectra, the respective chemical shift values of carbon atoms C1α and β, C6α and β showed glycoside formation and arylation (Table 8.12).

8.1.6.4 Synthesis of 4-*O*-(α-D-Glucopyranosyl-(1′→4)D-Glucopyranosyl)Vanillin Using Glucosidases by Response Surface Methodology

Design: Central Composite Rotatable Design (CCRD), 32 experiments, 5 variables at 5 levels (Sivakumar et al. 2006a)

Variables: Glucosidase concentration, vanillin concentration, incubation period, buffer volume and pH in case of both the glucosidases. Table 8.13 shows the coded and actual levels of the variables employed in the design matrix. Actual set of experiments undertaken as per CCRD with coded values and the maltosylation yields obtained are given in Table 8.14.

Equation: A second-order polynomial equation was developed to study the effects of the variables on the esterification yields in terms of linear, quadratic and cross product terms.

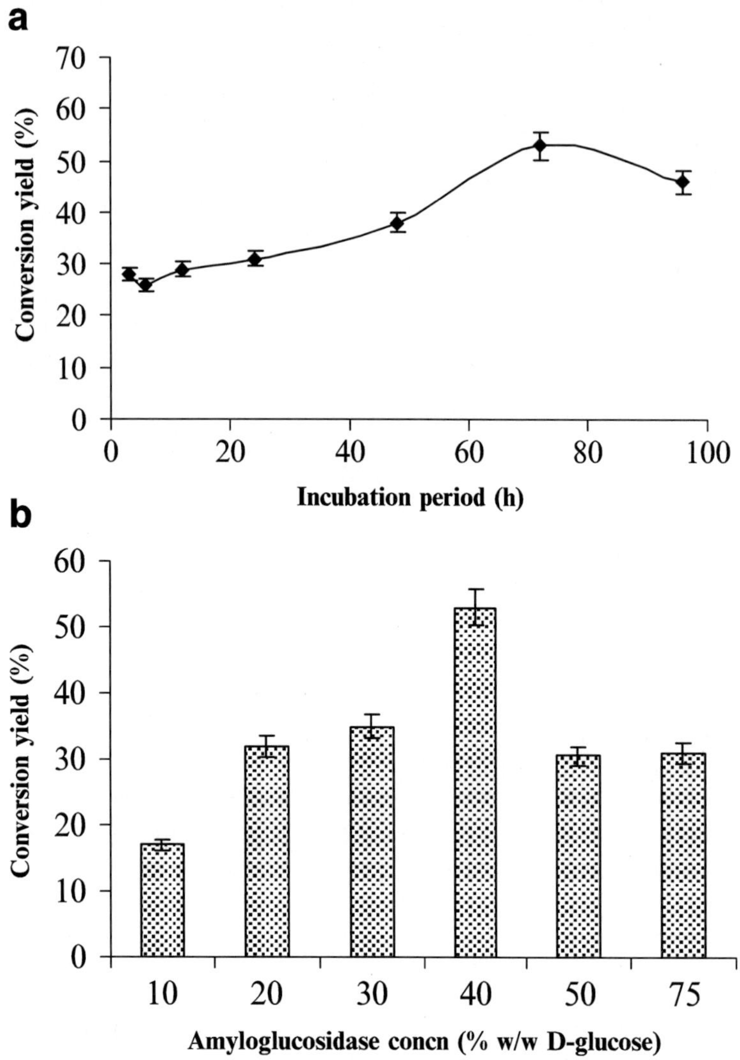

Fig. 8.7 (**a**) Reaction profile for 4-*O*-(D-glucopyranosyl) vanillin synthesis by the reflux method. Conversion yields were from HPLC with respect to 1 mmol of D-glucose. Reaction conditions: D-glucose, 1 mmol; vanillin, 1 mmol; amyloglucosidase, 40% (w/w D-glucose); 0.1-mM pH 4 acetate buffer; solvent, di-isopropyl ether; temperature, 68°C; and (**b**) effect of amyloglucosidase concentration for 4-*O*-(D-glucopyranosyl) vanillin synthesis. Reaction conditions: D-glucose, 1 mmol; vanillin, 1 mmol; 0.1-mM pH 4 acetate buffer; solvent, di-isopropyl ether; temperature, 68°C; and incubation period, 72 h

Coefficients: Microsoft Excel software, Version 5.0

Analysis of Variance (*ANOVA*): Microsoft Excel software ANOVA shows the model is significant at $P<0.01$.

Optimisation: Microsoft Regression software A second-order polynomial equation was developed to study the effect of the variables on the maltoside yields.

8.1.6.4.1 Amyloglucosidase-Catalysed Synthesis of 4-*O*-(α-D-Glucopyranosyl-(1′→4)D-Glucopyranosyl)Vanillin

The data obtained using amyloglucosidase were fitted to a second-order polynomial equation, and the predictive equation obtained with coefficients exhibited an R^2 value of 0.81.

Table 8.11 Optimisation of reaction conditions for the synthesis of 4-*O*-(D-glucopyranosyl)vanillin

Reaction conditions[a]	Variable parameter[b]	Yield (%)[c]
	Incubation period (h)	
Vanillin – 1 mmol	3	28
D-Glucose – 1 mmol	6	26
pH – 4	12	29
Buffer concentration – 0.1 mM (1 mL)	24	31
Enzyme – 40% w/w D-glucose	48	38
	72	53
	96	46
	pH (0.01 M)	
Vanillin – 1 mmol[a]	4	51
D-Glucose – 1 mmol	5	28
Enzyme – 40% w/w D-glucose	6	22
Buffer concentration – 0.1 mM (1 mL)	7	18
Incubation period – 72 h	8	17
	Buffer concentration (mM)	
Vanillin – 1 mmol	0.02	13
D-Glucose – 1 mmol	0.04	46
Enzyme – 40% w/w D-glucose	0.06	48
pH – 4	0.08	50
Incubation period – 72 h	0.1	51
	0.12	42
	Enzyme concentration (% w/w D-glucose)	
Vanillin – 1 mmol	10	17
D-Glucose – 1 mmol	20	32
pH – 4	30	35
Buffer concentration – 0.1 mM (1 mL)	40	53
Incubation period – 72 h	50	30
	75	31
	Vanillin (mmol)	
pH – 4	0.5	49
Buffer concentration – 0.1 mM (1 mL)	1	53
D-Glucose – 1 mmol	1.5	47
Enzyme – 40% w/w D-glucose	2	48
Incubation period – 72 h	2.5	46

[a]Initial reaction conditions
[b]Other variables are the same as under reaction conditions, except the specified ones
[c]HPLC yields expressed with respect to 1-mmol D-glucose employed

Equation:

$$Y = -3.7258\,X_1 + 25.3461\,X_5 + 0.0051\,X_3X_3 - -3.7238\,X_5X_5t + 0.6985\,X_1X_4 + 0.4896\,X_1X_5 - 0.0281X_2X_3 - 0.7730\,X_3X_4 + 5.7750\,X_4X_5$$

where X_1, enzyme concentration; X_2, vanillin concentration; X_3, incubation period; X_4, buffer volume; X_5, pH; and Y, yield. Table 8.14 shows the predicted yields obtained by using the reduced equation.

The effect of amyloglucosidase and buffer concentration on the extent of maltosylation at 1.5-mmol vanillin, pH 6 and 72-h incubation period

Table 8.15 Validation data for the amyloglucosidase-catalysed reactions at selected random conditions[a]

Expt. no.	Amyloglucosidase (% w/w maltose)	Vanillin (mmol)	Incubation period (h)	Buffer volume (mL)	pH	Predicted yield (%)	Experimental yield (%)
1	30	1.5	72	1.2	6	17.6	10.7
2	15	2	72	1.2	6	15.9	22.1
3	30	1	24	1.2	6	41.1	49.6
4	30	0.5	72	2	6	19.6	29.7
5	30	1.5	72	0.8	6	17.7	8.6
6	45	1.5	48	1.2	6	27	30.2
7	30	2	72	1.2	6	16.6	13.9
8	30	1.5	48	1.2	7	24.8	22.1
9	30	1.5	36	1.6	6	43.9	38.7
10	20	1.5	24	1.2	6	40.3	42.3

[a]Conversion yields obtained from HPLC with respect to 0.5 mmol of maltose

conditions showed that (Table 8.15) there appears to be a good correspondence between predicted and experimental yields at yields less than 40% and the correspondence appears to deviate a little at higher predictive yields.

8.1.6.4.2 β-Glucosidase-Catalysed Synthesis of 4-*O*-(α-D-Glucopyranosyl-(1′→4)β-D-Glucopyranosyl)Vanillin

Equation:

$$Y = -0.7408\,X_3 - 30.6961\,X_4 + 20.7320\,X_5 \\ +2.6099\,X_2X_2 - -1.4606\,X_5X_5 + 0.0092\,X_1X_3 \\ +0.9891\,X_1X_4 - 0.2937X_1X_5 + 0.0720\,X_3X_5$$

where X_1, enzyme concentration; X_2, vanillin concentration; X_3, incubation period; X_4, buffer volume; X_5, pH; and Y, yield. Table 8.14 shows the predicted yields obtained by using the reduced equation. Average absolute deviation between the experimental and predicted yields using this model is 22.3%.

Correlation Coefficient: $R^2 - 0.69$

Figure 8.10a shows the effect of β-glucosidase and vanillin concentrations on the maltoside yield at pH 6, 0.125-mM (1.25-mL) buffer concentration and 72-h incubation period. At all the β-glucosidase (10–50% w/w maltose) concentrations, the maltosylation yield increases with increase in vanillin concentration. However, increase in β-glucosidase concentration exhibited very little enhancement in yield. Maximum yield is depicted at 50% (w/w maltose) β-glucosidase concentration and 2–2.5 mmol of vanillin.

Effect of β-glucosidase and buffer concentration at pH 6, 1.5-mmol vanillin and 72-h incubation period on the maltosylation yield is shown in Fig. 8.10b. In the saddle-shaped surface obtained, a crossover point indicating reversal in the maltosylation behaviour is observed at 30% (w/w maltose) β-glucosidase concentration and 0.125-mM (1.25-mL) buffer concentration. Up to 30% (from 10% w/w maltose) β-glucosidase and 0.125-mM to 0.04-mM (1.25-mL to 0.4-mL) buffer concentration, the maltosylation yield decreases with increase in enzyme concentration. Above this crossover values in both the variables, the yield increases. Here also, a critical β-glucosidase to buffer concentration appears to influence the extent of maltosylation.

The maximum yield predicted based on the response model is 33.6% for the β-glucosidase-catalysed reaction under the following conditions: β-glucosidase, 50% (w/w of maltose); vanillin, 2.5 mmol; buffer concentration, 0.125 mM (1.25 mL) pH 6; and 72-h incubation period. The experiment conducted at the above optimum conditions gave 27.1% yield. Validation experiments performed at certain random selected conditions also showed good correspondence between experimental and predicted yields (Table 8.16).

Hence, the present RSM study has clearly showed the usefulness of CCRD technique in clearly bringing out the most salient features of this maltosylation reaction.

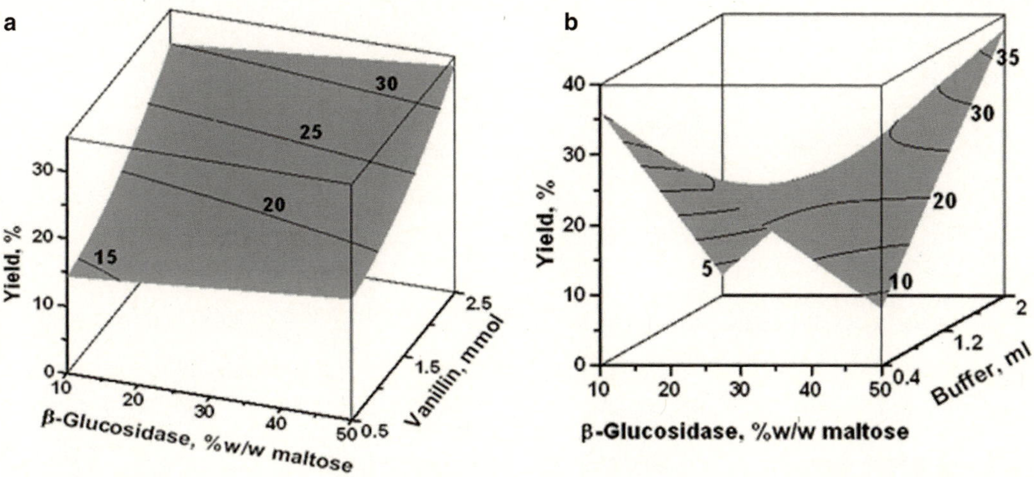

Fig. 8.10 Three-dimensional surface plots showing the effect of variables in the β-glucosidase-catalysed reaction: (**a**) vanillin and β-glucosidase concentrations on the extent of maltosylation yield (pH 6; buffer concentration, 0.125 mM/1.25 mL; incubation period, 72 h) and (**b**) β-glucosidase concentration and buffer concentration on the extent of maltosylation yield (pH 6; incubation period, 72 h; vanillin concentration, 1.5 mmol)

Table 8.16 Validation data for the β-glucosidase-catalysed reactions at selected random conditions[a]

Expt. no.	β-Glucosidase (% w/w maltose)	Vanillin (mmol)	Incubation period (h)	Buffer volume (mL)	pH	Predicted yield (%)	Experimental yield (%)
1	50	2.5	72	1.2	6	33.6	27.1
2	10	0.5	72	1.2	6	14.3	22.5
3	30	1.5	72	1.2	6	21.3	18.7
4	40	1.5	72	1.5	6	22.3	20.6
5	25	1.5	80	1.2	6	20.2	17.5
6	45	1.5	100	1.2	6	25.7	21.3
7	35	1.5	72	1.2	6	21.8	15.7
8	30	1.5	48	0.6	6	22.7	27.6
9	30	1.5	108	1.2	7	20.9	15.5
10	30	1.5	72	0.5	6	22.1	24.7

[a]Conversion yields obtained from HPLC with respect to 0.5 mmol of maltose

8.1.7 Syntheses of DL-Dopa Glycosides

DL-Dopa (DL-3,4-dihydroxy phenylalanine), an aromatic amino acid precursor of dopamine, is the most effective drug for Parkinson's disease (Yar 1993). Parkinson's disease is characterised by a severe and progressive degeneration of nigrostriatal dopamine (DA) neurons (Angerlacenci 2007) associated with the deficiency of catecholamine and dopamine (Shetty et al. 2002). It is generally accepted that after administration, L-dopa in Parkinson's disease is converted into dopamine by aromatic L-amino acid decarboxylase (AADC) within the serotonergic (5-HT) fibres in the striatum and *substantia nigra pars reticulate* (Yamada et al. 2007). Glucose is the brain source of energy, and this and other hexoses are also transferred across the blood–brain barrier (BBB) by the glucose carrier GLUT1. Such a transport will be facilitated if L-dopa is converted into the glycoside as the L-dopa converted product glucosyl dopamine is able to interact with the glucose transporter (GLUT1) and absorbed into the central nervous system (CNS) from the blood stream (Dalpiaz et al. 2007). During chronic treatment with

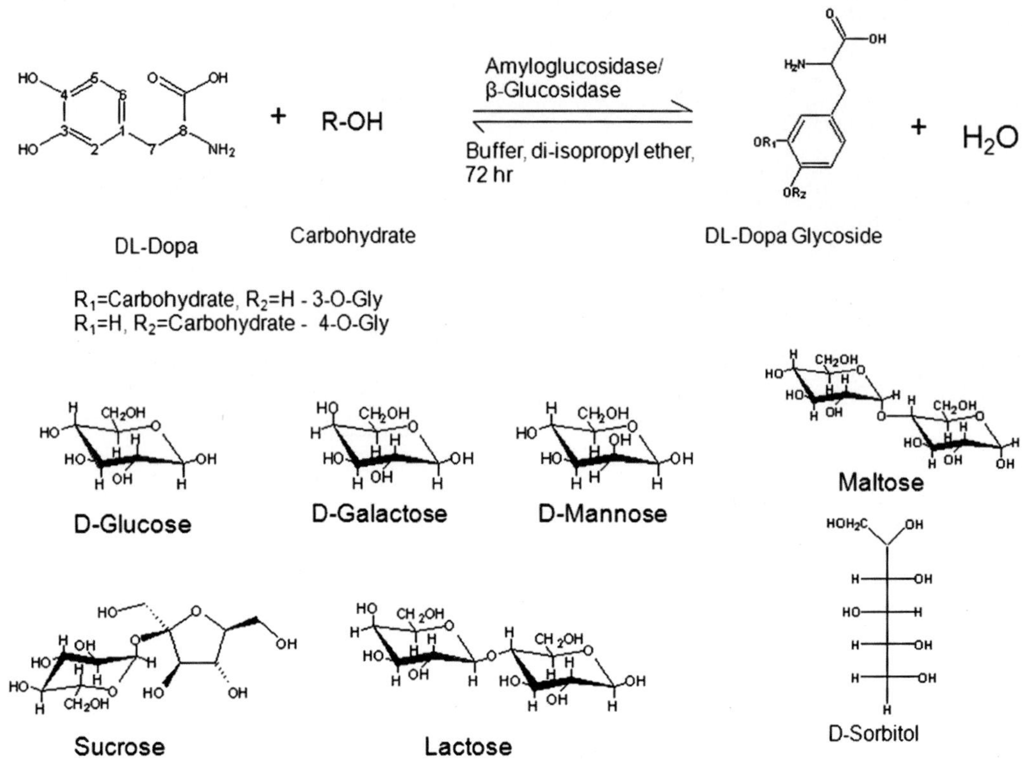

Scheme 8.5 Syntheses of DL-dopa glycosides

L-dopa, a variety of transport problems like involuntary movements occur which can be overcome by the employment of glucosyl derivatives (Pardridge 2002; Madrid et al. 1991). Side effects of the drugs can be reduced, and drug stability can be increased by modification of the aglycon molecule. For example, L-dopa and new dopaminergics can be modified to improve bioactivity properties (Pras et al. 1995; Giri et al. 2001).

L-Dopa was first isolated from *Vicia faba* (Guggenheim 1913; Vered et al. 1994) as a β-anomer (Nagasawa et al. 1961; Andrews and Pridham 1965). *Vicia faba* has been incorporated into dietary strategies to manage Parkinsonian motor oscillations (Kempster et al. 1993). A crude extract from the petals of *Mirabilis jalapa* was mixed with *cyclo*-Dopa and UDP-glucose to give *cyclo*-Dopa-5-*O*-glucoside by the action of glucosyl transferases (Sasaki et al. 2004; Wyler et al. 2004). However, no chemical or enzymatic methods have been reported yet for the synthesis of DL-dopa glycosides. Enzymatic method could be the alternative one which does not require protection and deprotection process (Fernandez et al. 2003; Roode et al. 2003) and provide milder reaction conditions, easy workup, less pollution, higher yields and selectivity. It can also give rise to stable dopa derivatives with enhanced stability and pharmacological activity (Suzuki et al. 1996; Vijayakumar and Divakar 2007). This section describes the synthesised DL-dopa glycosides using amyloglucosidase from *Rhizopus* mould and β-glucosidase from sweet almond in organic media (Scheme 8.5, Sivakumar et al. 2009).

HPLC analysis of the glycosides prepared exhibited the following retention times for the substrates and products: DL-dopa, 8.5 min; D-glucose, 6.2 min; DL-dopa-D-glucoside, 13 min; D-galactose, 7.1 min; DL-dopa-D-galactoside, 8.9 min; D-mannose, 6.7 min; DL-3-hydroxy-4-*O*-(β-D-mannopyranosyl) phenylalanine, 8.7 min; lactose, 9.3 min; DL-3-hydroxy-4-*O*-(β-D-galactopyranosyl-(1′→4)β-D-glucopyranosyl)phenylalanine, 1.9 min; D-sorbitol, 6.7 min; DL-3-hydroxy-4-*O*-(6-D-sorbitol)phenylalanine, 7.9 min; D-mannitol, 6.8 min; and DL-dopa-D-mannitol, 7.8 min.

8.1.7.1 Synthesis of DL-Dopa-D-Glucoside Using Amyloglucosidase

Glucosylation reaction between DL-dopa and D-glucose catalysed by amyloglucosidase from *Rhizopus* mould showed optimum conditions (Table 8.17) of incubation period, 72; pH 6.0, 0.06–0.14-mM (0.6–1.4-mL) buffer and yield in the range 62–59%. Up to 40% (w/w D-glucose) enzyme, the conversion yield more or less remained the same (yields 59%, 65%, 62% and 59% for 10%, 20%, 30% and 40% w/w d-glucose/enzyme, respectively). About 62% yield was obtained at 0.5-mmol DL-dopa.

8.1.7.2 Syntheses of DL-Dopa Glycosides of Other Carbohydrates Using Amyloglucosidase and β-Glucosidase and Their Spectral Characterisation

DL-Dopa glycosides with other eleven carbohydrates using amyloglucosidase and β-glucosidase (Scheme 8.5) gave bis products of glycosides and C6-O-arylated products.

The glycosides were characterised by UV, IR, mass, melting point, optical rotation and 2-D HSQCT, which provided good information on the nature and proportions of the products formed (Table 8.18).

UV spectra of DL-dopa glycosides showed $\sigma \rightarrow \sigma^*$ band ranging from 191.5 to 198.5 nm (199.5 nm for DL-dopa), $\sigma \rightarrow \pi^*$ band ranging from 221.5 to 226 nm (221 nm for DL-dopa) and $n \rightarrow \pi^*$ band from 295 nm to 297.5 nm (280 nm for DL-dopa). IR spectra showed 1,019–1,040-cm^{-1} band for the glycosidic C–O–C aryl–alkyl symmetrical stretching and 1,307-1,320-cm^{-1} band for the asymmetrical stretching frequencies. In 2-D Heteronuclear Single Quantum Coherence Transfer (HSQCT) spectra, the respective chemical shift values of carbon atoms C1α and β, C6α and β showed glycoside formation and arylation.

8.1.8 L-Dopa Glycosides

Syntheses of L-dopa glycosides were carried out under the optimum conditions of 0.5 mmol, carbohydrates - 1 mmol, amyloglucosidase/immobilised β-glucosidase - 10% w/w of carbohydrate, 0.1 mM (1 mL) of 0.01-M pH 6 phosphate buffer for an incubation period of 72 h in 100-mL diisopropyl ether (Scheme 8.6, Vadivelan and Divakar 2011a). Diverse carbohydrate molecules, aldohexoses (D-glucose, D-galactose and D-mannose), ketohexose (D-fructose), pentoses (D-ribose and D-arabinose), disaccharides (lactose, maltose, sucrose) and carbohydrate alcohol (D-mannitol), were employed in the presence of amyloglucosidase and immobilised β-glucosidase.

UV spectra of L-dopa glycosides (Tables 8.19 and 8.20) showed shift in the $\sigma \rightarrow \sigma^*$ band in the 198.5–205-nm (28.5 nm for L-dopa) range, shift in the $\pi \rightarrow \pi^*$ band in the 267.5–295-nm (281 nm for L-dopa) range and shift in the $n \rightarrow \pi^*$ band around 388 nm (390 nm for L-dopa). IR spectra showed shifts in the 1,015–1,047-cm^{-1} band for the glycosidic C–O–C aryl–alkyl symmetrical stretching and 1,206–1,312-cm^{-1} band for the asymmetrical stretching frequencies. Mass spectral data also confirmed product formation.

In 2-D Heteronuclear Single Quantum Coherence Transfer (HSQCT) spectra, the respective chemical shift values of carbon atoms C1α and β, C6α and β showed glycoside formation and arylation.

However, there are distinct differences in this enzymatic reaction by both amyloglucosidase and β-glucosidase between the two substrates L-dopa and DL-dopa as described below. The nature and proportion of the L-dopa glycosides are shown in Tables 8.19 and 8.20. Amyloglucosidase catalysed L-dopa reaction with all the eleven carbohydrate molecules employed. However, β-glucosidase catalysed the reaction with only D-glucose, D-galactose, D-ribose and lactose. Amyloglucosidase showed selectivity in the case of D-mannose, D-fructose, D-ribose, maltose, sucrose, D-mannitol and D-sorbitol. About 20 individual glycosides were synthesised enzymatically using the glucosidases, of which 16 are being reported for the first time. The

Table 8.17 Optimisation of reaction conditions for the synthesis of DL-dopa-D-glucoside using amyloglucosidase

Reaction conditions[a]	Variable parameter[b]	Conversion yields (%)[c]
	Incubation period (h)	
DL-Dopa – 0.5 mmol	3	42
D-Glucose – 1 mmol	6	48
pH – 6	12	58
Buffer concentration – 0.1 mM (1 mL)	24	60
Amyloglucosidase – 10% w/w D-glucose	48	63
	72	62
	96	57
	120	49
	pH (0.01 M)	
DL-Dopa – 0.5 mmol[a]	4	16
D-Glucose – 1 mmol	5	38
Amyloglucosidase – 30% w/w D-glucose	6	62
Buffer concentration – 0.1 mM (1 mL)	7	48
Incubation period – 72 h	8	22
	Buffer concentration (mM)	
DL-Dopa – 0.5 mmol	0.03	51
D-Glucose – 1 mmol	0.06	62
Amyloglucosidase – 30% w/w D-glucose	0.1	63
pH – 6	0.14	59
Incubation period – 72 h	0.18	52
	0.22	47
	Amyloglucosidase (% w/w D-glucose)	
DL-Dopa – 0.5 mmol	10	59
D-Glucose – 1 mmol	20	65
pH – 6	30	62
Buffer concentration – 0.1 mM (1 mL)	40	59
Incubation period – 72 h	50	47
	75	45
	DL-Dopa (mmol)	
pH – 6	0.2	22
Buffer concentration – 0.1 mM (1 mL)	0.4	41
D-Glucose – 1 mmol	0.5	62
Amyloglucosidase – 10% w/w D-glucose	0.8	53
Incubation period – 72 h	1.2	56
	1.6	55
	2	54

[a]Initial reaction conditions
[b]Other variables are the same as under reaction conditions, except the specified ones
[c]HPLC yields expressed with respect to 1-mmol D-glucose employed

new glycosides reported are L-dopa-D-glucoside, L-dopa-D-galactoside, L-dopa-mannoside, L-dopa-fructoside, L-dopa-riboside, L-dopa-arabinoside, L-dopa-lactoside, L-dopa-maltoside, L-dopa-sucroside and D-mannitolide.

Only monoglycosylated/arylated products were formed. No bis products with both the OH groups at 3rd and 4th positions glycosylated/arylated could be detected. This indicated that steric effects are responsible for the formation of only monoglycosylated products. Under the reaction conditions employed, L-dopa gave a mixture of 3-*O*- and 4-*O*-glycosylated products with many of the carbohydrates employed. C6-O-arylated

Table 8.18 Syntheses of DL-3,4-dihydroxyphenylalanine (DL-dopa) glycosides using amyloglucosidase and β-glucosidase[a]

Glycosides	Amyloglucosidase catalysis		β-Glucosidase catalysis	
	Product (% proportion)[b]	Yields (%)[c]	Product (% proportion)[b]	Yields (%)[c]
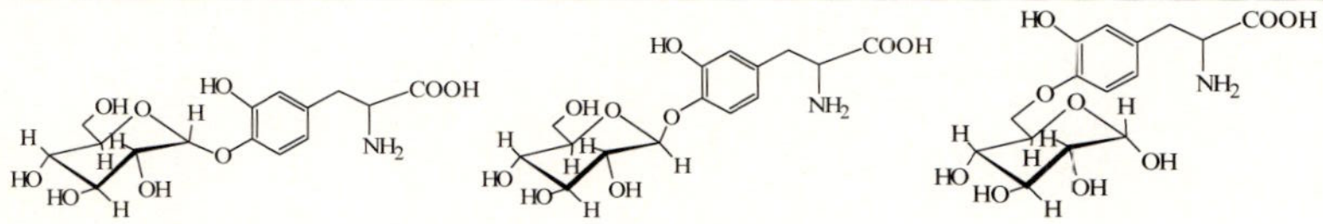 DL-3-Hydroxy-4-*O*-(α-D-glucopyranosyl)phenylalanine DL-3-Hydroxy-4-*O*-(β-D-glucopyranosyl)phenylalanine DL-3-Hydroxy-4-*O*-(6-D-glucopyranosyl)phenylalanine 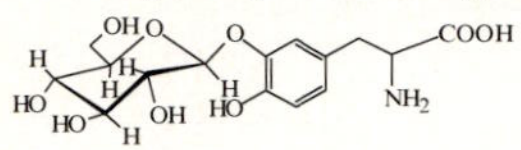 DL-4-Hydroxy-3-*O*-(β-D-glucopyranosyl)phenylalanine	4-*O*-C1α (24), 4-*O*-C1β (48), 4-*O*-C6-*O*-arylated (6), 3-*O*-C1β (22)	62	4-*O*-C1β (28), 4-*O*-C6-*O*-arylated (72)	33
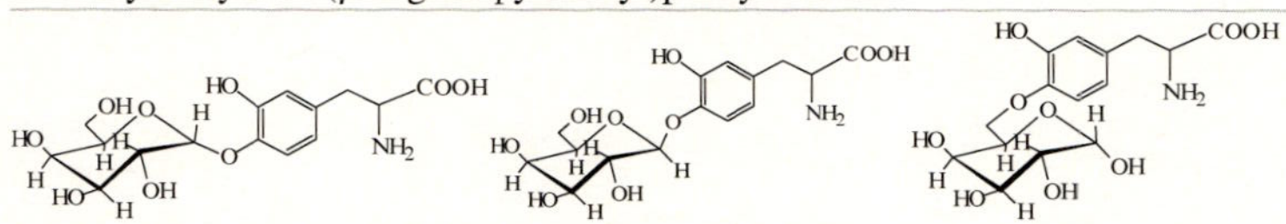 DL-3-Hydroxy-4-*O*-(α-D-galactopyranosyl)phenylalanine DL-3-Hydroxy-4-*O*-(β-D-galactopyranosyl)phenylalanine DL-3-Hydroxy-4-*O*-(6-D-galactopyranosyl)phenylalanine 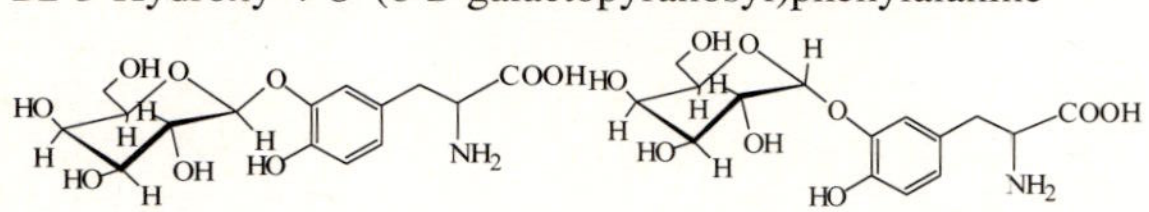 DL-4-Hydroxy-3-*O*-(α-D-galactopyranosyl)phenylalanine DL-4-Hydroxy-3-*O*-(β-D-galactopyranosyl)phenylalanine	4-*O*-C1α (27), 4-*O*-C1β (25), 4-*O*-C6-*O*-arylated (29), 3-*O*-C1α (10), 3-*O*-C1β (9)	46	4-*O*-C1α (17), 4-*O*-C1β (35), 4-*O*-C6-*O*-arylated (23), 3-*O*-C1α (21), 3-*O*-C1β (4)	31
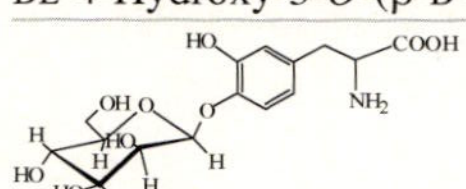 DL-3-Hydroxy-4-*O*-(β-D-mannopyranosyl)phenylalanine	4-*O*-C1β	61	4-*O*-C1β	32

(continued)

Table 8.18 (continued)

Glycosides	Amyloglucosidase catalysis		β-Glucosidase catalysis	
	Product (% proportion)[b]	Yields (%)[c]	Product (% proportion)[b]	Yields (%)[c]
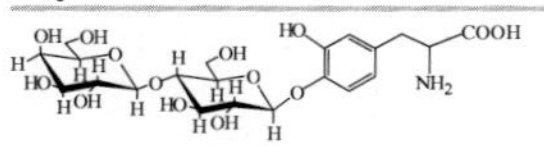 DL-3-Hydroxy-4-*O*-(β-D-galactopyranosyl-(1′→4)β-D- glucopyranosyl)phenylalanine	–	–	4-*O*-C1β	17
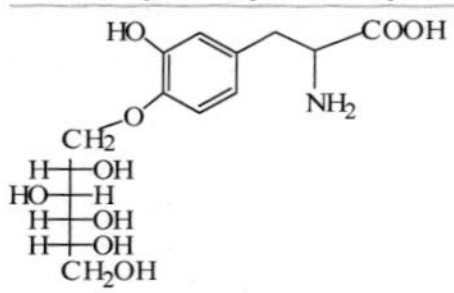 DL-3-Hydroxy-4-*O*-(6-D-sorbitol)phenylalanine	4-*O*-C6-*O*-arylated	12	–	–
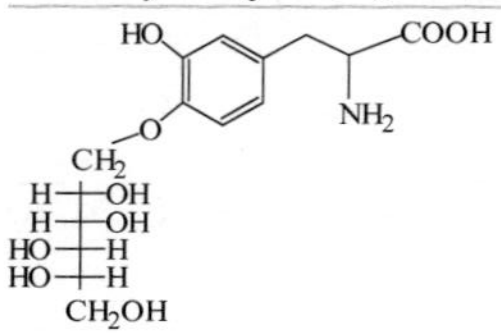 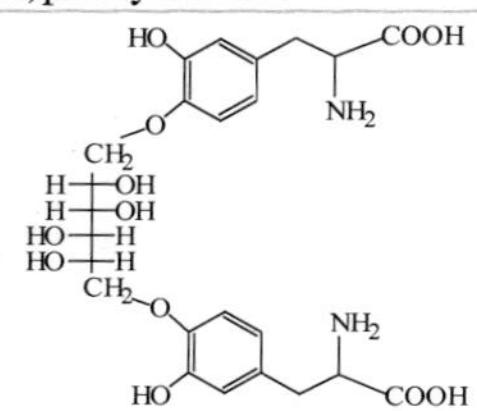 DL-3-Hydroxy-4-*O*-(1-D-mannitol)phenylalanine 1,6-*O*-(Bis DL-3-hydroxy-4-*O*-phenylalanine)D-mannitol	4-*O*-C1-*O*-arylated (70), Bis 4-*O*-C1,6-di-*O*-arylated (30)	20	–	–

[a]DL-3,4-Dihydroxyphenylalanine, 0.5 mmol; carbohydrate, 1 mmol; enzyme concentration, 10% w/w of carbohydrate; solvent, di-isopropyl ether; buffer, 0.1-mM (1-mL) pH 6 phosphate buffer; incubation period, 72 h

[b]The product proportions were determined from the area of respective $^{1}H/^{13}C$ signals

[c]Conversion yields were from HPLC with respect to free carbohydrate. Error in yield measurements is ±10%

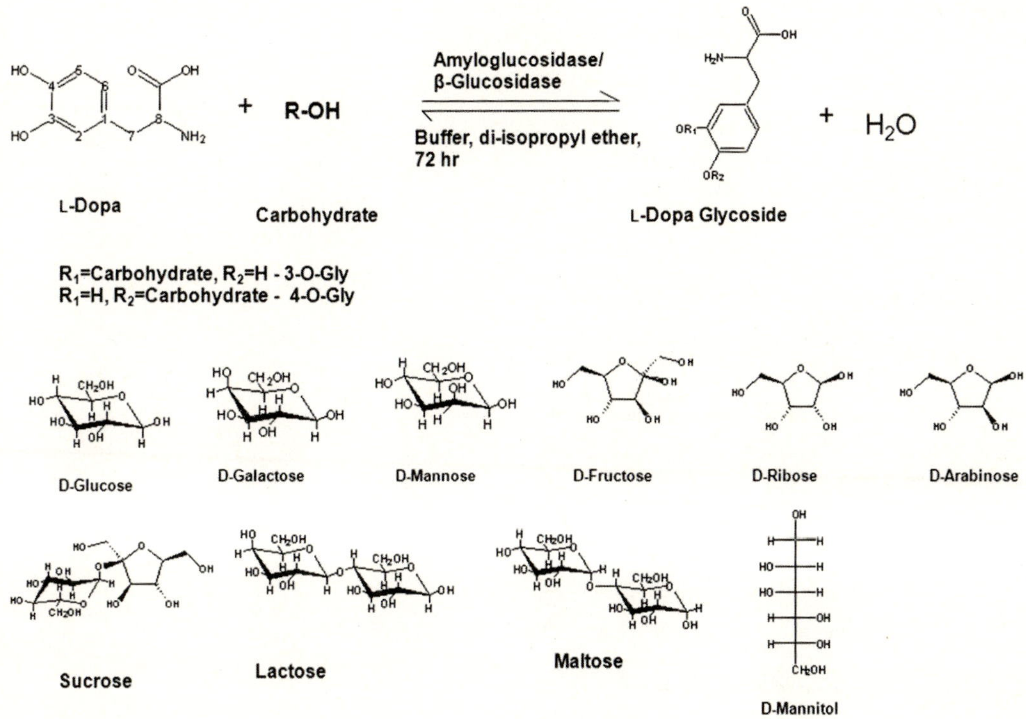

Scheme 8.6 Syntheses of L-dopa glycosides

compounds are observed with D-glucose and D-galactose. While D-fructose gave 2-O, sucroside and D-mannitol gave 1-O aryl derivatives.

D-Glucose exhibited 33.3% α and 66.7% β compared to the 60:40α:β-anomeric composition of free D-glucose employed. D-Galactose showed 42.3% α and 57.7% β compared to the 92:8α:β-anomeric composition of free D-galactose employed. Similarly D-arabinose showed 25.9% α and 74.1% β compared to the 95:5α:β-anomeric composition of the D-arabinose employed.

8.1.9 Syntheses of Dopamine Glycosides

Parkinson's disease is a neurodegenerative disease characterised by bradykinesia, tremors, rigidity and difficulty in walking (Geng et al. 2004, 2007). The usual treatment for this disorder is the use of L-dopa, which enters the central nervous system (CNS) through active transport and is enzymatically cleaved in the brain to release dopamine. To overcome the drawbacks of L-dopa treatment in transport problems across blood–brain barrier (BBB), new derivatives of dopamine able to penetrate the BBB, by making use of specific transport systems (Pardridge 2002; Audus et al. 1992; Madrid et al. 1991), are employed. Glucose is the brain's source of energy and other hexoses are transported across the BBB by the glucose carrier GLUT1. Glycosyl dopamine derivatives bearing the sugar moiety linked to either the amino group or the catechol ring of dopamine through amide, ester or glycosidic bonds were synthesised chemically with potent anti-Parkinsonian properties (Fernandez et al. 2003). The glucosyl dopamine derivative is able to interact with the glucose transporter (GLUT1) and absorbed into the CNS from the blood stream (Dalpiaz et al. 2007). Glycosylated form of dopamine is stable in the periphery and, after reaching the brain through GLUT1 carrier, gets hydrolyzed by the action of brain enzymes to release dopamine (Fernandez et al. 2003).

Table 8.19 Synthesis of L-dopa glycosides using amyloglucosidase and immobilised β-glucosidase

		Amyloglucosidase catalysis	
S. no.	Glycosides	% of product (proportion)	% of product (conversion)
1	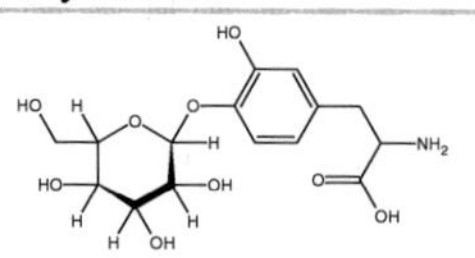 L-3-Hydroxy-4-O-(α-D-glucopyranosyl) phenylalanine	**12**	54
	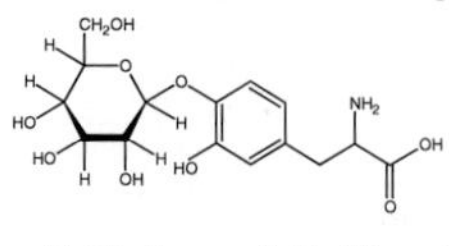 L-3-Hydroxy-4-O-(β-D-glucopyranosyl)phenylalanine	**42**	
	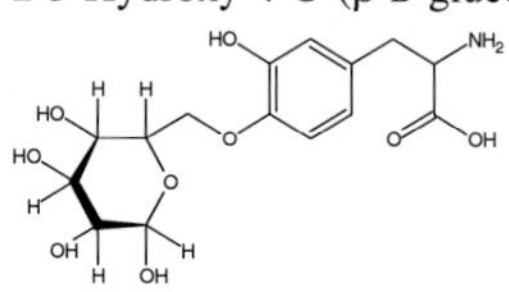 L-3-Hydroxy-4-O-(6-O- D-glucopyranosyl)phenylalanine	**14**	
	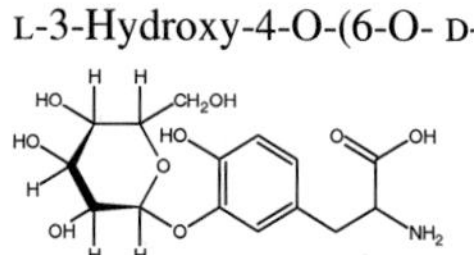 L-4-Hydroxy-3-O-(β-D-glucopyranosyl)phenylalanine	**22**	
2	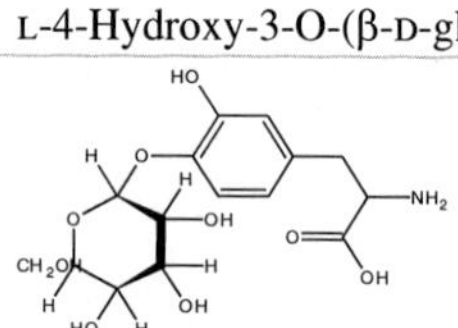 L-3-Hydroxy-4-O-(α-D-galactopyranosyl) phenylalanine	**22**	44
	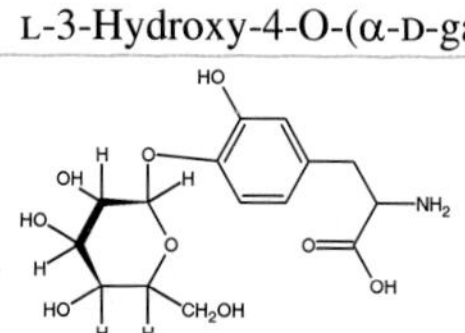 L-3-Hydroxy-4-O-(β-D-galactopyranosyl)phenylalanine	**30**	
	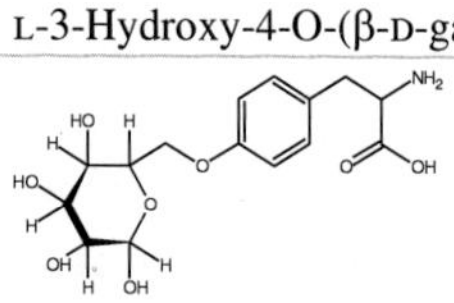 L-3-Hydroxy-4-O-(6-O-D-galatopyranosyl)phenylalanine	**28**	
	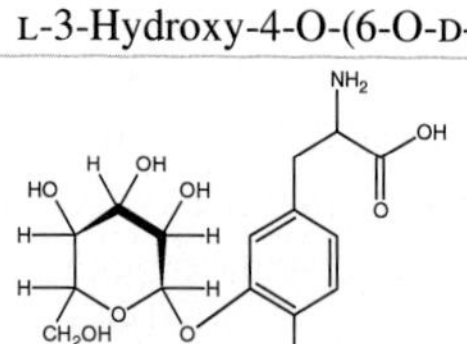 L-4-Hydroxy-3-O-(α-D-galactopyranosyl) phenylalanine	**20**	

(continued)

Table 8.19 (continued)

S. no.	Glycosides	Amyloglucosidase catalysis	
		% of product (proportion)	% of product (conversion)
3	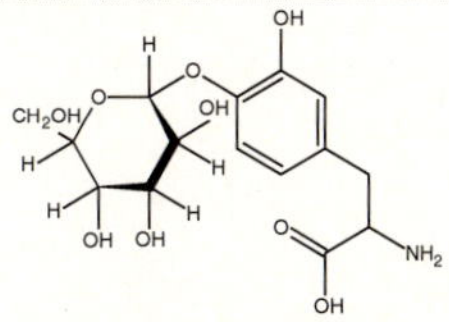 L-3-Hydroxy-4-O-(α-D-mannopyranosyl)phenylalanine	15	46.3
4	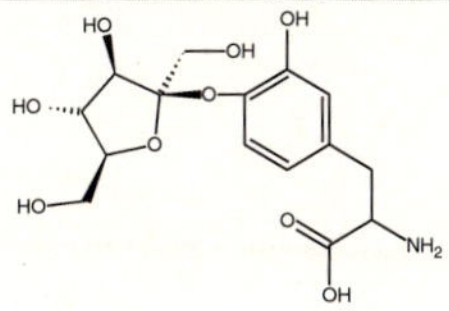 L-3-Hydroxy-4-O-(2-O-D-fructofuranosyl) phenylalanine	16	8.7
5	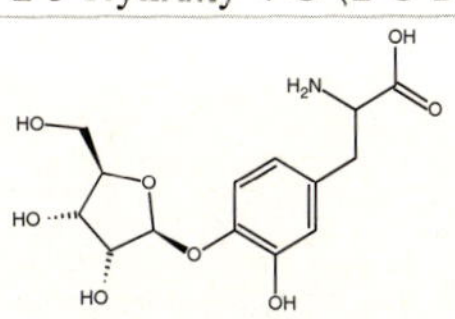 L-3-Hydroxy-4-O-(α-D-ribofuranosyl)phenylalanine	17	62.7
6	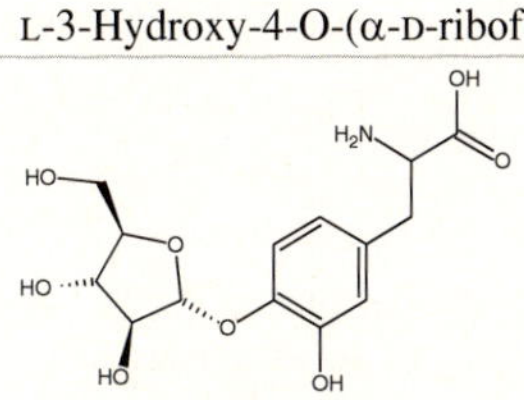 L-3-Hydroxy-4-O-(α-D-arabinofuranosyl) phenylalanine	25.9	63.8
	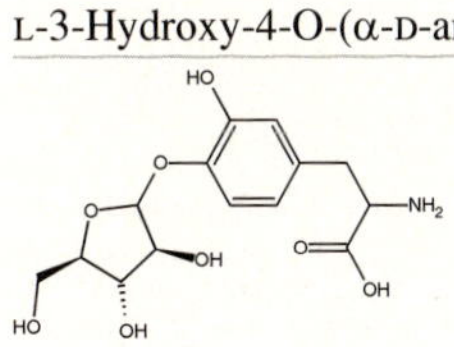 L-3-Hydroxy-4-O-(β-D-arabinofuranosyl) phenylalanine	74.1	
7	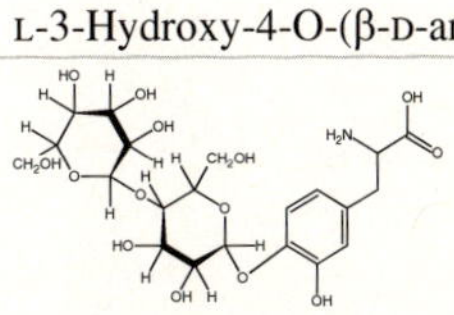 L-3-Hydroxy-4-O-(α-lactosyl) phenylalanine	41.6	29
	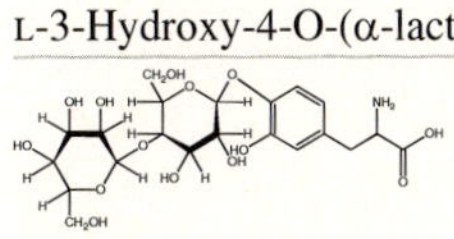 L-3-Hydroxy-4-O-(β-lactosyl) phenylalanine	58.4	
8	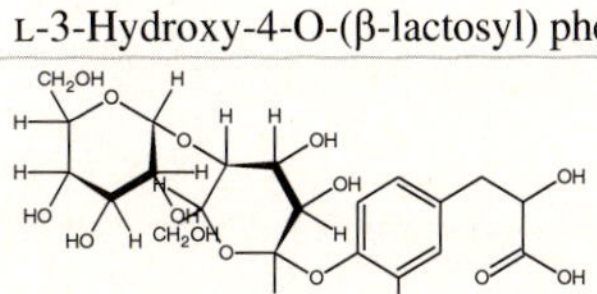 L-3-Hydroxy-4-O-(β-maltosyl) phenylalanine		35.7

(continued)

Table 8.19 (continued)

		Amyloglucosidase catalysis	
S. no.	Glycosides	% of product (proportion)	% of product (conversion)
9	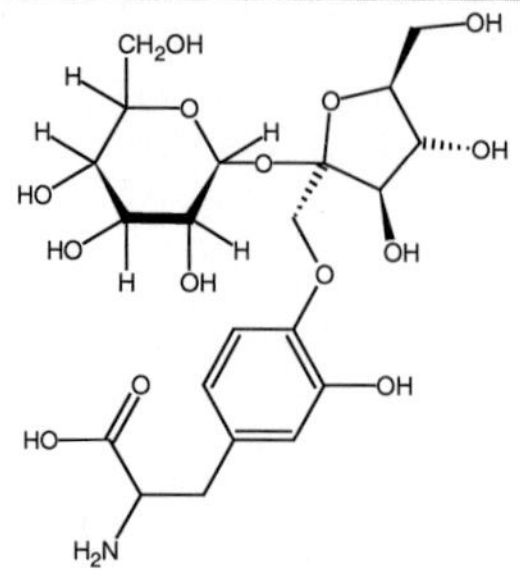 L-3-Hydroxy-4-O- (1-O-sucrose) phenylalanine		8.5
10	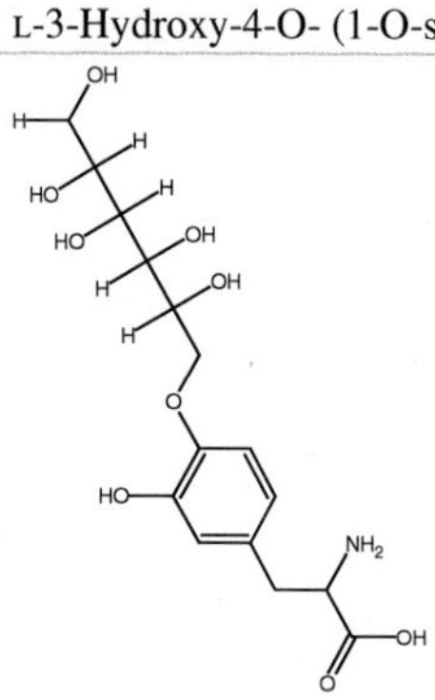 L-3-Hydroxy-4-O-(1-O-D-mannitolide) phenylalanine		5

Table 8.20 Synthesis of L-dopa glycosides using immobilised β-glucosidase

S. no.	Glycosides	% of product (proportion)	% of conversion
1	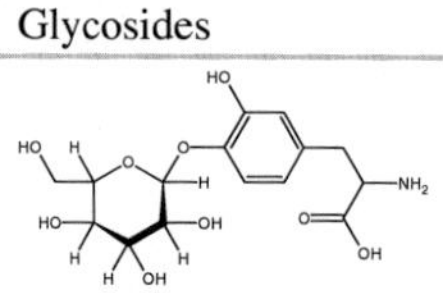 L-3-Hydroxy-4-O-(α-D-glucopyranosyl) phenylalanine	**24**	57.6
	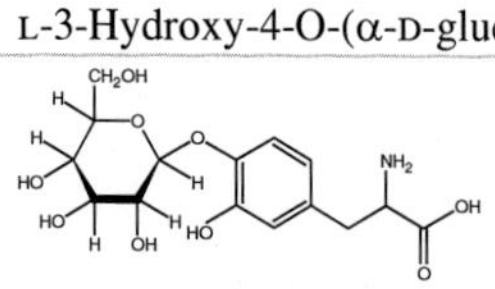 L-3-Hydroxy-4-O-(β-D-glucopyranosyl)phenylalanine	**48**	
	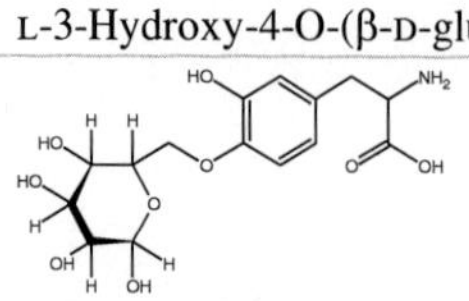 L-3-Hydroxy-4-O-(6-O-D-glucopyranosyl)phenylalanine	**6**	

(continued)

Table 8.20 (continued)

S. no.	Glycosides	% of product (proportion)	% of conversion
	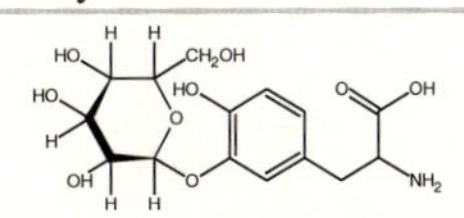 L-4-Hydroxy-3-O-(β-D-glucopyranosyl)phenylalanine	**22**	
2	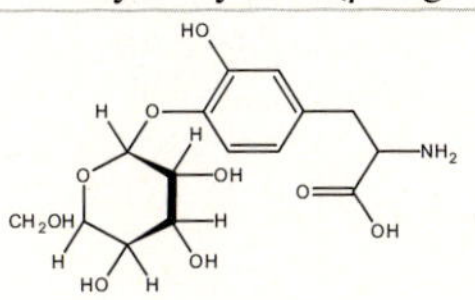 L-3,-Hydroxy-4-O-(α-D-galactopyranosyl)phenylalanine	43.9	55
	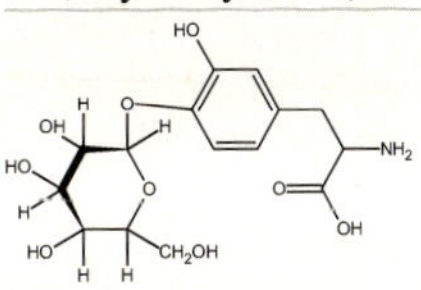 L-3-Hydroxy-4-O-(β-D-galactopyranosyl)phenylalanine	56.1	
3	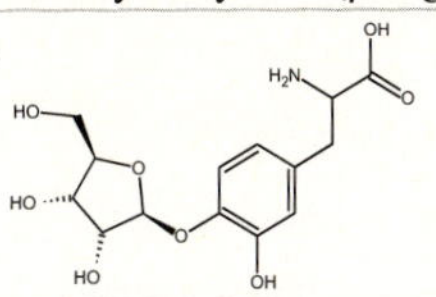 L-3-Hydroxy-4-O-(α-D-ribofuranosyl)phenylalanine	46.2	44
	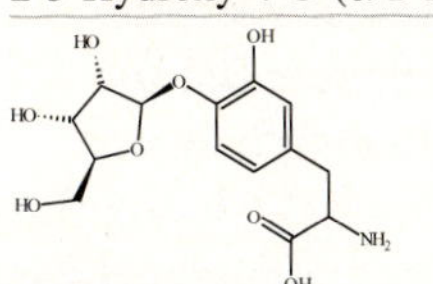 L-3-Hydroxy-4-O-(β-D-ribofuranosyl)phenylalanine	53.8	
4	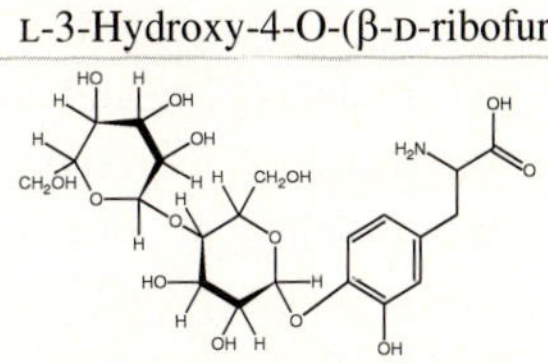 L-3-Hydroxy-4-O-(α-D-lactosyl) phenylalanine	8.5	25
	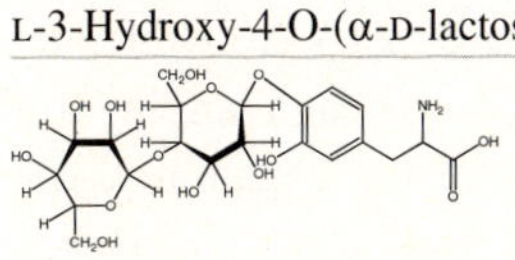 L-3-Hydroxy-4-O-(β-D-lactosyl) phenylalanine	87.5	

Employing chemical methods for the synthesis of dopamine glycosides require protection and deprotection steps (Fernandez et al. 2003). Enzymatic method is the alternative one which provides milder reaction conditions, easy workup, less pollution, higher yields and selectivity (Suzuki et al. 1996; Vic and Thomas 1992; Vijayakumar et al. 2007).

It is in this context that dopamine glycosides were prepared using amyloglucosidase from

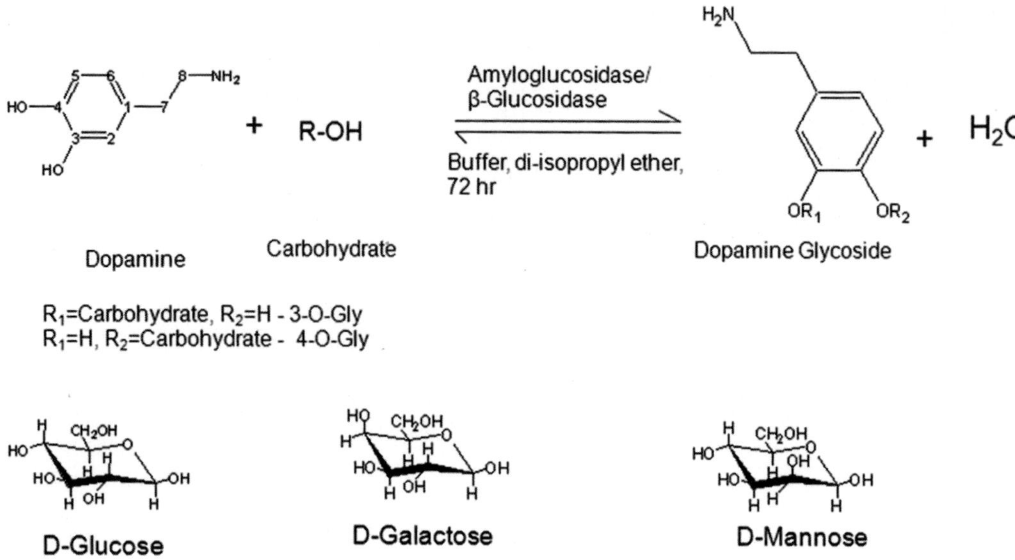

Scheme 8.7 Syntheses of dopamine glycosides

Rhizopus mould and immobilised β-glucosidase in organic media (Scheme 8.7, Sivakumar and Divakar 2009a).

8.1.9.1 Amyloglucosidase-Catalysed Glucosylation of Dopamine

Enzymatic glucosylation between dopamine and D-glucose was optimised in terms of incubation period, pH, buffer, enzyme and dopamine concentrations using both amyloglucosidase and immobilised β-glucosidase. Optimum conditions determined were as follows: 58% at 24-h incubation (Table 8.21) at pH 6.0 and 40% (w/w D-glucose) amyloglucosidase showed the highest yield of 45% at 0.5 mmol of dopamine.

8.1.9.2 Immobilised β-Glucosidase-Catalysed Glucosylation of Dopamine

Optimum conditions for the glucosylation of dopamine using immobilised β-glucosidase were worked out. Optimum conditions are 58% yield at 72 h in 0.04-mM (0.4-mL) pH 5.0 buffer and a yield of 65% with 50% (w/w D-glucose) immobilised β-glucosidase (Table 8.22) with 0.5-mmol dopamine.

8.1.9.3 Syntheses of Dopamine Glycosides of Various Carbohydrates Using Amyloglucosidase and Immobilised β-Glucosidase and Their Spectral Characterisation

Syntheses of the other dopamine glycosides were carried out under optimised conditions mentioned above, with dopamine and carbohydrates: D-glucose, D-galactose and D-mannose (Scheme 8.7) using amyloglucosidase and immobilised β-glucosidase.

Ultraviolet-visible spectra of dopamine glycosides showed $\sigma \rightarrow \sigma^*$ band ranging from 191.5 to 201 nm (199 nm for dopamine), $\sigma \rightarrow \pi^*$ band ranging from 217 to 225.5 nm (218.5 nm for dopamine) and $\pi \rightarrow \pi^*$ and $n \rightarrow \pi^*$ band from 276 to 286 nm (280 nm for dopamine). IR spectra showed 1,028–1,079-cm^{-1} band for the glycosidic C–O–C aryl–alkyl symmetrical stretching and 1,196–1,204-cm^{-1} band for the asymmetrical stretching frequencies. In 2-D Heteronuclear Single Quantum Coherence Transfer (HSQCT) spectra, the respective chemical shift values of carbon atoms C1α and β, C6α and β showed glycoside formation and arylation (Table 8.23).

Table 8.21 Optimisation of reaction conditions for the synthesis of dopamine-D-glucoside using amyloglucosidase

Reaction conditions	Variable parameter[b]	Conversion yields (%)[c]
	Incubation period (h)	
Dopamine – 0.5 mmol	3	44
D-Glucose – 1 mmol	6	45
pH – 6	12	47
Buffer concentration – 0.06 mM (0.6 mL)	24	58
Amyloglucosidase – 40% w/w D-glucose	48	48
	72	52
	96	48
	120	37
	pH (0.01 M)	
Dopamine – 0.5 mmol[a]	4	No yield
D-Glucose – 1 mmol	5	16
Amyloglucosidase – 30% w/w D-glucose	6	35
Buffer concentration – 1 mL (0.1 mM)	7	29
Incubation period – 72 h	8	23
	Buffer concentration (mM)	
Dopamine – 0.5 mmol	0.03	21
D-Glucose – 1 mmol	0.06	39
Amyloglucosidase – 30% w/w D-glucose	0.1	27
pH – 6	0.15	22
Incubation period – 72 h	0.2	9
	0.25	9
	Amyloglucosidase (% w/w D-glucose)	
Dopamine – 0.5 mmol	10	13
D-Glucose – 1 mmol	20	26
pH – 6	30	34
Buffer concentration – 0.06 mM (0.6 mL)	40	45
Incubation period – 72 h	50	19
	75	21
	Dopamine (mmol)	
pH – 6	0.25	33
Buffer concentration – 0.06 mM (0.6 mL)	0.5	50
D-Glucose – 1 mmol	0.75	48
Amyloglucosidase – 40% w/w D-glucose	1	31
Incubation period – 72 h	1.5	25
	2	16

[a]Initial reaction conditions
[b]Other variables are the same as under reaction conditions, except the specified ones
[c]HPLC yields expressed with respect to 1-mmol D-glucose employed

8.1.10 Serotonyl Glycosides

Serotonin is primarily found in the gastrointestinal tract and central nervous system of humans as well as animals. In addition to animals, it is also found in insects, fungi and plants (Kang et al. 2009). Serotonin is an important neurotransmitter in the central and peripheral nervous systems. It has been reported to modulate the regulation of a variety of major physiological functions including affective behaviour, memory and thermoregulation via the interaction at serotonin receptor subtypes (Hoyer and Martin 1997). Basically 80% of the human body's total serotonin is

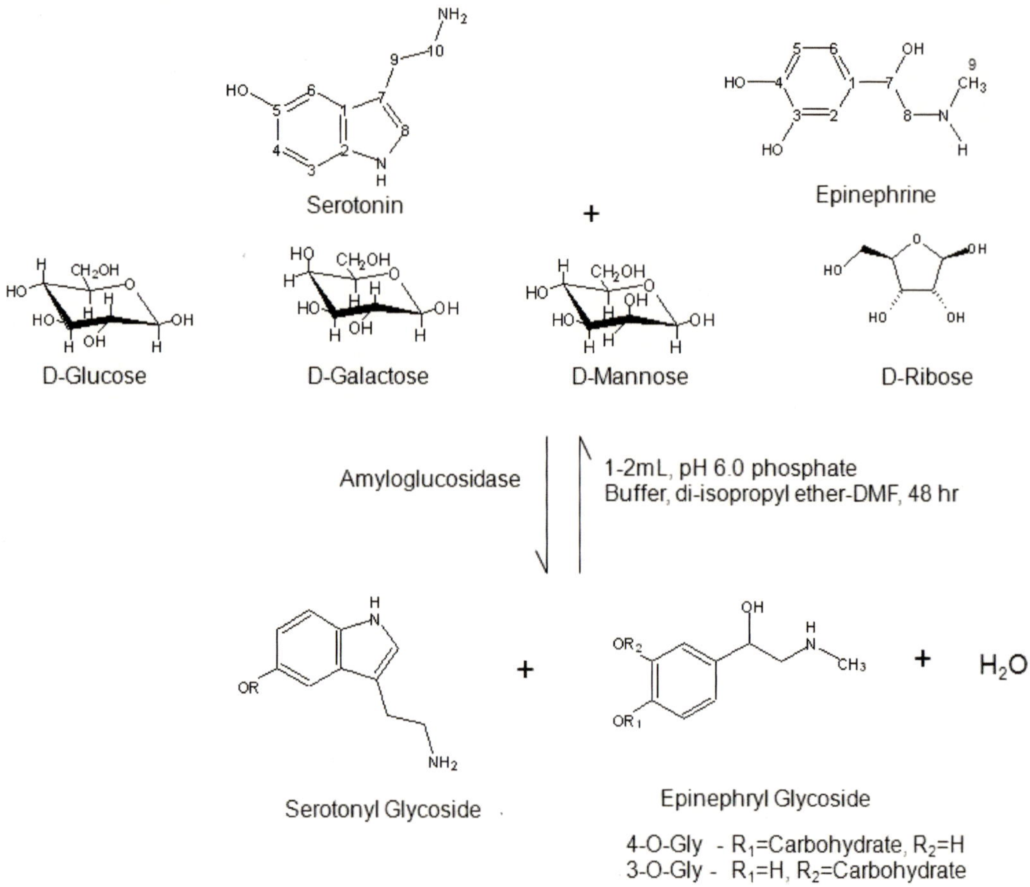

Scheme 8.8 Glycosylation of serotonin and epinephrine

Serotonin can be made more water-soluble (solubility of serotonin in water 2 g/100 mL at 27°C) and stable by derivatising. Serotonin supplements especially derivatised ones could be better substituents for external administration in diet which resulted in the preparation of serotonyl glycosides (Scheme 8.8, Vadivelan and Divakar 2011b).

The reaction did not take place without the use of the enzyme amyloglucosidase from *Rhizopus* mould under the reaction conditions employed. Serotonin (Scheme 8.8) contains phenolic –OH group at position 5. The carbohydrates employed were D-glucose, D-galactose, D-mannose, D-ribose, D-arabinose, D-fructose, D-maltose, lactose, sucrose, D-sorbitol and D-mannitol. The reaction did not occur at the primary and secondary amino groups, respectively. In case of serotonin, reaction took place with D-glucose, D-galactose, D-mannose and D-ribose only. The other carbohydrate molecules did not react. This could be due to competitive binding of serotonin and carbohydrates to the enzyme restricting reactions with certain carbohydrates as facile transfer of tightly bound serotonin to the lesser nucleophilic carbohydrates rendered impossible. Glycosylation yields were generally in the range 13–29% (Table 8.24).

UV spectra of serotonin glycosides showed UV_{max} between 205 and 215 nm (σ-σ^* transition), 274 and 296 nm (n-π^* transition) and around 367 nm for extended n-π^* transition (270 nm for free serotonin). IR C–O–C glycosidic aryl–alkyl symmetrical stretching frequencies in the 1,016–1,088-cm^{-1} range and asymmetrical stretching frequencies in the 1,200–1,208-cm^{-1} range indicated that serotonin has undergone glycosylation. Mass spectra also confirmed glycosylation through detection of the parent M^+ ion for the glycosides. In 2-D Heteronuclear Single Quantum Coherence Transfer (HSQCT) spectra, the respective chemical shift values of carbon atoms C1α and β, C6α and β showed glycoside formation and arylation.

Table 8.24 Amyloglucosidase-catalysed syntheses of serotonin and epinephrine glycosides

Glycoside	Product (% proportion)[a]	Yield (%)[b]
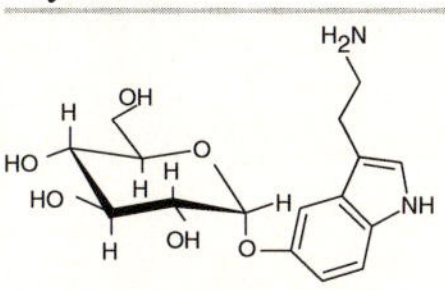 **7a**: Serotoninyl-5-*O*(–α-D-glucopyranoside) 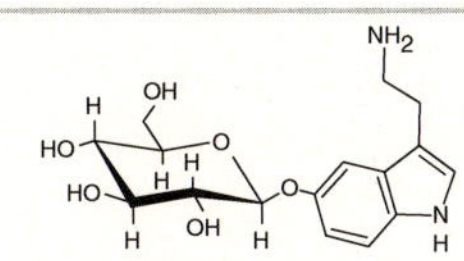 **7b**: Serotoninyl-5-*O* (–β-D-glucopyranoside) 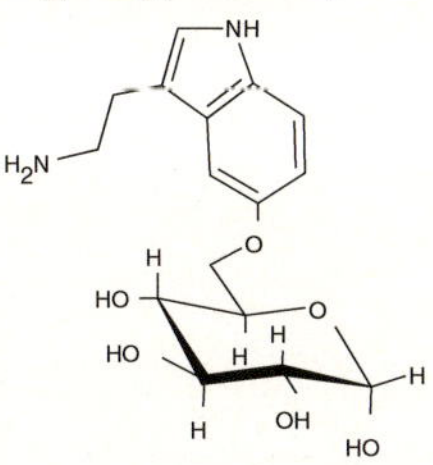 **7c**: Serotoninyl-5-*O*-(6-α-D-glucopyranoside) 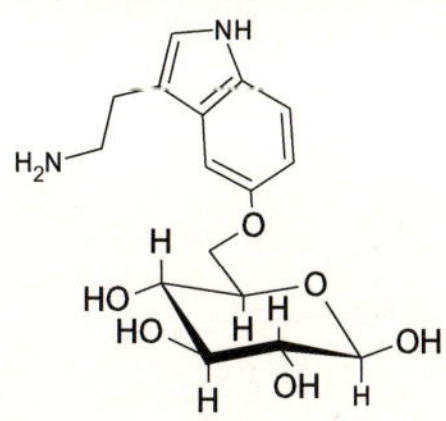 **7d** Serotoninyl-5-*O*-(6-β-D-glucopyranoside)	C1 α (26), C1 β (39), 6-*O*-α (17) and 6-*O*-β (19) monoglycosides	15
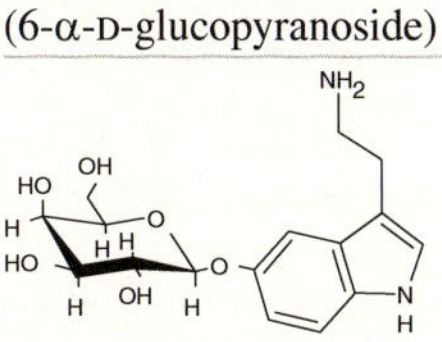 **8a**: Serotoninyl-5-*O* (–β-D-galactopyranoside) 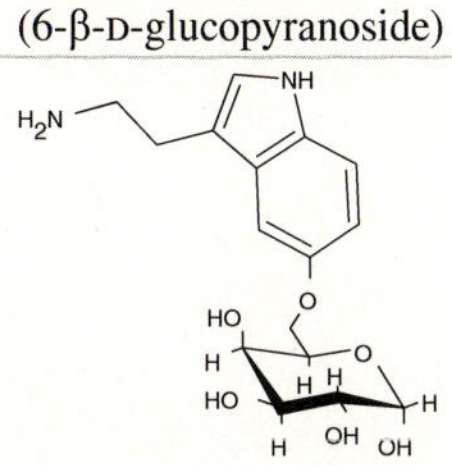 **8b**: Serotoninyl-5-*O*-(6-α-D-galactopyranoside) 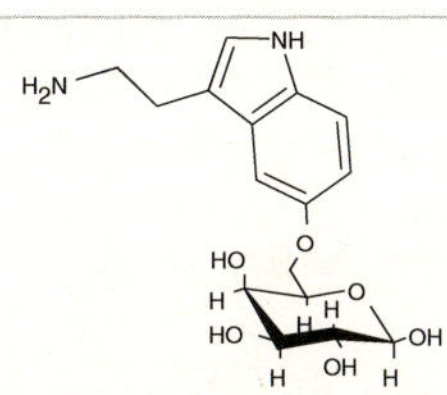 **8c**: Serotoninyl-5-*O*-(6-β-D-galactopyranoside)	C1α (49) and 6- *O*-α (25), 6-*O*-β (26) monogalactoside	13
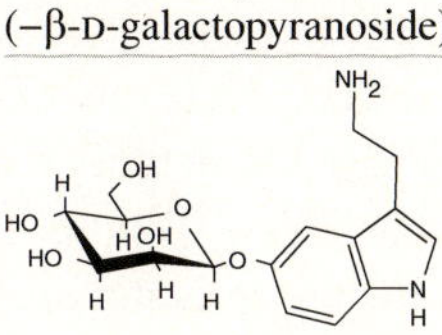 **9a** Serotoninyl-5-*O* (–β-D-mannopyranoside) 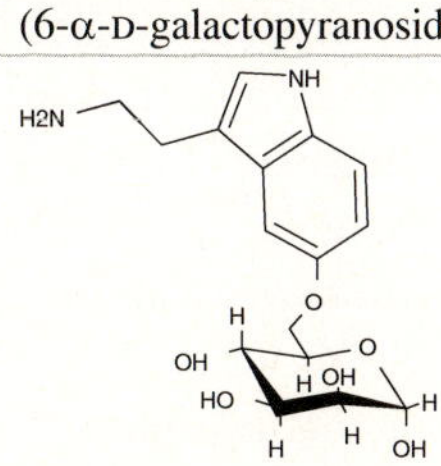 **9b**: Serotoninyl-5-*O*-(6-α-D-mannopyranoside) 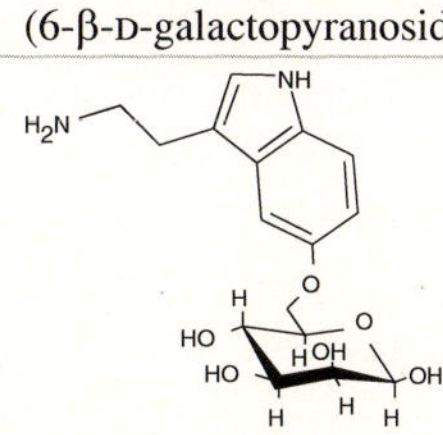 **9c**: Serotoninyl-5-*O*-(6-β-D-mannopyranoside)	C1β (48) and 6-*O*-α (22) and 6- *O*-β (20) mono-mannoside	18
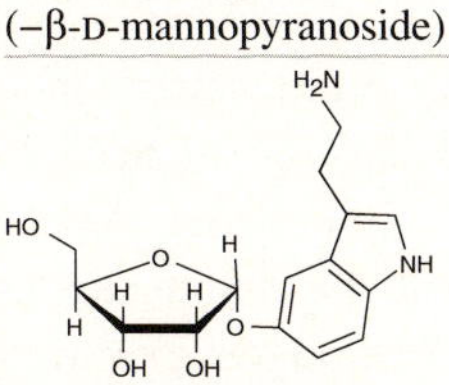 **10a**: Serotoninyl-5-*O* (–α-D-ribofuranoside) 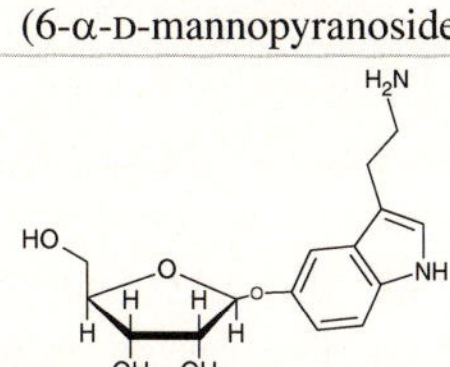 **10b**: Serotoninyl-5-*O* (–β-D-ribofuranoside)	C1α (72) and C1β (28) monoriboside	29

(continued)

Table 8.24 (continued)

Glycoside	Product (% proportion)[a]	Yield (%)[b]
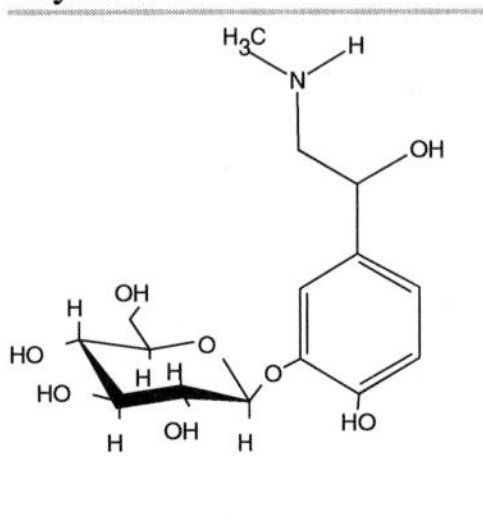 **11a:** Epinephrinyl-3-*O*-(−β-D-glucopyranoside) 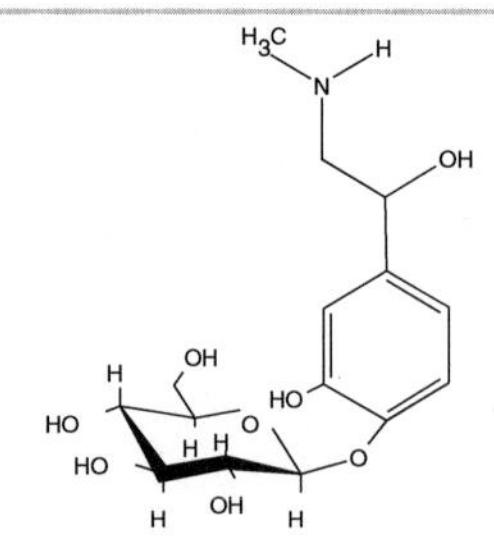 **11b**: Epinephrinyl-4-*O*-(−β-D-glucopyranoside)	3-*O*-C1β (63) and 4-*O*-C1β (37) monoglucoside	18
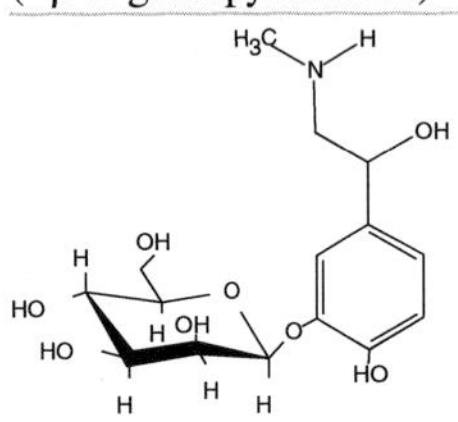 **12a:** Epinephrinyl-3-*O* (−β-D-mannopyranoside) 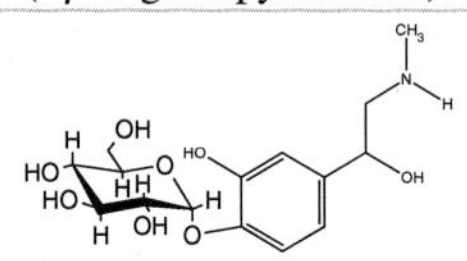 **12b:** Epinephrinyl-4-*O*(−α-D-mannopyranoside) 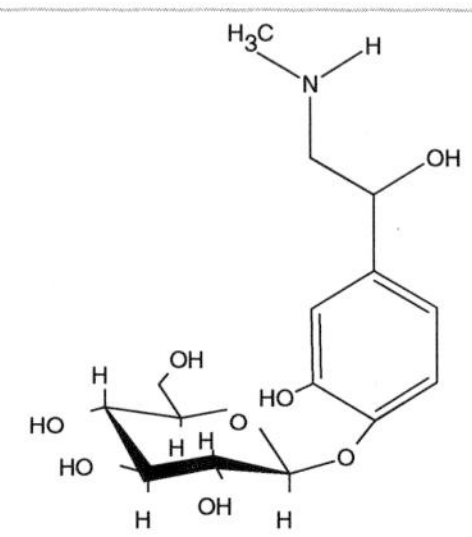 **12c:** Epinephrinyl-4-*O* (−β-D-mannopyranoside)	C1α (72) and C1β (28) monoriboside	29

[a]Product proportions were determined by ^{13}C 2-D HSQCT NMR C1/C6 peak areas or their cross peaks
[b]Conversion yields were from HPLC with errors in yield measurements of ±5–10%

Except for the serotoninyl glucosides, all the other glycosides and 6-O-derivatives are being reported for the first time. Since the phenolic group at position 5 in serotonin is not sterically hindered, all the four carbohydrates D-glucose, D-galactose, D-mannose and D-ribose underwent facile reaction with serotonin.

8.1.11 Epinephryl Glycosides

Epinephrine is a hormone and also a neurotransmitter, briefly called as adrenaline. As a hormone, it acts on nearly all body tissues. Epinephrine's various functions result from its binding to a variety of adrenergic receptors triggering a number of metabolic changes. Binding to α-adrenergic receptors inhibits insulin secretion, stimulates glycogenolysis in the liver and muscle and stimulates glycolysis in muscle (Sircar 2007). β-Adrenergic receptor binding triggers glucagon secretion in the pancreas, increased adrenocorticotropic hormone secretion by the pituitary gland and increased lipolysis by adipose tissue. All these effects lead to increased blood glucose and fatty acids, providing substrates for energy production within cells throughout the system (Sircar 2007). Due to its vasoconstrictive effects, epinephrine is the drug of choice for treating anaphylaxis and also useful in treating sepsis. It is also used as a bronchodilator for asthma as its specific β-agonists are unavailable or ineffective.

Epinephrine can be made more water-soluble (solubility of epinephrine 0.01 g/100 mL at 18°C) and stable by derivatising. Epinephrine supplements like glycosides could be better supplements, and hence epinephryl glycosides were prepared enzymatically (Scheme 8.8, Vadivelan and Divakar 2011b).

In case of epinephrine, out of the 11 carbohydrates employed, only D-glucose and D-mannose reacted effectively. The other carbohydrate molecules did not react. Structural comparison between serotonin and epinephrine clearly shows that epinephrine is more polar compared to serotonin and hence could bind much more strongly to the enzyme compared to serotonin, inhibiting the enzyme towards reaction with other carbohydrate molecules. Compared to the other carbohydrate molecules besides D-glucose and D-mannose, epinephrine could bind more strongly than those other carbohydrate molecules thereby inhibiting reaction.

While the reaction with glucose exhibited a conversion yield of 18%, D-mannose gave only 29% yield. Both 3-O- and 4-O-products were formed. Only phenolic –OH groups of epinephrine at position 3 and 4 reacted in a facile manner. Epinephrine contains two phenolic OH molecules and alcoholic OH group, all three capable of undergoing glycosylation. No product formation arising out of reaction at the secondary hydroxyl group at position 7 was detected.

UV spectra showed λ_{max} at 208 nm and 231 nm (σ-σ^* transition) for the reaction products of D-glucose and D-mannose, respectively, and another λ_{max} at 266 nm for D-glucoside and 285 nm (n-π^* transition) for D-mannoside (279 nm for epinephrine itself). Similarly IR, glycosidic aryl–alkyl C–O–C stretching frequencies at 1,048 cm^{-1} for D-glucoside and 1,021 cm^{-1} for D-mannoside and asymmetrical stretching frequencies at 1,212 cm^{-1} and 1,223 cm^{-1} for the respective glucoside and mannosides indicated glycosylation. 2-D HSQCT spectra conformed the product formation further from the respective chemical shift values (Table 8.24). The C3 and C4 carbons of epinephrine showed shifts at around 144–145 ppm in their respective derivatives indicating reaction involving these phenolic hydroxyl groups.

Presence of hydrophobic/hydrophilic phenolic –OH group epinephrine bestowed excellent nucleophilicity to these molecules promoting reaction with few diverse carbohydrate molecules. Only monoglycosylated/arylated products were detected. However, loss of regiospecificity in case of epinephrine could be due to employment of large concentration of the enzyme. Reaction with only few carbohydrate molecules indicated that both serotonin and epinephrine possess the propensity to inhibit the enzyme strongly. The glycosides prepared are more water-soluble and stable than their respective aglycons, hence portending excellent pharmaceutical properties for the prepared glycosides.

8.2 Vitamins

Vitamins are organic compounds that are required in small amounts for the normal functioning of the body and maintenance of metabolic integrity. As with many other biologically active compounds, glycoside derivatives of vitamins have been identified and their properties investigated. The formation of glycosylated derivatives of vitamins, whether natural in plants, animals or microorganisms or by intentional chemical modifications, represents a process that may cause dramatic changes in chemical, nutritional and metabolic properties of vitamins.

Hence, water-soluble vitamins were glycosylated to improve their water solubility, and fat-soluble vitamins were glycosylated also to improve their water solubility. Both classes of vitamins on glycosylation could possess better nutritional and pharmacological properties that are very much desired.

Enzymatic syntheses of more water-soluble thiamin (vitamin B1), riboflavin (vitamin B2) and pyridoxine (vitamin B6) and fat-soluble retinol (vitamin A), ergocalciferol (vitamin D2), cholecalciferol (vitamin D3) and α-tocopherol (vitamin E) with carbohydrates (D-glucose, D-galactose, D-mannose, D-fructose, D-ribose, D-arabinose, maltose, sucrose, lactose, D-sorbitol and D-mannitol) were carried out using amyloglucosidase from *Rhizopus* mould and β-glucosidase isolated from sweet almond. The usual 11 carbohydrate molecules were employed (Sivakumar and Divakar 2009b; Ponrasu et al. 2008, 2009; Einstein Charles et al. 2009a, b; Manohar and Divakar 2010).

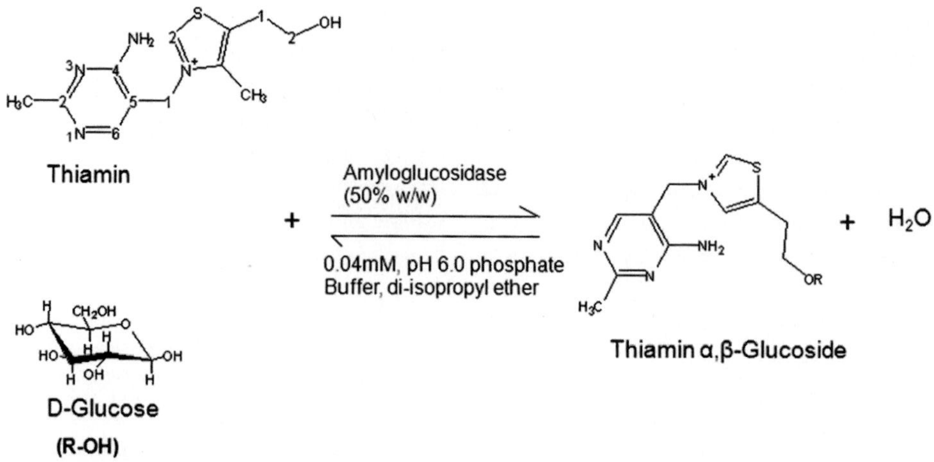

Scheme 8.9 Synthesis of thiamin glucoside

8.2.1 Glucosylation of Thiamin

Thiamin (3-[(4′amino-2′-methyl-pyrimidine-5′-yl-methyl)-5-(2-hydroxy-ethyl)]-4-methylthiazolium chloride hydrochloride, vitamin B_1), a water-soluble vitamin belonging to the B complex group, is an important cofactor of decarboxylase, transketolase and carboxylase (Kren et al. 1998). The characteristic odour and a strong tongue-pricking taste of thiamin could be reduced by preparing derivatives of this vitamin (Suzuki and Uchida 1994).

Enzymatic synthesis of O-α-glucosylthiamin has been reported by a consecutive action of cyclomaltodextrin glucanotransferase from *Rhizopus stearothermophilus* and glucoamylase from *Rhizopus sp.* (Uchida and Suzuki 1968). Glucosylation of thiamin (vitamin B1) is discussed here (Scheme 8.9, Ponrasu et al. 2009).

8.2.1.1 Response Surface Methodology

Design: Central Composite Rotatable Design (CCRD), 32 experiments, 5 variables at 5 levels (Ponrasu et al. 2009)

Variables: Five variables were employed. Actual set of experiments undertaken and the glucosylation yields obtained are shown in Table 8.25.

Equation: A second-order polynomial equation was developed to study the effects of the variables on the esterification yields in terms of linear, quadratic and cross product terms.

$$Y = 279.21 - 1.93 * Xe + 33.99 * Xc - 2.37 * Xt - 58.46 * Xv - 37.13 * Xp + 0.007 * Xe * Xe + 57.88 * Xc * Xc + 0.007 * Xt * Xt + 6.15 * Xv * Xv + 0.74 * Xp * Xp - 0.05 * Xe * Xc - 0.0005 * Xe * Xt - 0.02 * Xe * Xv + 0.20 * Xe * Xp + 0.11 * Xc * Xt - 33.50 * Xc * Xv - 15.42 * Xc * Xp + 0.12 * Xt * Xv + 0.21 * Xt * Xp + 11.92 * Xv * Xp$$

where Xe is the enzyme concentration, Xc is the thiamin concentration, Xt is the incubation period, Xv is the buffer concentration, Xp is the pH and Y is the yield.

Coefficients: Kyplot Software Version 2.0 Beta

Analysis of Variance (ANOVA): Kyplot, Software. R^2 value of 0.74. ANOVA shows the model is significant at $P<0.01$.

Optimisation: Microsoft Regression software

Contour plots (Fig. 8.11a–e) clearly brought out the glucosylation behaviour of the enzyme under the reaction conditions employed.

Glucosylation reaction gave a mixture of α- and β-glucosides. β-Glucosidase is a retaining

Table 8.25 Experimental design with experimental and predictive yields of glucosylation based on response surface methodology[a]

Expt. no.	Immobilised β-glucosidase (% w/w D-glucose)	Thiamin(mmol)	Incubation period (h)	Buffer concentration (mM)	pH (0.01 M)	Experimental yield (%)	Predicted yield (%)
1	40	0.5	48	0.08	7.0	11.9	13.2
2	40	0.5	48	0.16	5.0	31.3	32.6
3	40	0.5	96	0.08	5.0	20.6	20.6
4	40	0.5	96	0.16	7.0	57.9	52.3
5	40	1.0	48	0.08	5.0	41.0	39.3
6	40	1.0	48	0.16	7.0	24.3	17.1
7	40	1.0	96	0.08	7.0	31.0	22.3
8	40	1.0	96	0.16	5.0	37.0	28.4
9	80	0.5	48	0.08	5.0	17.2	24.5
10	80	0.5	48	0.16	7.0	45.0	46.7
11	80	0.5	96	0.08	7.0	35.2	35.4
12	80	0.5	96	0.16	5.0	22.0	22.3
13	80	1.0	48	0.08	7.0	20.0	18.6
14	80	1.0	48	0.16	5.0	24.2	22.9
15	80	1.0	96	0.08	5.0	29.3	26.6
16	80	1.0	96	0.16	7.0	53.4	45.2
17	20	0.75	72	0.12	6.0	16.2	26.5
18	100	0.75	72	0.12	6.0	32.8	30.6
19	60	0.25	72	0.12	6.0	42.4	34.9
20	60	1.25	72	0.12	6.0	12.5	28.1
21	60	0.75	24	0.12	6.0	32.5	28.3
22	60	0.75	120	0.12	6.0	25.6	37.9
23	60	0.75	72	0.04	6.0	14.0	12.6
24	60	0.75	72	0.20	6.0	19.8	29.3
25	60	0.75	72	0.12	4.0	17.4	15.8
26	60	0.75	72	0.12	8.0	14.5	24.2
27	60	0.75	72	0.12	6.0	18.4	17.0
28	60	0.75	72	0.12	6.0	17.9	17.0
29	60	0.75	72	0.12	6.0	18.0	17.0
30	60	0.75	72	0.12	6.0	19.7	17.0
31	60	0.75	72	0.12	6.0	17.7	17.0
32	60	0.75	72	0.12	6.0	18.8	17.0

[a]Conversion yields obtained from HPLC with respect to 1.0 mmol of glucose. Error in yield measurement will be ±5%. This applies to all the yields given in the subsequent tables also. Data are an average from two measurements

enzyme which gives rise to β-glucosides exclusively. Formation of a mixture of α- and β-glucosides indicated that the β-glucosidase prepared by the tannin precipitation method could contain contaminant α-glucosidase responsible for such products. This was confirmed by detecting an α-glucosidase activity of 305.3 AU (μmol/min·mg of enzyme preparation) for the native enzyme and 101.3 AU (μmol/min·mg of enzyme preparation) for the immobilised enzyme.

NMR data clearly confirmed the formation of two glucosides: 2-*O*-(α-D-glucopyranosyl)thiamin and 2-*O*-(β-D-glucopyranosyl)thiamin (Scheme 8.9). A doublet at 4.86 ppm with an equatorial–equatorial coupling constant value of 2.6 Hz corresponded to the α-glucoside, and another at 4.37 ppm with an axial–equatorial

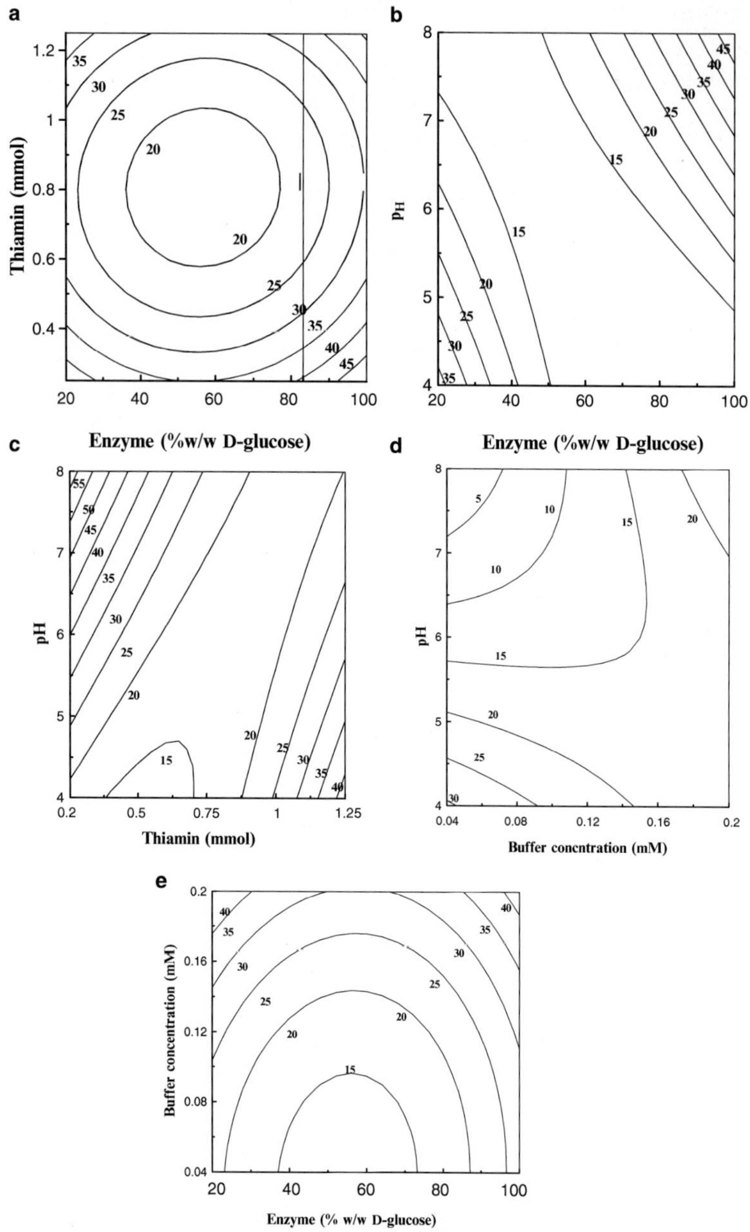

Fig. 8.11 Contour plots showing the effect of (**a**) concentration of thiamin and enzyme, (**b**) pH and enzyme concentration, (**c**) pH and thiamin concentration, (**d**) pH and buffer concentration and (**e**) buffer and enzyme concentration on the glycoside yields

coupling constant value of 6.4 Hz corresponded to the β-anomer. Carbon-13 chemical shift values also confirmed the chemical shift value of the α-anomer at 98 ppm and β-anomer at 103 ppm.

Effect of immobilised β-glucosidase and thiamin concentration on the extent of glucosylation (Fig. 8.11a), enzyme concentration and pH on the extent of glucosylation (Fig. 8.11b) and buffer concentration and pH on the extent of glucosylation (Fig. 8.11d) showed reversal in glucosylation at 60% (w/w D-glucose) enzyme concentration, 0.12-mM buffer concentration and pH 6. Below this crossover point, the yields gradually decreased which later increased beyond this

Table 8.26 Validation data for the immobilised β-glucosidase-catalysed reactions at selected random conditions[a]

Expt. no.	Immobilised β-glucosidase (% w/w D-glucose)	Thiamin (mmol)	Incubation period (h)	Buffer concentration (mM)	pH	Predicted yield (%)	Experimental yield (%)
1	50	0.75	72	0.12	6.0	17.0	27.0
2	20	0.75	72	0.12	4.0	41.0	31.0
3	20	0.25	72	0.12	6.0	43.0	49.4
4	30	1.0	72	0.10	5.0	35.0	33.0
5	40	0.40	72	0.10	5.0	23.6	21.5
6	30	0.25	60	0.11	7.0	37.6	34.0
7	40	0.75	60	0.08	5.0	26.0	21.0
8	60	0.50	60	0.14	5.0	21.3	22.6
9	100	0.75	72	0.12	6.0	31.0	32.3
10	60	0.50	72	0.12	6.0	23.0	24.0

[a]Conversion yields obtained from HPLC with respect to 1.0 mmol of D-glucose. Data are an average from two measurements

point. Such a crossover point indicated that under the experimental conditions corresponding to this point, the enzyme behaviour enhanced the yields thermodynamically and kinetically. An enzyme concentration of 60% (w/w D-glucose) and 0.75 mmol of thiamin corresponded to a complete binding of thiamin to the enzyme. Similarly a buffer concentration of 0.12 mM at pH 6 also could favourably dispose the enzyme for enhanced conversion through a favourable conformational change.

Figure 8.11c illustrating the effect of thiamin concentration and pH and Fig. 8.11d that of buffer concentration and pH bring out the effect of thiamin ionisation on the extent of glucosylation. Thiamin possesses two ionisable moieties – a thiazolium group with a pK of 16.9–18.9 and an amino group with a pK of 7.0–8.5. At a buffer concentration of 0.12 mM, lower concentrations of thiamin could result in complete neutralisation of amine hydrochloride as the pH and buffer concentrations are increased, leading to higher conversions (Fig. 8.11c, d). Below pH 6, positively charged thiamin could bind strongly to the enzyme, thus acting as an efficient inhibitor, whereas above pH 6, it could act as an efficient acceptor molecule for D-glucose due to loss of most of its charge on the free amino group. However, the thiazolium moieties with a pK of 16.9–18.9 will be least affected at pH 6. Figure 8.11e illustrates the effect of enzyme and buffer concentration on the extent of glucosylation at 0.75-mmol thiamin concentration, 72-h incubation period and pH 6. Higher buffer concentrations favoured enhanced glucosylated yields at all the enzyme concentrations.

A maximum conversion yield of 52% predicted by the response model for the glucosylation of thiamin at 40% (w/w D-glucose) immobilised β-glucosidase, 0.5-mmol thiamin, 96-h incubation period, 0.16-mM buffer concentration and pH 7 was found to yield 58% under the above experimental conditions. Validation experiments performed at certain selected random conditions gave moderate to good correspondence between experimental and predicted yields (Table 8.26). Thus the present response surface methodology study has brought the enzymatic behaviour of β-glucosidase clearly in this glucosylation reaction.

8.2.2 Syntheses of Riboflavinyl Glycosides

Riboflavin [1-deoxy-1-(3,4-dihydro-7,8-dimethyl-2,4-dioxobenzo{g}pteridine-10(2 H)-yl)-D-ribitol], vitamin B_2, is a constituent member of the vitamin B complex (LeBlane et al. 2005). Riboflavin is a necessary growth factor important for erythrocyte production (Bates 1993). As a prosthetic group of oxido-reductive flavoenzymes, it functions as a flavin nucleotide in the process of electron transport, leading to

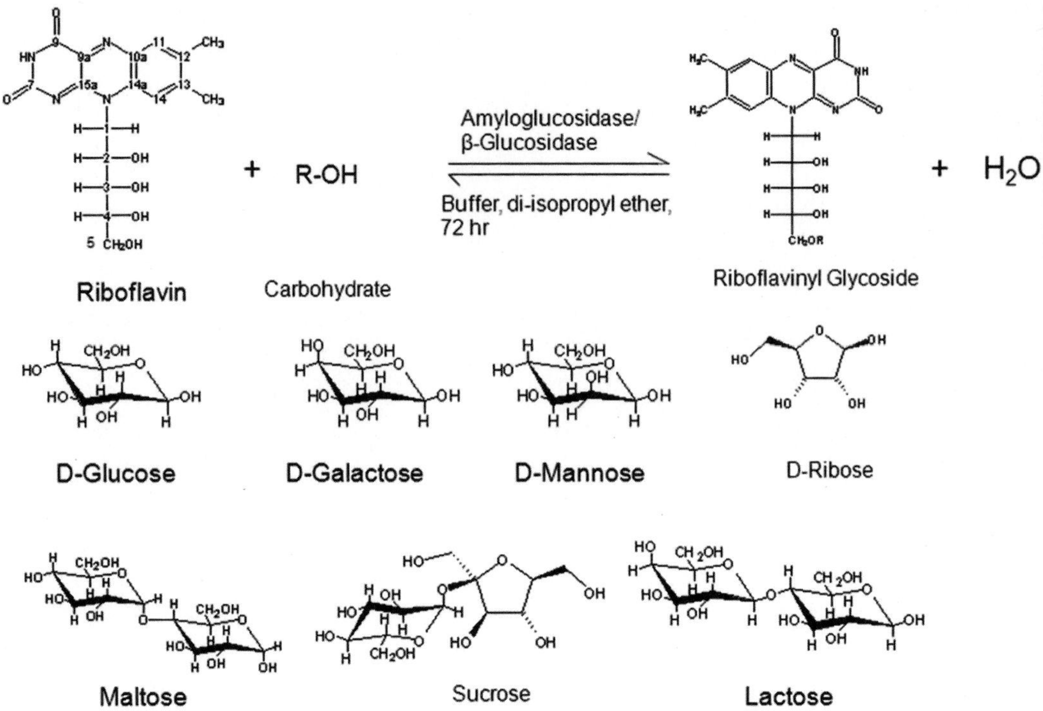

Scheme 8.10 Syntheses of riboflavinyl glycosides

oxidative degradation of pyruvate, fatty acids and amino acids (Werner et al. 2005). Riboflavin is a potent antioxidant, protective against many diseases including cancer (Sengodan et al. 2003). Also it degrades nicotine into a pharmacologically inactive substance (Dickerson et al. 2004). Recent researches focus on the role of plasma riboflavin in determining the homocysteine concentration, which is a risk factor in cardiovascular diseases (Powers 2003). Because of its limited gastrointestinal absorption, large doses of riboflavin remain unabsorbed.

Riboflavin is synthesised in plants and few microorganisms (Bacher et al. 2000). It is also obtained from animal source (Whitby 1971). As early as 1952, Whitby reported 5′-D-riboflavinyl-D-glucopyranoside from riboflavin by using homogenates of rat liver. Enzymes from different microorganisms have also been employed (Suzuki and Uchida 1969) for the preparation of 5′-D-riboflavinyl-α-D-glucopyranoside. Uchida and Suzuki (1968) and Suzuki and Uchida (1983) have reported preparation of riboflavin glucoside in growing cultures of *Ashbya gossypii* and *Eremothecium ashbyii* and in germinating barley seeds. Tachibana (1955) has reported preparation of riboflavinyl-β-D-galactoside in cell suspension cultures of *Aspergillus oryzae*. In 1971, Tachibana again showed riboflavin glucoside and other oligosaccharides from *Escherichia coli, Clostridium acetobutyricum, Leuconostoc mesenteroides* and cotyledons of *pumpkin Cucurbita pepo* and of sugar beet *Beta vulgaris*. However, on the following pages, an exclusive enzymatic synthesis of riboflavinyl glycosides is discussed (Scheme 8.10, Sivakumar and Divakar 2009b).

HPLC retention times for the substrates and products (Fig. 8.12) are as follows: riboflavin, 5.5 min; D-glucose, 6.5 min; 5-*O*-(D-glucopyranosyl) riboflavin, 8.2 min; D-galactose, 7.1 min; 5-*O*-(D-galactopyranosyl)riboflavin, 12.4 min; D-mannose, 6.7 min; 5-*O*-(D-mannopyranosyl)riboflavin, 13.4 min; D-ribose, 7.4 min; 5-*O*-(D-ribofuranosyl) riboflavin, 9.1 min; maltose, 11.5 min; 5-*O*-(α-D-

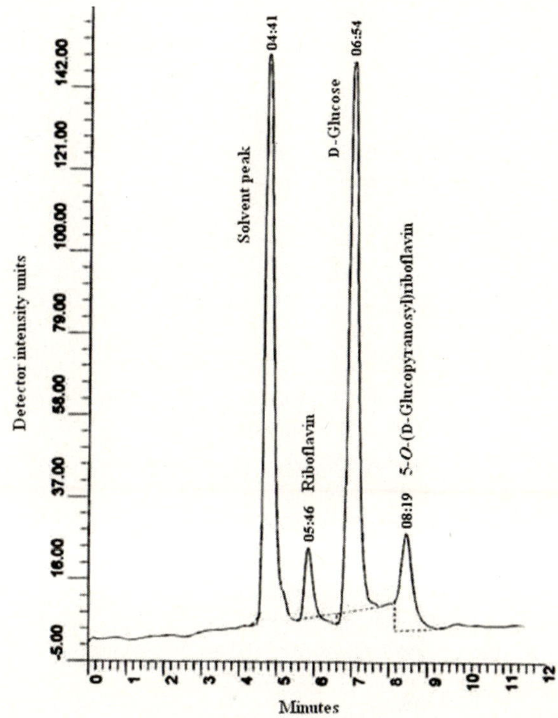

Fig. 8.12 HPLC chromatogram for the reaction mixture of 5-*O*-(D-glucopyranosyl) riboflavin. HPLC conditions: aminopropyl column (10 μm, 300 mm × 3.9 mm); solvent, CH_3CN: H_2O (70:30 v/v); flow rate, 1 mL min^{-1}; RI detector. Peak retention times: solvent peak, 4.4 min; riboflavin, 5.5 min; D-glucose, 6.5 min; and 5-*O*-(D-glucopyranosyl) riboflavin, 8.2 min

glucopyranosyl-(1′ → 4)D-glucopyranosyl) riboflavin, 17.2 min; sucrose, 9.6 min; 5-*O*-(1-D-fructofuranosyl-(2 → 1′)α-D-glucopyranosyl) riboflavin, 16.1 min; lactose, 9.3 min; and 5-*O*-(β-D-galactopyranosyl-(1′ → 4)β-D-glucopyranosyl) riboflavin, 15.1 min.

8.2.2.1 Synthesis of 5-*O*-(D-Glucopyranosyl)Riboflavin Using Amyloglucosidase

Optimum conditions for the synthesis of 5-*O*-(D-glucopyranosyl)riboflavin (Table 8.27) are: maximum yield - 31% between 12 and 72 h (Table 8.27, Fig. 8.13), pH-7, Amyloglucosidase - 50% (w/w D -glucose), riboflavin - 1 mmol and Buffer- 0.1-mM (1-mL).

8.2.2.2 Synthesis of 5-*O*-(β-D-Glucopyranosyl)Riboflavin Using β-Glucosidase

5-*O*-(β-D-Glucopyranosyl)riboflavin using β-glucosidase isolated from sweet almond was also synthesised (Scheme 8.10 and Table 8.28 and Fig. 8.14).

Determination of water solubility of 5-*O*-(D-glucopyranosyl)riboflavin showed that it is soluble to the extent of 8.2 g L^{-1} at room temperature (25°C). Hence, 5-*O*-(D-glucopyranosyl)riboflavin was found to be more soluble than riboflavin itself (0.2 g L^{-1} at 25°C) in water under identical conditions.

8.2.2.3 Syntheses of Riboflavinyl Glycosides of Other Carbohydrates Using Amyloglucosidase and β-Glucosidase and Their Spectral Characterisation

Syntheses of riboflavinyl glycosides with other carbohydrates (D-glucose, D-galactose, D-mannose, D-ribose, maltose and sucrose) were attempted under (Scheme 8.10) optimum conditions using amyloglucosidase and β-glucosidase (Table 8.29).

Spectral characterisation showed the following features. Ultraviolet-visible spectra of riboflavinyl glycosides showed shifts in $\sigma \rightarrow \sigma^*$ band in the 193–196.5-nm (191 nm for free riboflavin) range, $\sigma \rightarrow \pi^*$ band in the 223–230.5-nm (224 nm for free riboflavin) range and $\pi \rightarrow \pi^*$

Table 8.27 Optimisation of reaction conditions for the synthesis of 5-*O*-(D-glucopyranosyl) riboflavin using amyloglucosidase

Reaction conditions	Variable parameter[b]	Yield (%)[c]
	Incubation period (h)	
Riboflavin – 0.5 mmol	3	18
D-Glucose – 1 mmol	6	17
pH – 7	12	25
Buffer concentration – 0.1 mM (1 mL)	24	25
Amyloglucosidase – 50% w/w D-glucose	48	23
	72	25
	96	8
	pH (0.01 M)	
Riboflavin – 0.5 mmol[a]	4	11
D-Glucose – 1 mmol	5	14
Amyloglucosidase – 40% w/w D-glucose	6	15
Buffer concentration – 0.1 mM (1 mL)	7	17
Incubation period – 72 h	8	14
	Buffer concentration (mM)	
Riboflavin – 0.5 mmol	0.05	17
D-Glucose – 1 mmol	0.1	19
Amyloglucosidase – 40% w/w D-glucose	0.15	6
pH – 7	0.2	5
Incubation period – 72 h	0.25	2
	Amyloglucosidase concentration (% w/w D-glucose)	
Riboflavin – 0.5 mmol	10	18
D-Glucose – 1 mmol	20	19
pH – 7	30	23
Buffer concentration – 0.1 mM (1 mL)	40	24
Incubation period – 72 h	50	25
	75	21
	Riboflavin (mmol)	
pH – 7	0.25	19
Buffer concentration – 0.1 mM (1 mL)	0.5	25
D-Glucose – 1 mmol	1	31
Amyloglucosidase – 50% w/w D-glucose	1.5	25
Incubation period – 72 h	2	25

[a]Initial reaction conditions
[b]Other variables are the same as under reaction conditions, except the specified ones
[c]HPLC yields expressed with respect to 1-mmol D-glucose employed

band in the 253.5–283.5-nm (266 nm for free riboflavin) range and IR glycosidic C–O–C symmetrical stretching frequencies in the 1,031–1,071-cm^{-1} range and glycosidic C–O–C asymmetrical stretching frequencies in the 1,148–1,298-cm^{-1} range indicating that riboflavin had undergone glycosylation. From the 2-D HSQCT spectra of the riboflavinyl glycosides, the glycoside formation was confirmed from their respective ^{1}H and ^{13}C chemical shift values of carbon atoms C1α, C1β and C6-*O*-arylated compounds. Mass spectra also confirmed the formation of the glycosides. Two-dimensional HSQCT data clearly indicated that the glycosylation has occurred at the 5-CH_2OH of ribitol moiety of riboflavin.

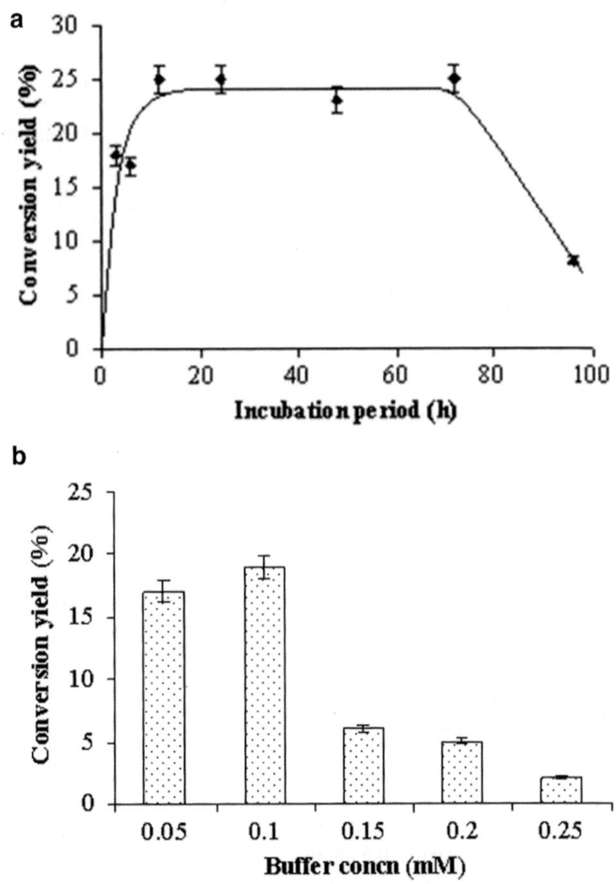

Fig. 8.13 (**a**) Reaction profile for 5-*O*-(D-glucopyranosyl)riboflavin synthesis by the reflux method. Conversion yields were from HPLC with respect to 1 mmol of D-glucose. Reaction conditions: D-glucose, 1 mmol; riboflavin, 0.5 mmol; amyloglucosidase, 50% (w/w D-glucose); 0.1-mM (1-mL) pH 7 phosphate buffer; solvent, di-isopropyl ether; and temperature, 68°C. (**b**) Effect of buffer concentration for 5-*O*-(D-glucopyranosyl) riboflavin synthesis. Reaction conditions: D-glucose, 1 mmol; riboflavin, 0.5 mmol; amyloglucosidase, 40% (w/w D-glucose); pH 7 phosphate buffer; solvent, di-isopropyl ether; temperature, 68°C; and incubation period, 72 h

8.2.3 Pyridoxine Glycosides

Vitamin B_6 is a collective term for a group of three related compounds, pyridoxine, pyridoxal, pyridoxamine and their phosphorylated derivatives. Pyridoxal 5′-phosphate is an important cofactor in a wide range of biochemical processes including amino acid metabolism and antibiotic biosynthesis (Studart et al. 2005) and as regulatory molecules in signal transduction.

Pyridoxine is a light and heat sensitive water-soluble vitamin. However, pyridoxine glycosides are less heat and light sensitive than pyridoxine (Kawai et al. 1971a, b). Synthesis of pyridoxine-7-*O*-α-D-glucoside by transglycosylation using *Coriolus* and *Verticillium* cultured cells (Asano and Wada 2003) and enzyme from the marine mollusc *Aplysia fasciata* (Andreotti et al. 2006) was reported. In this section a systematic study was undertaken to prepare pyridoxine glycosides using β-glucosidase isolated from sweet almond in di-isopropyl ether nonpolar medium (Scheme 8.11, Einstein Charles and Divakar 2009).

Under optimum conditions of 40% (w/w D-glucose), 0.18-mM (1.8-ml) pH 5 acetate buffer for 72-h incubation period β-glucosidase gave a maximum yield of 36 for pyridoxine-D-glucoside (Sivakumar et al. 2006a, b; Ponrasu et al. 2008). Table 8.30 describes the optimum yields obtained.

Syntheses of the other pyridoxine glycosides were carried out at the optimised conditions, with D-glucose, D-galactose, D-mannose, D-fructose, D-ribose, D-arabinose, maltose, lactose, sucrose, D-sorbitol and D-mannitol, and the reaction occurred with only the three aldohexoses employed (Table 8.31) with yields in the range 23–40%. Non-reactivity of the other carbohydrate

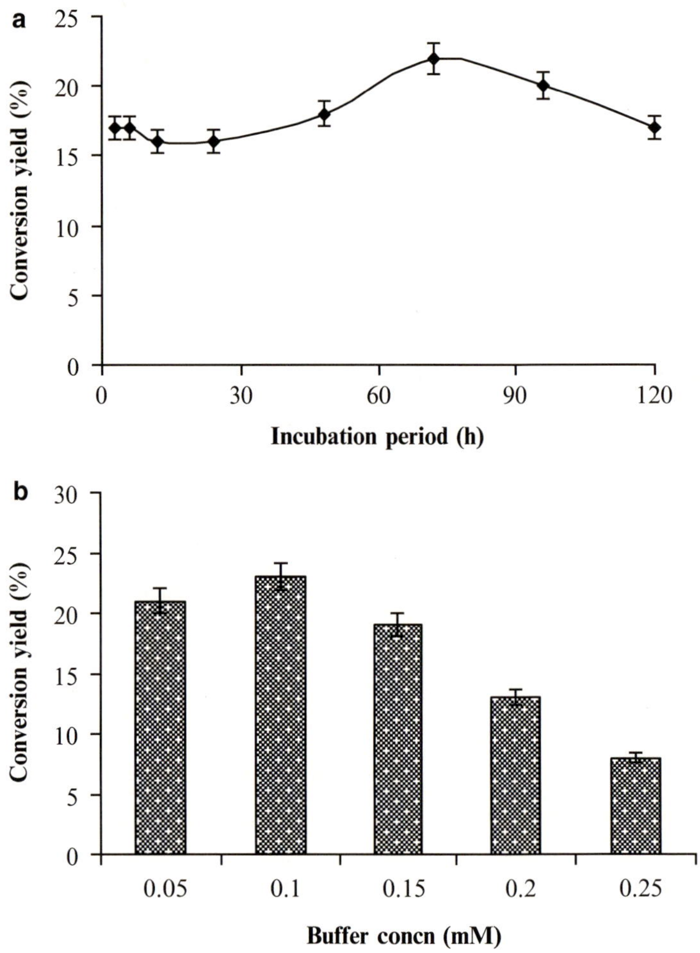

Fig. 8.19 (**a**) Reaction profile for 6-*O*-(β-D-glucopyranosyl)α-tocopherol synthesis by the reflux method. Conversion yields were from HPLC with respect to 0.5 mmol of D-glucose. Reaction conditions: D-glucose, 0.5 mmol; α-tocopherol, 0.5 mmol; β-glucosidase, 40% (w/w D-glucose); 0.1-mM (1-mL) pH 6 phosphate buffer; solvent, di-isopropyl ether; and temperature, 68°C. (**b**) Effect of buffer concentration for 6-*O*-(β-D-glucopyranosyl)α-tocopherol synthesis. Reaction conditions: D-glucose, 0.5 mmol; α-tocopherol, 0.5 mmol; β-glucosidase, 40% (w/w D-glucose); pH 6 phosphate buffer; solvent, di-isopropyl ether; temperature, 68°C; and incubation period, 72 h

was prepared in the present work from water-insoluble α-tocopherol.

8.2.7.3 Syntheses of α-Tocopheryl Glycosides of Other Carbohydrates Using β-Glucosidase and Their Spectral Characterisation

Syntheses of the other α-tocopheryl glycosides were carried out at the above-determined optimised conditions, with α-tocopherol and carbohydrates: D-glucose, D-galactose and D-mannose (Table 8.38).

Ultraviolet-visible spectra of α-tocopheryl glycosides showed $\sigma\rightarrow\sigma^*$ band in the 193–198.5-nm (199 nm for α-tocopherol) range, $\sigma\rightarrow\pi^*$ band in the 223–224.5-nm (228 nm for α-tocopherol) range and $\pi\rightarrow\pi^*$ band in the 270.5–273.5-nm (292 nm for α-tocopherol) range; IR spectra showed 1,028–1,084-cm^{-1} range band for the glycosidic C–O–C aryl–alkyl symmetrical stretching and 1,259–1,260-cm^{-1} range band for the asymmetrical stretching frequencies indicating that α-tocopherol had undergone glycosylation. NMR and Mass data confirmed the product formation. The phenolic carbon chemical shift value at 150.5 ppm (145.4 ppm for free α-tocopherol) indicated that glucosylation occurred at the phenolic OH group of α-tocopherol.

Table 8.38 Syntheses of α-tocopheryl glycosides using β-glucosidase

Glycosides	β-Glucosidase catalysis[a]	
	Product (% proportion)[b]	Yields (%)[c]
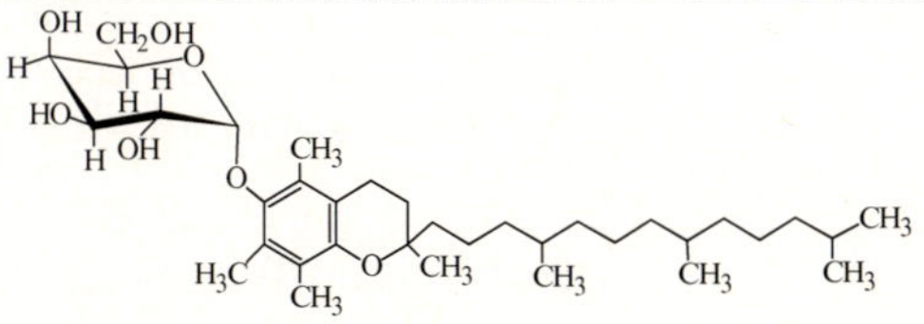 6-*O*-(β-D-Glucopyranosyl)α-tocopherol	C1β-glucoside	23
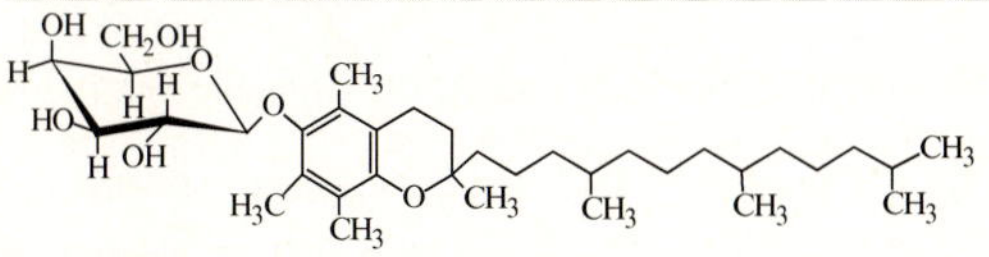 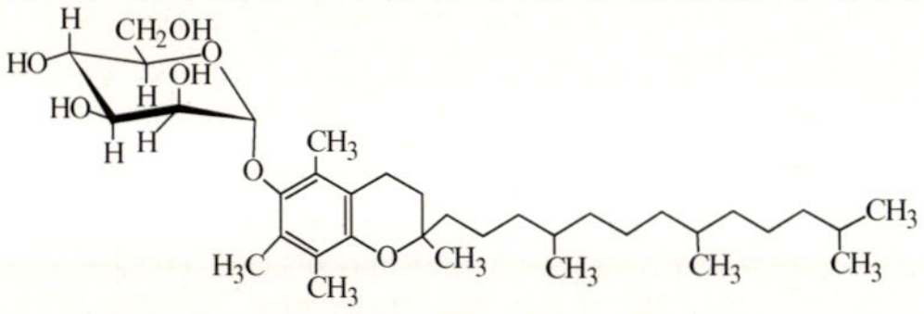 6-*O*-(α-D-Galactopyranosyl)α-tocopherol 6-*O*-(β-D-Galactopyranosyl)α-tocopherol	C1α-galactosides (41), C1β-galactosides (59)	11
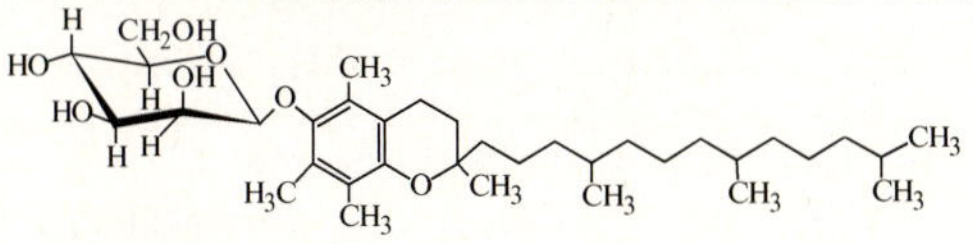 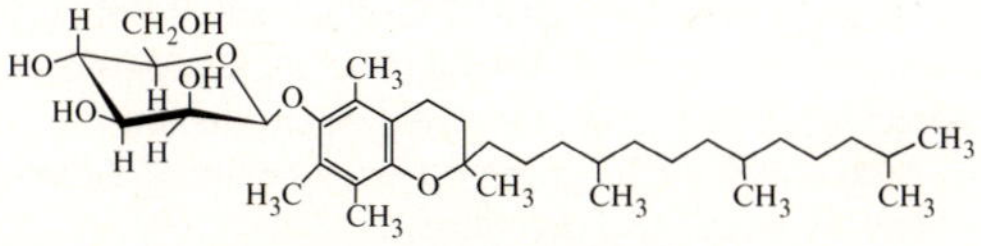 6-*O*-(α-D-Mannopyranosyl)α-tocopherol 6-*O*-(β-D-Mannopyranosyl)α-tocopherol	C1α-mannoside (46), C1β-mannoside (54)	18

[a]α-Tocopherol, 0.5 mmol; carbohydrate, 0.5 mmol; β-glucosidase concentration 40% (w/w D-glucose); solvent, di-isopropyl ether; buffer, 0.1-mM (1-mL) pH 6 phosphate buffer; incubation period, 72 h

[b]The product proportions were determined from the area of respective $^{1}H/^{13}C$ signals

[c]Conversion yields were from HPLC with respect to free carbohydrate. Error in yield measurements is ±5–10%

References

Akao T, Yoshino T, Kobashi K, Hatlori M (2002) Evaluation of salicin as an antipyretic prodrug that does not cause gastric injury. Planta Med 68:714–718

Andreotti G, Giordano A, Tramice A, Mollo E, Trincone A (2006) Hydrolyses and transglycosylations performed by purified α-D-glucosidase of the marine mollusk Aplysia fasciata. J Biotechnol 122:274–284

Andrews RS, Pridham JB (1965) Structure of dopa glucoside from *Vicia faba*. Nature 205:1213–1214

Angerlacenci M (2007) Dopamine dysregulation of movement control in L-dopa induced dyskinesia. Trends Neurosci 30(5):236–243

AOAC Official Methods of Analysis (1995) Chapter 46. 16th edn, vol II, p 12

Asano Y, Wada K (2003) Regioselective glucosylation of pyridoxine by microorganisms. Biosci Biotechnol Biochem 67:499–507

Audus KL, Chikhale PJ, Miller WD, Thompson SE, Borchardt RT (1992) Brain uptake of drugs. The influence of chemicals and biological factors. Adv Drug Res 23:1–64

Bacher A, Eberhardt S, Fischer M, Kis K, Richter G (2000) Biosynthesis of vitamin B2 (riboflavin). Annu Rev Nutr 20:153–167

Bates C (1993) Riboflavin. Int J Vitam Nutr Res 63:274–277

Berger M, Gray JA, Roth C (2009) The expanded biology of serotonin. Ann Rev Med 60:355–366

Bhabadesh C, Somnath M, Dehdas B, De Amit K (1996) Capsaicin a unique antioxidant, antifungal, anti-inflammatory analgesic compound with antifungal activity against dermatophytes. Med Sci Res 24:669–670

Burri J, Graf M, Lambelet P, Loliger J (1999) Vanillin more than a flavoring agent – a potential antioxidant. J Sci Food Agric 48:49–56

Burton GW, Joyce A, Ingold KU (1983) Is vitamin E the only lipid soluble, chain-breaking antioxidant in human blood plasma and erythrocyte membrane. Arch Biochem Biophys 221:281–290

Chahid Z, Montet D, Pina M, Graille J (1992) Effect of water activity on enzymatic synthesis of alkylglycosides. Biotechnol Lett 14(4):281–284

Chen CW, Ou-Yang C-C, Yeh C-W (2003) Synthesis of galactooligosaccharides and transgalactosylation modeling in reverse micelles. Enzyme Microb Technol 33:497–507

Chen IJ, Yang JM, Yeh JL, Wu BN, Lo YC, Chen SJ (1992) Hypotensive and antinociceptive effects of ether linked and relatively non pungent analogs of N-nonanoyl-vanillylamide. Eur J Med Chem 27:187–192

Chiba S (1997) Molecular mechanism in α-glucosidase and glucoamylase. Biosci Biotech Biochem 61:1233–1239

Clark AM (1996) Natural products as a resource for new drugs. Pharm Res 13:1133–1141

Dalpiaz A, Filosa R, de Caprariis P, Conte G, Bortolotti F, Biondi C, Scatturin A, Prasad PD, Pavanc B (2007) Molecular mechanism involved in the transport of a prodrug dopamine glycosyl conjugate. Int J Pharm 336:133–139

Dickerson TJ, Yamamoto N, Janda Kim D (2004) Antibody-catalysed oxidative degradation of nicotine using riboflavin. Bio Org Med Chem 12:4981–4987

Du Y, Wei G, Linhardt RJ (2004) Total synthesis of quercetin 3-sophorotrioside. J Org Chem 69:2206–2209

Einstein Charles R, Divakar S (2009) β-Glucosidase catalyzed syntheses of pyridoxine glycosides. Biosci Biotech Biochem 73:233–236

Einstein Charles R, Ponrasu T, Udaya Sankar K, Divakar S (2009a) Syntheses of retinol glycosides using β-glucosidase in SCCO2 media. Appl Biochem Biotech 159:199–207

Einstein Charles R, Ponrasu T, Sivakumar R, Divakar S (2009b) Angiotensin converting enzyme inhibitory and antioxidant activities of enzymatically synthesized phenolic and vitamin glycosides. Biotech Appl Biochem 52:177–184

Feldman D, Xiao-Yan Z, Krishnan AV (2000) Vitamin D and prostate cancer. Endocrinology 141:5–9

Fernandez C, Nieto O, Fontenla JA, Rivas E, de Ceballos ML, Fernandez-Mayoralas A (2003) Synthesis of glycosyl derivatives as dopamine prodrugs Interaction with glucose carrier GLUT-1. Org Biomol Chem 1:767–771

Fitzgerald DJ, Stratford M, Gasson MJ, Narbad A (2005) Structure function analysis of the vanillin molecule and its antifungal properties. J Agric Food Chem 53:1769–1775

Frandsen TP, Dupont C, Lehmbeck J, Stoffer B, Sierks MR, Honzatko RB, Svensson B (1994) Site-directed mutagenesis of the catalytic base Glutamic acid 400 in glucoamylase from *Aspergillus niger* and of Tyrosine 48 and Glutamine 401, both hydrogen bonded to the gamma-carboxylate group of Glutamic acid 400. Biochemistry 33:13808–13816

Gannett PM, Nagel DL, Reilly PJ, Lawson T, Sharpe J, Toth B (1988) The capsaicinoids their separation, synthesis and mutagenicity. J Org Chem 53:1064–1071

Geng X, Song L, Pu X, Tu P (2004) Neuroprotective effects of phenylethanoid glycosides from *Cistanches salsa* against 1-Methyl-4-phenyl-1,2,3,6-tetrahydropyridine (MPTP)-induced dopaminergic toxicity in C57 Mice. Biol Pharm Bull 27(6):797–801

Geng X, Tian X, Tu P, Pu X (2007) Neuroprotective effects of echinacoside in the mouse MPTP model of Parkinson's disease. Eur J Pharmacol 564:66–74

Giri A, Dhingra V, Giri CC, Singh A, Ward OP, Narasu ML (2001) Biotransformations using plant cells, organ cultures and enzyme systems current trends and future prospects. Biotechnol Adv 19:175–199

Gomes DCF, Alegrio LV, Leon LL, de Lima MEF (2002) Total synthesis and anti-leishmanial activity of some curcumin analogues. Arzneim-Forsch 52:695–698

Guggenheim M (1913) Dioxyphenylalanine neue Aminosaureaus *Vicia faba*. Hoppe-Seylers's Zeitschr Physiol Chem 88:276–284

Hamada H, Tomi R, Asada Y, Furuya T (2001) Phytoremediation of bisphenol A by cultured suspension cells of *Eucalyptus perriniana* regioselective hydroxylation and glycosylation. Tetrahedron Lett 43:4087–4089

Hamada H, Nishida K, Furuya T, Ishihara K, Nakajima N (2002) Preparation of a new pepper chemoenzymatic synthesis of capsaicin oligosaccharide and 8-nordihydrocapsaicin. J Mol Catal B: Enzym 16:115–119

Hamada H, Ohiwa S, Nishida T, Katssuragi H, Takeda T, Hamada H, Nakajima N, Ishihara K (2003) One step glucosylation of capsaicinoids by cultured cells of *Phytolacca americana*. Plant Biotechnol 20:253–255

Hausslera MR, Wasserman RH, McCaina TA, Peterlik M, Bursaca KM, Hughesa MR (1976) 1,25-Dihydroxyvitamin D3-glycosides Identification of a calcinogenic principle of *Solanum malacoxylon*. Life Sci 18(10):1049–1056

Heyes Z, Szabo A, Nemeth J, Jakab B, Pinter E, Banvolgyl A, Kereskai L, Keri G, Szolcsanyl J (2004) Antiiflammatory and analgesic effects of somatostatin released from chronic arthritis model in the rat. Arthritis Rheum 50:1677–1685

Holick MF (2004) Sunlight and vitamin D for bone health and prevention of autoimmune diseases, cancers, and cardiovascular disease. Am J Clin Nutr 6:1678S–1688S

Houghton LA, Vieth R (2006) The care against ergocalciferol (vitamin D2) as a vitamin supplement. Am J Clin Nutr 84:694–697

Hoyer D, Martin G (1997) 5-HT receptor classification and nomenclature towards a harmonization with the human genome. Neuropharmacology 36(4–5):419–428

Ichikawa H, Yoshida N, Takano H, Ishikawa T, Handa O, Takagi T, Nito Y, Murase H, Yoshikawa T (2003) A novel water-soluble vitamin E derivative (TMG) protects against gastric mucosal damage induced by ischemic and reperfusion in rats. Dig Sci 48(1):54–58

Iorizzi M, Lanzotti V, De Marino S, Zollo F, Blanco-Molina M, Macho A, Munoz E (2001) New glycosides from *Capsicum annum* L. Var. *accuminatum*. Isolation, structure determination and biological activity. J Agric Food Chem 49:2022–2029

Isozaki Y, Yoshida N, Ichikawa H, Kuroda M, Kokura S, Naito Y, Okanoue T, Yoshikawa T (2005) A novel water-soluble vitamin E derivative protect against aspirin-induced gastric mucosal injury in rats. Int J Mol Med 16(6):1035–1040

James SY, Williams MA, Newland AC, Colston KW (1999) Leukemia cell differentiation cellular and molecular interactions of retinoids and vitamin D. Gen Pharmacol 32:143–154

Jones BJ, Blackburn TP (2002) The medical benefit of 5-HT research. Pharmcol Biochem Behav 71(4):555–681

Kaljuzhin OV, Shkalev MV (2000) Immunomodulator and pharmaceutical compositions with antitumor properties, and a food additive. Patent EP1038532 (CA 129 335732)

Kaminaga Y, Nagatsu A, Akiyama T, Sugimoto N, Yamazaki T, Maitani T, Mizukami H (2003) Production of unnatural glucosides of curcumin with drastically enhanced water solubility by cell suspension cultures of *Catharanthus roseus*. FEBS Lett 555:311–316

Kang K, Park S, Kim YS, Lee S, Back K (2009) Biosynthesis and biotechnological production of serotonin derivatives. Appl Micorb Biotechnol 83(1):27–34

Katrien T, Miloslav P, Sophie S, Guy E, Yvette M (1997) New antioxidant mixture for long term stability of serotonin, dopamine and their metabolites in automated microbore liquid chromatography with dual electrochemical detection. J Chromatograph B Biomed Sci Appl 694(2):297–303

Kawada T, Watanabae T, Takahashi T, Tanaka T, Iwai K (1986) Effects of capsaicin on lipid metabolism in rats fed a high fat diet. J Nutr 116:1272–1278

Kawai F, Yamada H, Ogata K (1971a) Properties of pyridoxine glucoside. J Vitaminol 17:121–124

Kawai F, Yamada H, Ogata K (1971b) Studies on transglycosidation to vitamin B6 by microorganisms Part IV. Purification of a bacterial enzyme catalyzing pyridoxine glucoside synthesis. Agric Biol Chem 35:184–190

Kempster PA, Bogetic Z, Secombei JW, Martin HD, Balazs NDH, Wahlquist MC (1993) Motor effect of broad beans *Vicia faba* in Parkinson's disease Single dose studies. Asia Pacific J Clin Nutr 2:85–89

Kometani T, Tanimoto H, Nishimura T, Kanbara I, Okada S (1993a) Glucosylation of capsaicin by cell suspension cultures of *Coffea arabica*. Biosci Biotech Biochem 57:2192–2193

Kometani T, Tanimoto H, Nishimura T, Okada S (1993b) Glucosylation of vanillin by cultured plant cells. Biosci Biotech Biochem 57:1290–1293

Kondo Y, Shimoda K, Takimura J, Hamada H, Hamada H (2006) Glycosylation of vitamin E homologue by cultured plant cells. Chem Lett 35:324–326

Kren V (2001) Chemical biology and biomedicine of glycosylated natural compounds. In: Fraser-Reid B, Tatsuta K, Thiem J (eds) Glycoscience chemistry and chemical biology. Springer, Berlin, 3: 2471-2529

Kren V, Martinkova L (2001) Glycosides in medicine: the role of glycosidic residue in biological activity. Curr Med Chem 8:1303–1328

Kren V, Hunkova Z, Halada P, Suzuki Y (1998) Transglycosylation of thiamine by fungal β-N-Acetylhexosaminidases. Biosci Biotechnol Biochem 62(12):2415–2417

Lahmann M, Thiem J (1997) Synthesis of α-tocopheryl oligosaccharides. Carbohydr Res 299:23–31

Lan T, Dalip R, Errol MY (2004) Isolation and purification of the hypoglycemic principle present in Cap. frutescens. Phytotherap Res 18:95–96

LeBlane JG, Burgees C, Sesma F, de Giori GS, Van Sinderen D (2005) *Lactococcus lactis* is capable of improving the riboflavin status in deficient rats. British J Nutr 94:262–267

Lee Y, Howard LR, Villalon B (1995) Flavonoids and antioxidant activity of fresh pepper (*C. annum*) cultivars. J Food Sci 60:473–476

Lehninger LA (1977) Biochemistry, 3rd edn. Worth Publishers Inc., New York, pp 355–357

Lopez-Malo A, Alzamora SM, Argaiz A (1998) Vanillin and pH synergistic effects on mold growth. J Food Sci 63:143–146

Madrid Y, Langer LF, Brem H, Langer R (1991) New directions in the delivery of drugs and other substances to the central nervous system. Adv Pharmacol 22:299–324

Manohar B, Divakar S (2010) Enzymatic synthesis of cholecalciferol glycosides. J Food Sci Technol 46(5):469–475

Marieb EN (2009) Essentials of human anatomy and physiology, 8th edn. Pearson/Benjamin Cummings, San Fransisco, p 336. ISBN 321-51342-8

Marty A, Chulalaksananukul W, Condoret JS, Willemont RM, Durand G (1990) Comparison of lipase-catalyzed esterification in supercritical carbon dioxide and n-hexane. Biotechnol Lett 12(1):11–16

Matondo RB, Punt C, Homberg J, Toussaint MJ, Kisjes R, Korporaal SJ, Akkerman JW, Cuppen E, de Bruin A (2009) Deletion of the serotonin transporter in rats disturbs serotonin homeostasis without impairing liver regeneration. Am J Physiol Gastrointest Liver Physiol 296(4):G963–G968

Maurizio C, Kati O, Maria O, Sabrina P, Bruno S (2007) Vitamin D in rheumatoid arthritis. Autoimmun Rev 7:59–64

Mello JRB (2003) Calcinosis-calcinogenic plants. Toxicon 41:1–12

Mohri K, Watanabe Y, Yoshida Y, Satoh M, Isobe K, Sugimoto N, Tsuda Y (2003) Synthesis of glycosylcurcuminoids. Chem Pharm Bull 51:1268–1272

Moon JH, Terao J (1998) Antioxidant activity of caffeic acid and dihydrocaffeic acid in Lard and human low-density lipoprotein. J Agri Food Chem 46: 5062–5065

Morre DJ, Chueh PJ, Morre DM (1995) Capsaicin inhibits preferentially the NADH oxidase and growth of transformed cells in culture. Proc Natl Acad Sci USA 92:1831–1835

Murase H, Yamauchi R, Kato K, Kunieda T, Terao J (1997) Synthesis of a novel vitamin E derivative, 2-(α-D-glucopyranosyl) methyl-2,5,7,8-tetramethyl chroma- 6-ol transglycosylation. Lipids 32:73–78

Nagasawa T, Takagi H, Kawakami K, Suzuki T, Shahaghi Y (1961) The browning compounds of bean III isolation of dopa-O-β-D-glucoside and the enzymic mechanism for the color change of broad bean. Agric Biol Chem 25:441–447

Nsoukpoe-Kossi CN, Sedaghat-Herati R, Ragi C, Hotchandani S, Tajmir-Riahi HA (2007) Retinol and retinoic acid bind human serum albumin: stability and structural features. Int J Biol Macromol 40:484–490

Ochiai J, Takano H, Ichikawa H, Naito Y, Yoshida N, Yanagisawa R, Yoshino S, Murase H, Yoshikawa T (2002) A novel water-soluble vitamin E derivative, 2-(α-D-glucopyranosyl)methyl-2,5,7,8-tetramethylchroman-6-ol, protects against acute lung injury and mortality in endotoxemic rats. Int J Mol Med 18(6):580–584

Ohnuki K, Haramizu S, Oki K, Watanabe T, Yazawa S, Fushiki T (2001) Administration of capsiate, a non-pungent capsaicin analog, promotes energy metabolism and suppresses body fat accumulation in mice. Biosci Biotechnol Biochem 65:2735–2740

Ong DE, Newcomer ME, Chytil F (1994) Biology chemistry and medicine. In: Sporn MB, Roberts AB, Goodman DS (eds) The retinoids. Raven Press Ltd., New York, pp 283–318

Ooi Y, Hashimoto T, Mitsuo N, Satoh T (1985) Enzymatic formation of β-galactosidase from *Aspergillus oryzae* and its application to the synthesis of chemically unstable cardiac glycosides. Chem Pharm Bull 33:1808–1814

Orihara Y, Furuya T, Hashimoto N, Deguchi Y, Tokoro K, Kanisawa T (1992) Biotransformation of isoeugenol and eugenol by cultured cells of *Eucalyptus perriniana*. Phytochemistry 31:827–831

Pardridge WM (2002) Targeting neurotherapeutic agents through the blood–brain barrier. Arch Neurol 59:35–40

Peehl DM, Krishnan AV, Feldman D (2003) Pathways mediating the growth-inhibitory actions of vitamin D in prostate cancer. J Nutr 133:2461–2469

Ponrasu T, Charles RE, Sivakumar R, Divakar S (2008) Syntheses of α-tocopheryl glycosides by glucosidases. Biotechnol Lett 30:1431–1439

Ponrasu T, Manohar B, Divakar S (2009) A response surface methodological study on prediction of glucosylation yields of thiamin using immobilized β- glucosidase. Process Biochem 44:251–255

Powers HJ (2003) Riboflavin and health. Am J Clin Nutr 77:1352–1360

Pras N, Woerdenbag J, Van Uden W (1995) Bioconversion potential of plant enzymes for the production of pharmaceuticals. Plant Cell Tissue Org Cult 43:117–121

Rashid MH, Inoue M, Kondo S, Kawashima T, Bakoshi S, Ueda H (2003) Novel expression of vanilloid receptor I on capsaicin-insensitive fibres accounts for the analgesic effect of capsaicin in neuropathic pain. J Pharmacol Exp Ther 304:940–948

Reichel L, Sckickle R (1943) Synthesen neuer methoxylierter Chalkon-Flavanon-glucoside unter physiologischen Bedingungen. Chemie und Biochemie der Pflanzenstoffe, XII. Mitteilung Plant Chem 76B:1134–1137

Robyt JF (1998) Essentials of carbohydrate chemistry. Springer, New York, pp 64–68

Roode BM, Franssen MCR, van der Padt A, Boom RM (2003) Perspective for the industrial enzymatic production of glycosides. Biotechnol Prog 19:1391–1402

Russell AJ, Beckman EJ (1991) Should the high diffusivity of a supercritical fluid increase the rate of an enzyme-catalyzed reaction? Enzyme Microbiol Technol 13(12):1007–1007

Sasaki N, Adachi T, Koda T, Ozeki Y (2004) Detection of UDP-glucose Cyclo-dopa 5-O-glucosyltransferase activity in four O' clocks (*Mirabilis jalapa L.*). FEBS Lett 568:159–162

Sato T, Takeuchi H, Takahashi K, Kurosu J, Yoshida K, Tsugane T, Shimura S, Kino K, Kirimura K (2003) Selective α-glucosylation of eugenol by α-glucosyl transfer enzyme of *Xanthomonas campestris* WU-9701. J Biosci Bioeng 96:199–202

Satoh T, Miyataka H, Yamamoto K, Hirano K (2001) Synthesis and physiological activity of novel tocopheryl glycosides. Chem Pharm Bull 49(8):948–953

Schmid B, Kotter I, Heide L (2001) Pharmacokinetics of salicin after oral administration of a standard willow bark extract. Eur J Clinic Pharmacol 57:387–391

Scholz D, Mebel M, Topelmann J, Grossmann I, Scholze J, Mrochen H (1983) Prevention of osteonecrosis following renal transplantation by using vitamin D2 (ergocalciferol). Proc Eur Dial Transplant Assoc 20:331–337

Sengodan S, Palanivel S, Panchanadham S (2003) Curative effect of riboflavin, niacin and ascorbic acid on tamoxifen mediated endometrial carcinoma bearing Sprage-Dawley rats with reference to lipid peroxidation and antioxidant status. J Clin Biochem Nutr 33:39–45

Shetty P, Atallah MT, Shetty K (2002) Effects of UV treatment on the proline-linked pentose phosphate pathway for phenolics and L-DOPA synthesis in dark germinated *Vicia faba*. Process Biochem 37:1285–1295

Shimoda K, Konda Y, Abe K, Hamada H, Hamada H (2006) Formation of water-soluble vitamin derivatives from lipophilic vitamins by cultured plant cells. Tetrahedron Lett 47:2695–2698

Sierks MR, Ford C, Reilly PJ, Svensson B (1990) Catalytic mechanism of fungal glucoamylases as defined by mutagenesis of Asp 176, Glu179, and Glu180 in the enzyme from *Aspergillus awamori*. Protein Eng 3:193–198

Sircar S (2007) Medical physiology. Thieme Publishing Group, Stuttgart, p 536. ISBN 3-13-144061-9

Sivakumar R, Divakar S (2006) Glycosylation of vanillin by amyloglucosidase in organic media. Tetrahedron Lett 47:695–699

Sivakumar R, Divakar S (2007) Syntheses of N-vanillyl-nonanamide glycosides using amyloglucosidase from *Rhizopus* mold and β-glucosidase from sweet almond. Biotechnol Lett 29:1537–1548

Sivakumar R, Divakar S (2009a) Enzymatic syntheses of dopamine glycosides. Enzyme Microb Technol 44:33–39

Sivakumar R, Divakar S (2009b) Amyloglucosidase from *Rhizopus* mold and β- glucosidase from sweet almond catalysed syntheses of riboflavinyl glycosides. Indian J Biotechnol 8:400–407

Sivakumar R, Manohar B, Divakar S (2006a) Synthesis of vanillyl-maltoside using glucosidases by response surface methodology. Eur Food Res Technol 226:255–263

Sivakumar R, Vijayakumar GR, Manohar B, Divakar S (2006b) Competitive substrate inhibition of amyloglucosidase from *Rhizopus mold* by vanillin and curcumin in respective glucosylation reactions. Biocatal Biotrans 24:299–305

Sivakumar R, Ponrasu T, Divakar S (2009) Syntheses of DL-dopa glycosides using glucosidases. Glycoconj J 26:199–209

Smith DC, Johnson CS, Freeman CC, Muindi J, Wilson JW, Trump DL (1999) A phase I trial of calcitriol (1,25-dihydroxycholecalciferol) in patients with advanced malignancy. Clin Cancer Res 5:1339–1345

Sommer J, Schroeder C, Stockigt J (1997) *In vivo* formation of vanillin glucoside. Plant Cell Tiss Org Cult 50:119–123

Sophie L, Robert G, Pierre K (2004) A new method of solvent free O – and N – glycosilation using activated carbon fibers (ACF) as a promoter. Applications to the synthesis of saponin and nucleoside analogues. Chem Commun 5:586–587

Stephan D, Peter K (2003) Vanillins – a novel family of DNA-PK inhibitors. Nucleic Acid Res 31:5501–5512

Studart MT, Raschle T, Forster G, Amrhein N, Fitzpatrick TB (2005) Vitmin B6 biosynthesis in higher plants. Proc Natl Acad Sci USA 102:13687–13692

Suzuki Y, Uchida K (1969) Biosynthesis of riboflavin-α-glucoside by plant grains. Arch Biochem Biophys 130(1):683–684

Suzuki Y, Uchida K (1983) Formation of glucosyl riboflavin in germinating barley seeds. Nogaku Kenkyu 9:285–292

Suzuki Y, Uchida K (1994) Enzymatic formation of a new derivative of thiamin, β-galactosylthiamin. Biosci Biotech Biochem 58:1273–1276

Suzuki Y, Ki YH, Uchida K, Takami M (1996) Enzymatic synthesis of glucosylated and phosphatidylated biologically active compounds. J Appl Glycosci 43:273–282

Tachibana S (1955) Sugar compounds of riboflavin II Biosynthesis of riboflavinyl galactoside by *Aspergillus oryzae*. Vitamins 9:119–124

Tachibana S (1971) Isolation of riboflavin glycosides. In: Vitamin and coenzymes. McCormick CDB, Wright LD (eds) Methods in enzymology. Academic Press, New York, 18B: 413–416

Tai CY, Huang SC, Huang MS, Liu HS (2001) Hydrolysis of amylopectin by isoamylase under supercritical carbon dioxide. J Chin Inst Chem Eng 32(3):269–275

Tapavicza SV, Bell D, Kopp-Holtwiesche B (2000) Plant growth enhancement against phytopathogenic fungi and/or soil borne pests. Patent WO 0002451 (CA 132)

The Merck Index (2006) In: Neil MJO (ed) An encyclopedia of chemicals, drugs and biologicals, 14th edn. Merck and Co., Whitehouse Station

Tietze LF, Griesbach U, Schuberth I, Bothe U, Marra A, Dondoni A (2003) Novel carboranyl C-glycosides for the treatment of cancer by boron neutron capture therapy. Chem Eur J 9:1296–1302

Uchida K, Suzuki Y (1968) Production of riboflavin glucoside in growing cultures of *Ashbya gossypii* and *Eremothecium ashbyii*. Denpon Nogei Kagaku Kaishi 42(4):233–237

Vadivelan G, Divakar S (2011a) Enzymatic synthesis of L-Dopa glycosides. Indian J Chem Sect B 50B:1779–1785

Vadivelan G, Divakar S (2011b) Enzymatic Synthesis of serotonyl and epinephryl glycosides. Int J Fund Appl Life Sci 1(2):15–21

Vered Y, Rabey JM, Paleveitch D, Grosskopf I, Harsat A, Yanowski A, Shabtai H, Graff E (1994) Bioavailability of levodopa after consumption of *Vicia faba* seedlings by Parkinsonian patients and control subjects. Clin Neuropharmacol 17:138–146

Vic G, Thomas D (1992) Enzyme-catalyzed synthesis of alkyl-β-D-glucosides in organic media. Tetrahedron Lett 33:4567–4570

Vic G, Thomas D, Crout DHG (1997) Solvent effect on enzyme-catalyzed synthesis of β-D-glucosides using the reverse hydrolysis method application to the preparative-scale synthesis of 2-hydroxybenzyl and octyl β-D-glucopyronosides. Enzyme Microb Technol 20:597–603

Vijayakumar GR, Divakar S (2005) Synthesis of guaiacol-α-D-glucoside and curcumin- bis-α-D-glucoside by an amyloglucosidase from *Rhizopus*. Biotechnol Lett 27:1411–1415

Vijayakumar GR, Divakar S (2007) Amyloglucosidase catalyzed synthesis of eugenyl and curcuminyl glycosides. Biotechnol Lett 29:575–584

Vijayakumar GR, Manohar B, Divakar S (2006) Amyloglucosidase catalyzed synthesis of curcumin-bis-α-D-glucoside-A response surface methodological study. Eur Food Res Technol 223:725–730

Vijayakumar GR, George C, Divakar S (2007) Synthesis of n-alkyl glucosides by amyloglucosidase. Ind J Chem B 46B:314–319

Walker J, Alshafie G, Abou-Issa H, Curley R Jr (2002) An improved synthesis of the C-linked glucuronide of N-(4-hydroxyphenyl) retinamide. Bioorg Med Chem Lett 12:2447–2450

Watanabae T, Kawada T, Kato T, Harada T, Iwai K (1994) Effects of capsaicin analogs on adrenalin catecholamine secretion in rats. Life Sci 54:369–374

Werner R, Manthey KC, Griffin JB, Zempleni J (2005) HepG2 cells develop signs of riboflavin deficiency within 4 days of culture in riboflavin-deficient medium. J Nutr Biochem 16:617–624

Whitby LG (1971)Glycosides or riboflavin. In: Vitamin and coenzymes. McCormick CDB, Wright LD (eds) Methods in enzymology. Academic Press, New York, 18B:404–413

Wieder R, Novick SC, Hollis BW, Bryan M, Chanel SM, Owusu K, Camastra D, Saunders T, Pliner L, Harrison J, Bonate P, Williams T, Soignet S (2003) Pharmacokinetics and safety of ILX23-7553, a non-calcemic-vitamin D3 analogue, in a phase I study of patients with advanced malignancies. Invest New Drugs 21:445–452

Wyler H, Meuer U, Bauer J, Stravs-Mombelli L (2004) Cyclodopaglucoside((2S)-5-(β-D-glucopyranosyloxy)-6-hydroxyindoline-2-carboxylicacid) and its occurrence in red beet (*beta Vulgaris Varnn rubra L.*). Helvetica Chim Acta 67(5)):1348–1355

Yamada H, Aimi Y, Nagatsu I, Taki K, Kudo M, Arai R (2007) Immunohistochemical detection of L-dopa derived dopamine within serotonergic fibres in the striatum and the substantia nigra pars reticulata in Parkinsonian model rats. Neurosci Res 59:1–7

Yar MD (1993) Parkinson's disease, the L-dopa era. Adv Neurobiol 60:11–17

Zhou JH (2000) Herbal sweetening and preservative composition comprising licorice extract and mogrosides obtained from plants belonging to cucurbitaceae and/or momordica. Patent US 6103240 (CA 133 168393)

9 Glycosylation of Phenols and Vitamins: An Overview

Abstract

The reactions were discussed in this chapter in terms of incubation period, pH, buffer, enzyme and substrate concentrations, regio- and stereoselectivity.

9.1 General

Glycosylation potentialities of amyloglucosidase from *Rhizopus* sp., β-glucosidase from sweet almond and calcium alginate bead-entrapped β-glucosidase were explored in detail in the synthesis of a wide variety of carbohydrate acceptor molecules such as alcohols (alcohols of carbon chain length C1–C18), phenols (guaiacol, eugenol, curcumin, vanillin, N-vanillyl-nonanamide, capsaicin, L-dopa, DL-dopa, dopamine, serotonin and epinephrine) and vitamins (retinol (vitamin A), thiamin (vitamin B1), riboflavin (vitamin B2), pyridoxine (vitamin B6), ergocalciferol (vitamin D2), cholecalciferol (vitamin D3) and α-tocopherol (vitamin E)). The donor carbohydrates include aldohexoses (D-glucose, D-galactose and D-mannose), ketohexose (D-fructose), pentoses (D-arabinose and D-ribose), disaccharides (maltose, sucrose and lactose) and carbohydrate alcohols (D-mannitol and D-sorbitol).

A novel experimental set-up was developed for even large-scale synthesis of glycosides using lesser enzymes and larger concentrations of substrates to give higher yields. This set-up involved refluxing the reaction mixture containing the alcohol/phenol/vitamin and carbohydrate with appropriate concentrations of amyloglucosidase and β-glucosidase (native/immobilised) in the presence of specified pH and buffer concentrations in 100 mL of di-isopropyl ether solvent at 68°C for a specified incubation period. This set-up gave higher yields and selectivity than the conventional shake-flask experiments.

Optimisation of reaction parameters for the synthesis of above-mentioned phenolic and vitamin glucosides was carried out in terms of incubation period, pH, buffer, enzyme and phenol/vitamin concentrations.

The glycosylation yields were better with phenols compared to those with the vitamins employed. Vitamins are not better nucleophiles compared to the phenols employed. Amyloglucosidase from *Rhizopus* mould gave higher yield with lesser selectivity, and β-glucosidase from sweet almond gave lesser yield and greater selectivity. In general, conversion yields ranging from 5 to 62% for amyloglucosidase catalyses and 7 to 65% for β-glucosidase (native/immobilised) catalyses were obtained for various phenols/vitamins. Invariably DL-dopa and dopamine gave high yields with less regioselectivity with both the glucosidases employed. Loss of regioselectivity in many of the glycosylation/arylation

S. Divakar, *Enzymatic Transformation*,
DOI 10.1007/978-81-322-0873-0_9, © Springer India 2013

reactions could be due to the employment of large amount of the enzymes. This is inevitable, as the reversible reaction requires such large concentrations of enzymes for conversions.

Phenols employed for the glycosylation possesses 3,4-dihydroxy phenyl derivatives, where a substituent *para* to the hydroxyl group at position 4 of the phenyl ring is substituted by -CHO or -CH═CH- or $-CH_2$ and the 3rd position substituted with $-OCH_3$ and -OH. All the phenols employed in this study possess one or two phenolic groups – one invariably at the 4th position and another as such as in DL-dopa and dopamine or modified to a OCH_3 group (vanillin, N-vanillyl-nonanamide and curcumin). However, both the OH groups have reacted giving monoglycosides and arylated products, but no bis derivatives in case of DL-dopa, L-dopa, dopamine and epinephrine were observed. In curcumin both the phenolic OH groups on either side reacted to give a bis products. Hence, the presence of bulky groups *para* to 4-OH did not hinder formation of glycoside. A reasonable extent of selectivity exhibited by these enzymes has been utilised for the preparation of the glycosides, thereby eliminating the need for elaborate protective and deprotective strategies.

Effect of incubation period showed that the yield increased up to a time period of 72 h and then showed a remarkable drop at still higher incubation periods, which could be due to partial hydrolysis of the glycosides formed after 72 h. Since the enzymes could not be recovered, they were not reused again. Hence, the activity of the enzyme after the reaction could not be determined. Refluxing di-isopropyl ether for a specified incubation period did not produce any peroxides in these glycosylation reactions.

Glycosylation described in the present work did not occur without the use of enzyme. Glycosidase reactions occur only in the presence of certain amount of water (Vic and Crout 1995; Vijayakumar et al. 2007), whose presence in the reaction mixture could be regulated carefully to get a good glycosylation yield. Besides imparting 'pH memory', added water is essential for maintaining the integrity of the three-dimensional structure of the enzyme molecule and therefore its activity (Dordick 1989) in a nonpolar solvent like di-isopropyl ether (Vijayakumar and Divakar 2005). Water has been added in the form of 10 mM buffer. When buffer concentration (buffer volume) was varied, the lower and higher buffer concentrations (buffer volume) resulted in lesser conversion yields. A lower buffer concentration may not be sufficient to keep the active conformation of the enzyme, and a higher buffer concentration could result in hydrolysis of the product.

Beyond a critical water concentration, glycosylation decreased due to the size of the water layer formed around the enzyme retarding the transfer of the glycosyl donor to the active site of the enzyme (Humeau et al. 1998; Camacho-Paez et al. 2003a, b). Besides it renders the water layer surrounding the enzymes more flexible by forming multiple H bonds and interacting with organic solvent causing denaturation (Valiveti et al. 1991). The protonation of the ionisable groups during the addition of buffer on the enzyme could be controlled by the type and availability of the counter ions as well as hydrogen ions resulting in 'pH memory', and the increase in ionic strength could play a favourable role in glycosylation (Pardridge 2002). The carbohydrate molecule could also reduce the water content of the reaction mixture. Adachi and Kobayashi (2005) have reported that the hexose, which is more hydrated, decreased the water activity in the system and shifts the equilibrium towards synthesis. The effect of pH showed that pH 4 for vanillin (Table 8.11) and curcumin (Table 8.6), pH 7 for N-vanillyl-nonanamide (Table 8.8), pH 6 for DL-dopa (Table 8.17) and pH 5 for dopamine (Table 8.21) are the best for obtaining maximum conversion. The three-dimensional structure of the enzyme up to pH 7 may still retain a highly active conformation. However, pH 8 was not the optimum one for any of the phenols employed.

At lesser enzyme concentrations, for a given amount of substrates (enzyme/substrate ratio low), rapid exchange between bound and unbound forms of both the substrates (carbohydrates and the aglycon) with the enzyme (on a weighted average based on binding constant values of both the substrates) leaves substantial number of unbound substrate molecules at the start of the reaction which decrease progressively as conversion takes place (Romero et al. 2003; Marty et al. 1992). This becomes more so, if one of them binds more firmly to the enzyme than the other (higher binding constant value) as the respective enzyme/substrate ratios keep changing (during the course of the reaction) unevenly till the conversion stops due to complete binding (inhibition) of the predominant substrate. At intermediatory enzyme concentrations, such a competitive binding results in favourable proportions of bound and unbound substrates to effect quite a good conversion. At higher enzyme concentrations, most of the substrates would be in the bound form leading to inhibition and lesser conversion (higher enzyme/substrate ratios). Also, the glycosylation reaction requires larger amount of enzyme compared to hydrolysis. Effect of enzyme concentration at a fixed D-glucose concentration showed a maximum conversion at around 40% (w/w D-glucose) in many cases. At higher concentrations of both glucosidases, conversion yields decreased probably due to inhibition of the enzyme by the phenols employed. While this leads to lesser selectivity, they also give rise to varying bound and unbound substrate concentrations till the conversion ends. For a given amount of enzyme and substrates, there is no increase in conversion beyond 72–120 h. Longer incubation periods of especially lesser enzyme concentrations could also result in partial enzyme inactivation. However, not all the enzyme is inactivated before the end of the reaction.

Amyloglucosidase-catalysed synthesis of n-octyl-α-D-glucoside, guaiacyl-α-D-glucoside, eugenyl maltoside, curcuminyl-bis-α-D-glucoside, DL-dopa glucoside, dopamine glucoside, riboflavinyl glucoside and tocopheryl-α-D-glucoside was optimised in terms of incubation period, buffer pH, buffer concentration, enzyme concentration and substrate concentrations. n-Octanol reacted with D-glucose, maltose and sucrose; guaiacol reacted with D-glucose and D-galactose; eugenol and curcumin reacted with D-glucose, D-mannose, maltose, sucrose and D-mannitol, and α-tocopherol reacted with only D-glucose. Phenols/vitamins underwent glycosylation/arylation with the following respective carbohydrate molecules. A few examples are vanillin (D-glucose, D-galactose, D-mannose, maltose, sucrose, lactose and D-sorbitol), N-vanillyl-nonanamide (D-glucose, D-galactose, D-mannose, D-ribose, maltose and lactose), curcumin (D-glucose, D-galactose, D-mannose and lactose), DL-dopa (D-glucose, D-galactose, D-mannose, D-sorbitol and D-mannitol), dopamine (D-glucose, D-galactose and D-mannose), riboflavin (D-glucose, D-galactose, D-mannose, D-ribose , maltose, sucrose and lactose) and ergocalciferol only with D-glucose and α-tocopherol (D-glucose, D-galactose and D-mannose). All the phenols employed invariably reacted with aldohexoses (D-glucose, D-galactose and D-mannose) employed to yield respective glycosides except in the reaction between N-vanillyl-nonanamide and D-mannose. Dopamine showed exclusively glycosylation only with aldohexoses.

Thus water-insoluble N-vanillyl-nonanamide, curcumin, capsaicin, retinol, ergocalciferol, cholecalciferol and α-tocopherol and less water-soluble vanillin and riboflavin were converted to more water-soluble glycosides thereby improving their potential bioavailability and pharmacological properties. Water solubility of the curcuminyl-bis-α-D-glucoside prepared was found to be 14 g L^{-1}, whereas curcumin itself is practically insoluble in water. Curcuminyl-bis-α-D-glucoside exhibited a total colour of 10.8 in DMSO.

Lack of stereo- and regiospecificity could be due to employment of larger amounts of the enzymes. This is inevitable, as this reversible reaction required such large concentrations of the enzymes to effect glycosylation.

Amyloglucosidase, besides effecting glycosylation, also facilitated the hydrolysis of maltose, lactose and sucrose. Lactose did not hydrolyze at all. Only in case of sucrose, the resultant glucose formed underwent trans-glucosylation with vanillin in the presence of amyloglucosidase to yield C1β glucosylated and C6-*O*-arylated product.

Both amyloglucosidase and β-glucosidase (native/immobilised) did not catalyse the reaction with D-fructose and D-arabinose with any of the phenols employed. Also, while β-glucosidase (native/immobilised) catalysed reaction with lactose, amyloglucosidase did not. Hence, D-fructose and D-arabinose could not result in not-so-facile formation of the required oxocarbenium ion intermediate (Chiba 1997) during the catalytic action of the enzyme, which is an essential requirement for glycosylation. In both the glucosidase-catalysed reactions, other than D-fructose and D-arabinose, the remaining above-mentioned carbohydrates underwent glycosylation/arylation with different phenols/vitamins to different extents. Such selective reactions could be due to stronger binding of the phenols/vitamins to the active site of enzymes than the respective unreacted carbohydrate molecules, thereby preventing facile transfer of the carbohydrate molecules to the nucleophilic phenolic or vitamin OH groups. During hydrolysis of amylase, the α-1,4 linked glucose units get hydrolysed which give rise to β-D-glucose by amyloglucosidase. This clearly showed that amyloglucosidase possesses 'inverting' potentiality. The same behaviour was observed in the glycosylation reaction also. In general amyloglucosidase catalyses gave C1α- and β-glycosides along with C6-*O*-aryl derivatives. However, β-glucosidase (native/immobilised) catalyses gave exclusively C1β-glycosides in several cases, indicating its capability to exhibit excellent regioselectivity in this glycosylation with the carbohydrate molecules.

In most cases, C1 glycosylated products were detected. Only few carbohydrate molecules showed C1-*O*-/C6-*O*-arylation. D-Sorbitol and D-mannitol gave arylated products by reacting only to the primary OH groups. No reaction occurred at the secondary hydroxyl groups of the carbohydrate molecules. Only monoglycosylated or monoarylated products were detected. No carbohydrate molecule gave bis products. Among the phenols employed, only curcumin showed bis-glycosylated products by reacting at both the hydroxyl moieties of the feruloyl units. Even DL-dopa, L-dopa, dopamine and epinephrine did not give bis products. β-Glucosidase showed generally β-glycosides and in very few cases C1-*O*-/C6-*O*-arylated products. However, amyloglucosidase on the other hand showed both C1α- and C1β-glycosylated and/or C6-arylated products. Phenolic OH at the 4th position readily reacted with the carbohydrate molecules employed and wherever possible DL-dopa. L-dopa, dopamine and epinephrine underwent reaction at the 3rd phenolic OH also. Totally hydrophobic acceptor molecules like curcumin, capsaicin N-vanillyl-nonanamide, retinol, ergocalciferol, cholecalciferol and α-tocopherol reacted with only very few carbohydrate molecules like D-glucose, D-galactose, D-mannose and lactose where as the remaining hydrophilic phenols and vitamins employed vanillin, L-dopa, DL-dopa, dopamine and riboflavin reacted with large number of carbohydrate molecules.

Out of 99 individual phenolic and vitamin glycosides synthesised enzymatically all the synthesised C1-glycosides and C6-O-alkylated/arylated of the carbohydrates, about 89 new compounds are reported in the present work: n-octyl-sucrose, guaiacyl-α-D-glucoside, guaiacyl-β-D-glucoside, guaiacyl-α-D-galactoside, eugenyl-α-D-mannoside, eugenyl maltoside, eugenyl sucrose, eugenyl-D-mannitol, 1,7-*O*-(bis-β-D-glucopyranosyl) curcumin, 1,7-*O*-(bis-α-D-galactopyranosyl)curcumin, 1,7-*O*-(bis-β-D-galactopyranosyl)curcumin, 1,7-*O*-(bis-α-D-mannopyranosyl) curcumin, 1,7-*O*-(bis-β-D-mannopyranosyl)curcumin, 1,7-*O*-(bis-β-D-galactopyranosyl-(1′ → 4)α-D-glucopyranosyl)curcumin, 1,7-*O*-(bis-β-D-galactopyranosyl-(1′ → 4) β-D-glucopyranosyl) curcumin, curcuminyl-bis-α-D-mannoside, curcuminyl-bis-maltoside, curcuminyl-bis-sucrose and curcuminyl-bis-D-mannitol, 4-*O*-(D-galactopyranosyl)vanillin, 4-*O*-(D-mannopyranosyl)vanillin, 4-*O*-(α-D-glucopyranosyl-(1′ → 4)D-glucopyranosyl) vanillin, 4-*O*-(D-fructofuranosyl-(2 → 1′) α-D-glucopyranosyl)vanillin, 4-*O*-(β-D-galactopyranosyl-(1′ → 4)β-D-glucopyranosyl) vanillin, 4-*O*-(D-sorbitol) vanillin, 4-*O*-(D-galactopyranosyl)N-vanillyl-nonanamide, 4-*O*-(β-D-mannopyranosyl)N-vanillyl-nonanamide, 4-*O*-(D-ribofuranosyl)N-vanillyl-nonanamide, 4-*O*-(α-D-glucopyranosyl-(1′ → 4)D-glucopyranosyl)N-vanillyl-nonanamide, 4-*O*-(β-D-galactopyranosyl-(1′ → 4)β-D-glucopyranosyl)N-vanillyl-nonanamide, 1,7-*O*-(bis-β-D-galacto-pyranosyl) curcumin, 1,7-*O*-(bis-β-D-mannopyranosyl) curcumin, 1,7-*O*-(bis-β-D-galactopyranosyl-

(1′→4)α-D-glucopyranosyl)curcumin, DL-Dopa-D-galactoside, DL-3-hyDroxy-4-*O*-(β-D-mannopyranosyl)phenylalanine, DL-3-hydroxy-4-*O*-(β-D-galactopyranosyl-(1′→4)β-D-glucopyranosyl)phenylalanine, DL-3-hyDroxy-4-*O*-(6-D-sorbitol) phenylalanine, DL-Dopa-D-mannitol, dopamine-D-galactoside, dopamine-D-mannoside, serotoninyl-5-O-α and β D-glucopyranoside, serotoninyl-5-O-6-O-α and β D-glucopyranoside, serotoninyl-5-O-β-D-galactopyranoside, serotoninyl-5-O-6-O- α and β D-galactopyranoside, serotoninyl-5-O-β D-mannopyranoside, serotoninyl-5-O-6-O-α and β D-mannopyranoside, serotoninyl-5-O-α and β D-ribopfuranoside, epinephryl 3-O and 4-O β-D-glucopyranoside, epinephryl 3-O-β-D-mannopyranoside, 4-O-α and β-D-mannopyranoside, 18-*O*-(α-D-glucopyranosyl)retinol, 18-*O*-(β-D-glucopyranosyl)retinol, 18-*O*-(α-D-galacto pyranosyl)retinol, 18-*O*-(β-D-galactopyranosyl) retinol, 18-*O*-(β-D-mannopyranosyl)retinol, 18-*O*-(1-D-fructofuranosyl)retinol, 18-*O*-(6-D-fructofuranosyl)retinol, 18-*O*-(1-D-sorbitol) retinol, 7-*O*-(α-D-glucopyranosyl)pyridoxine, 7-*O*-(β-D-glucopyranosyl)pyridoxine, 6-*O*-(α-D-glucopyranosyl)pyridoxine, 7-*O*-(α-D-galactopy ranosyl)pyridoxine, 7-*O*-(β-D-galactopyranosyl) pyridoxine, 7-*O*-(α-D-galactopyranosyl)pyridoxine, 7-*O*-(α-D-mannopyranosyl)pyridoxine, 7-*O*-(β-D-mannopyranosyl)pyridoxine , 6-*O*-(α-D-mannopyranosyl)pyridoxine 5-*O*-(α-D-galacto pyranosyl)riboflavin,5-*O*-(β-D-galactopyranosyl) riboflavin, 5-*O*-(α-D-mannopyranosyl)riboflavin, 5-*O*-(β-D-mannopyranosyl)riboflavin, 5-*O*-(α-D-ribofuranosyl)riboflavin,5-*O*-(β-D-ribofuranosyl) riboflavin, 5-*O*-(α-D-glucopyranosyl-(1′→4) α-D-glucopyranosyl)riboflavin, 5-*O*-(α-D-glucopyranosyl-(1′ → 4)6-D-glucopyranosyl) riboflavin, 5-*O*-(α-D-glucopyranosyl-(1′→4) 6′-D-glucopyranosyl)riboflavin, 5-*O*-(1-D-fructofuranosyl-(2 → 1′)α-D-glucopyranosyl) riboflavin, 5-*O*-(β-D-galactopyranosyl-(1′ → 4)β-glucopyranosyl)riboflavin, 6-*O*-(α-D-galactopyranosyl)α-tocopherol, 6-*O*-(β-D-galactopyranosyl)α-tocopherol, 6-*O*-(α-D-mannopyranosyl)α-tocopherol,6-*O*-(β-D-mannopyranosyl)α-tocopherol, 17-*O*-(α -D-galactopyranosyl)cholecalciferol, 17-*O*-(β-D-galactopyranosyl)cholecalciferol, 17-*O*-(6-D-galactopyranosyl)cholecalciferol, 17-*O*-(α-D-mannopyranosyl)cholecalciferol, 17-*O*-(β-D-mannopyranosyl)cholecalciferol, 17-*O*-(1-D-fructofuranosyl)cholecalciferol and 17-*O*-(β-D-fructofuranosyl)cholecalciferol.

Response surface methodological studies, useful for scaling up, were carried out for the optimisation of n-octyl-α-D-glucoside, curcuminyl glucoside, vanillyl maltoside and thiamin glucoside synthesis by shake-flask method and curcuminyl-α-D-glucoside synthesis by reflux method

9.2 n-Alkyl Glycosides

The present work (Chap. 7), especially the reflux method with amyloglucosidase, showed that the glucoside yields were much higher (methanol 25%, ethyl alcohol 44%, n-butanol 28%, n-hexanol 36% and n-octanol 46% with respect to D-glucose concentration employed) than those reported compared to those reported by Vic and Thomas (1992) – 13.1% for methanol, 9.8% for ethyl alcohol, 6.6% for n-butanol, 4.9% for n-hexanol and 3.6% for n-octanol in the reactions carried out with almond β-glucosidase. Among the nucleophilic straight-chain alcohols employed, while shake flask favoured medium-chain-length alcohols, the reflux method at pH 4.0 favoured shorter-chain-length alcohols and pH 5.0 more or less favoured almost all the chain length. This clearly shows that since pH affected ionisation states of the surface amino acid residues in amyloglucosidase between pH 4.0 and pH 5.0, the enzyme should be favouring a more open active site at pH 5.0, accommodating alcohols of different chain length favourably for glycosylation.

9.3 Curcuminyl-bis-Glycosides

Although β-glucosidase catalysis (Sect. 8.1.3) does not exhibit an inversion, it had significantly altered the α,β composition: D-galactose, 39% α-D-galactoside and 61% β-D-galactoside (compared to 92:8 α:β for free D-galactose), and D-mannose,

35% α-D-mannoside and 65% β-D-mannoside (compared to 27:73 α:β for free D-mannose). Also, curcumin showed bis-glycosylated products with the carbohydrates with which it reacted.

D-Galactose and D-mannose gave very low conversions (≤12%). Hence, they could possess very low binding potentiality (low binding constant value) to the enzyme. Presence of hydrophobic propanoid group *para* to the phenolic OH bestows good nucleophilicity in these molecules promoting reaction with carbohydrate molecules.

9.4 N-Vanillyl-Nonanamide Glycosides

All these glycosides were soluble in water to different degrees and also less pungent. They could hence be used in pharmaceutical applications. Both amyloglucosidase and β-glucosidase did not catalyse the reaction with D-fructose, D-arabinose, sucrose, D-sorbitol and D-mannitol (Sect. 8.1.4). Also, while β-glucosidase catalysed reaction with lactose, amyloglucosidase did not. N-Vanillyl-nonanamide could bind to these enzymes more strongly than the above carbohydrate molecules to the nucleophilic phenolic OH of N-vanillyl-nonanamide.

Amyloglucosidase clearly exhibited its 'inverting' potentiality in the glycosylation giving rise to the α-D-glucoside (70% of the α-glucoside without the β component compared to the 40:60 α:β anomeric composition of D-glucose employed), α-D-galactoside (42% α and 58% β compared to the 92:8 α:β anomeric composition of D-galactose employed) and α-D-riboside (33% α and 67% β compared to the 34:66 α:β anomeric composition of D-ribose employed). Both the glucosidases gave low conversions with D-ribose and maltose (≤10%). However, β-glucosidase catalysis gave exclusively C1β-glycosides with the exception of D-ribose, indicating its capability to exhibit excellent regioselectivity in this glycosylation with the carbohydrate molecules D-glucose, D-galactose, D-mannose, maltose and lactose. Even the presence of hydrophobic bulky alkyl side chain in N-vanillyl-nonanamide did not pose much of a steric hindrance when the carbohydrate molecules were transferred to its phenolic OH group.

9.5 Vanillyl Glycosides

D-Glucose employed was an α,β anomeric mixture (40:60); the glycosides formed showed predominant proportions of the α-anomer (>80%), indicating the potential for 'inverting' amyloglucosidase (from *Rhizopus* mould) to convert the majority of β-D-glucose into its respective α-D-glucoside (Sect. 8.1.6). In hydrolysis, amyloglucosidase hydrolyses α-1,4 linked glucose units in amylose to give β-glucose.

A maximum yield of 53% was obtained for a mixture of three mono-glucosides of 4-*O*-(D-glucopyranosyl)vanillin (Table 8.12). Also, amyloglucosidase catalysis gave C1α- and β-glycosides along with arylated derivatives in many cases except D-galactose and D-mannose. However, β-glucosidase catalysis gave exclusively C1β-glycosides with the exception of D-galactose and D-mannose, indicating its capability to exhibit excellent regioselectivity in this glycosylation with the carbohydrate molecules D-glucose, maltose and lactose. In case of β-glucosidase, it significantly altered the anomeric composition of D-galactose (23% α-D-galactoside and 77% β-D-galactoside compared to the 92:8 α:β anomeric composition of D-galactose employed) and D-mannose (44% α-D-mannoside and 56% β-D-mannoside compared to the 27:73 α:β anomeric composition of D-mannose employed). Among the carbohydrates employed, sucrose and D-sorbitol gave C1-*O*- and C6-*O*-arylated products with vanillin.

In case of sucrose, the C6-*O*-arylated product proportion was more compared to the C1-*O*-arylated product, which could be due to the steric hindrance offered by the C2 position of the fructose moiety when it is transferred to such phenolic nucleophiles. Even the presence of a hydrophobic aldehydic group in vanillin did not pose much of a steric hindrance when the carbohydrate molecules were transferred to its phenolic OH group. Both amyloglucosidase and β-glucosidase did not catalyse D-fructose, D-arabinose, D-ribose and D-mannitol. Vanillin could be a better inhibitor to these enzymes compared to these carbohydrate molecules, binding strongly to the enzyme, thereby blocking the facile transfer of the carbohydrate molecule to the nucleophilic phenolic OH of vanillin.

9.6 DL-Dopa Glycosides

Under the reaction conditions employed, DL-dopa gave a mixture of 3-*O*- and 4-*O*-glycosylated/arylated products with many of the carbohydrates employed (Sect. 8.1.7). Amyloglucosidase catalysed the reaction with D-glucose, D-galactose, D-mannose, D-sorbitol and D-mannitol. β-Glucosidase catalysed the reaction with D-glucose, D-galactose, D-mannose and lactose.

Amyloglucosidase gave selectivity with D-mannose to give 4-*O*-C1β and D-sorbitol to give 4-*O*-C6-*O*-arylated product. β-Glucosidase gave selectivity with D-mannose to give 4-*O*-C1β and lactose to give 4-*O*-C1β product. Glycosylated/arylated products at 3-*O*- and 4-*O*- positions were detected from D-glucose and D-galactose in the catalysis with both the enzymes. No other reacted carbohydrate molecule has formed products at the 3rd OH position.

Glucosidases employed did not catalyse the reaction with D-fructose, D-arabinose, D-ribose, maltose and sucrose. Amyloglucosidase exhibited 'inverting' potentiality in reacting with D-galactose. D-Galactose showed 66% α and 34% β compared to the 92:8 α:β anomeric composition of free D-galactose employed.

9.7 Dopamine Glycosides

Only monoglycosides were detected (Sect. 8.1.9). Bis glycosides involving both the 4-OH and 3-OH phenolic groups simultaneously were not detected. Under the reaction conditions employed, immobilised β-glucosidase showed selectivity with D-glucose to give rise to the 4-*O*-C1β product. Dopamine gave a mixture of 3-*O*- and 4-*O*-glycosylated products with D-galactose and D-mannose with both the enzymes and 4-*O*-C6-*O*-arylated product detected only with D-galactose.

Amyloglucosidase exhibited its 'inverting' potentiality in the glycosylation of D-galactose giving rise to 64% 4-*O*-products and 36% exclusively 3-*O*-C1β product compared to the 92:8 α:β anomeric composition of the D-galactose employed. Both amyloglucosidase and immobilised β-glucosidase did not catalyse the reaction with D-fructose, D-arabinose, D-ribose, maltose, sucrose, lactose, D-sorbitol and D-mannitol under the conditions employed. Dopamine could be a better inhibitor to these enzymes compared to the carbohydrate molecules, thereby blocking the transfer of the carbohydrate molecules to the phenolic OH group of dopamine.

9.8 Riboflavinyl Glycosides

Riboflavinyl glycosides have been formed from many of the carbohydrates employed (Sect. 8.2.2). The yields are shown in Table 8.29. In case of amyloglucosidase catalysis, 5-*O*-(D-ribofuranosyl) riboflavin was found to give the highest yield (40%) and 5-*O*-(α-D-glucopyranosyl-(1′→4) D-glucopyranosyl)riboflavin gave the lowest yield of 5% (Table 8.29). This clearly showed that D-ribose being a smaller carbohydrate molecule acts as a better acceptor than the bulkier maltose. β-Glucosidase gave the highest yield of 24% for 5-*O*-(β-D-glucopyranosyl)riboflavin and the lowest yield of 7% for 5-*O*-(D-mannopyranosyl) riboflavin. Also, while β-glucosidase favoured lactoside formation, amyloglucosidase did not.

In spite of possessing a ribitol group, riboflavin is soluble to the extent of only 0.2 g L^{-1} (Whitby 1971), which is due to the strong nonpolar nature of the aglycon molecule. However, attachment of a monosaccharide unit to ribitol improved the water solubility of riboflavin to 8.2 g L^{-1}. This shows that glycosylation is capable of counteracting the nonpolar aglycon characteristics to a great extent. Riboflavin showed glycosylation with many of the carbohydrates compared to the other vitamins employed (ergocalciferol and α-tocopherol) in spite of the bulky acceptor molecule. This could be due to the presence of primary OH present at the ribitol moiety in riboflavin, nucleophilic enough to serve as an efficient acceptor towards certain carbohydrate molecules which get glycosylated.

Amyloglucosidase gave the following α:β proportions: D-glucose (66% of α-D-glucoside and 44% β-D-glucoside compared to the 40:60 α:β anomeric composition of free D-glucose), D-galactose (52% α-D-galactoside and 48% β-D-galactoside compared to the 92:8 α:β of free D-galactose) and D-ribose (23% α-D-riboside and 77% β-D-riboside compared to the 34:66 α:β of free D-ribose). Although β-glucosidase in general does not exhibit inversion, it had significantly altered the α-β composition: D-galactose (47% α-D-galactoside and 53% β-D-galactoside) and D-mannose (59% α-D-mannoside and 41% β-D-mannoside comparable to 27:73 α:β of free D-mannose).

Amyloglucosidase catalysis showed selectivity in case of D-mannose by yielding the C1α-D-mannoside and with sucrose yielding C1-*O*-arylated product. It gave C1α- and β-glycosides in case of D-glucose, D-galactose and D-ribose. With D-glucose and maltose, C6-*O*-arylated products were also formed. β-Glucosidase showed selectivity with D-glucose yielding C1β-D-glucoside and with lactose yielding C1β-lactoside. However, it gave a mixture of C1α- and β-glycosides in case of D-galactose and D-mannose.

9.9 Retinyl Glycosides

β-Glucosidase did not catalyse the reaction with D-ribose, D-arabinose, maltose, sucrose, lactose and D-mannitol (Sect. 8.2.4). This could be stronger due to binding of retinol to the active site of β-glucosidase than the above-mentioned carbohydrate molecules, thereby preventing the facile transfer of these carbohydrate molecules to the nucleophilic olefinic OH of retinol. Our work on the preparation of several glycosides with different aglycons showed that diverse reactivity of the carbohydrate molecules was with respect to the aglycons employed (Sivakumar 2009; Vijayakumar 2007). With different aglycons, different reactivities were observed even under conventional reaction conditions. In most of the reactions carried out in our laboratory, glycosylation did not take place with D-ribose, D-arabinose, lactose and sucrose, however, with very few exceptions. This type of selectivity is not due to the $SCCO_2$ conditions employed.

A one-step synthesis involved milder reaction conditions, eliminating the need of protective and deprotective strategies to convert light and air sensitive, fat-soluble retinol into more stable and water-soluble glycosyl derivative, to be useful as a pharmacologically and therapeutically active water-soluble component of vitamin A. Such derivatisation of vitamin A was possible due to the potentiality of β-glucosidase to act in $SCCO_2$ media.

9.10 Ergocalciferyl Glycosides

Ergocalciferyl-D-glucosides (20-*O*-(α-D-glucopyranosyl)ergocalciferol, 20-*O*-(β-D-glucopyranosyl) ergocalciferol and 20-*O*-(6-D-glucopyranosyl) ergocalciferol) were detected in the reaction when amyloglucosidase was employed (Sect. 8.2.5, Table 8.34). However, amyloglucosidase did not show much regioselectivity in this reaction. Amyloglucosidase did not catalyse the reaction with the other carbohydrate molecules. This could be due to either steric hindrance caused by the bulkier hydrophobic vitamin D2 molecule when the carbohydrate molecules were transferred to its acyclic OH group or due to the stronger binding of vitamin D2 compared to the other carbohydrate molecules employed. The maximum conversion was found to 42% for 20-*O*-(D-glucopyranosyl)ergocalciferol. Being 'inverting' in nature, amyloglucosidase gave the following α:β proportions – 48% of α-D-glucoside and 52% β-glucoside compared to the 40:60 α:β anomeric composition of the D-glucose employed.

At a boiling temperature of 68 °C, the enzymes were stable, losing only about 20% of their activity during an incubation period of 50 h. Refluxing di-isopropyl ether during the incubation periods up to 50 h did not also produce peroxides in these glycosylation reactions.

β-Glucosidase isolated from sweet almond was entrapped into calcium alginate beads and used in the glycosylation reaction to facilitate recovery and reuse. In the reaction mixture, the enzymes dissolved in small volume of the buffer employed. They did not get precipitated by the di-isopropyl ether solvent. During workup the solvent in the reaction mixture was distilled off leaving a residue consisting of the enzyme, product and the unreacted substrates. Hence, the amount of the enzyme leached out into the buffer or the di-isopropyl ether solvent could not be determined. Reuse of the calcium alginate bead-entrapped β-glucosidase showed 30% reduction in activity, which could be due to loss of the enzyme.

9.11 Cholecalciferol Glycosides

Synthesis of other cholecalciferol glycosides (Table 8.36) showed that apart from the three aldohexoses and aldopentose employed, D-ribose, D-arabinose, maltose, sucrose, lactose, D-sorbitol and D-mannitol did not undergo glycosylation under the conditions employed. Although the reactions were conducted with disaccharides maltose, lactose and sucrose, transglycosylation did not occur under the conditions employed (Sect. 8.2.6). However, the disaccharides were hydrolyzed by the enzymes, the reason for which is not known.

In case of D-glucose and D-galactose, C-6 derivatives were also detected. Anomeric proportions were found to be greatly altered. In D-glucose the proportion of 52:48 is different from that of free D-glucose (60:40). However, in D-galactose β-anomer has been formed (45:56) compared to 92:8 of free D-galactose. In D-mannose, more of α-glycoside is formed (46:54) compared to (27:73) free D-mannose. The formation of α and β anomeric mixture is due to the α-glucosidase present in sweet almond isolate, which exhibited an activity of 359.7 AU (μmol $(\text{min·mg})^{-1}$ enzyme preparation). Thus through glycosylation, a water-insoluble cholecalciferol (vitamin D_3) was transformed into a water-soluble cholecalciferol glycoside in this study.

9.12 α-Tocopheryl Glycosides

α-Tocopheryl glycosides have been synthesised using β-glucosidase (Sect. 8.2.7). The yields and product proportions are shown in Table 8.37. β-Glucosidase 40% (w/w D-glucose) gave a maximum conversion yield of 23% for 6-*O*-(β-D-glucopyranosyl)-α-tocopherol at 0.1 mM (1 mL) of pH 6 phosphate buffer, indicated excellent regioselectivity.

The α and β forms of glycosides were determined based on the ^{1}H and ^{13}C chemical shift values of the anomeric proton and carbon signals, respectively. Besides, the coupling constant values of the anomeric ^{1}H protons were also useful in arriving at the anomeric configuration. In an α configuration, the anomeric carbon exhibited chemical shift values between 92 and 94 ppm. In a β configuration, such values are between 96 and 104 ppm. In an α configuration, the coupling between the anomeric and C2 proton is equatorial–axial corresponding to coupling constant values of 2–4 Hz. In case of a β configuration, the coupling between the anomeric and C2 proton is axial–axial corresponding to a value of 6–8 Hz. However, in D-mannose, since the OH at C2 position is axial, the coupling for the C2 and the anomeric protons is equatorial–equatorial (1–2 Hz) for the α configuration and axial–equatorial (2–4 Hz) for the β configuration.

Under the reaction conditions, 11% 6-*O*-(α-D-galactopyranosyl)-α-tocopherol, 6-*O*-(β-D-galactopyranosyl)α-tocopherol, 18% 6-*O*-(α-D-mannopyranosyl)α-tocopherol and 6-*O*-(β-D-mannopyranosyl)α-tocopherol were also obtained with β-glucosidase (Table 8.38). In case of D-galactose, the α/β proportions were found to be 41:59 (α:β) with respect to α/β proportions of 92:8, for the free D-galactose, thus favouring a slight excess of β-D-galactoside formation. However, β-glucosidase only marginally favoured α-D-mannoside formation as evidenced from the α/β proportions of 46:54 (α:β) for the glycosides compared to an α/β proportion of 27:73 for free D-mannose.

Syntheses of the other α-tocopheryl glycosides with carbohydrate molecules showed that except for the three aldohexoses employed, D-fructose, D-arabinose, D-ribose, maltose, sucrose, lactose, D-sorbitol and D-mannitol did not undergo glycosylation under the conditions employed. β-Glucosidase is not an inverting enzyme. In the present work, β-glucosidase gave β-glucoside and α/β anomeric mixture of glycosides with D-galactose and D-mannose. In the oxocarbenium ion mechanism (Chiba 1997), a planar carbenium ion centre formed with D-galactose and D-mannose could be available for attack by the nucleophilic α-tocopherol phenol from both above and below the plane giving rise to a mixture of α/β anomeric products. With D-glucose, β-glucosidase gave its natural β-glucoside. Since D-galactose and D-mannose are not natural products of β-glucosidase, the same selectivity could not be achieved by this enzyme.

There are not many reports on the glycosylating potential of amyloglucosidase in the literature. The results from this investigation have shown conclusively that amyloglucosidase and β-glucosidase (native/immobilised) from sweet almond are excellent enzymes for carrying out effective glucosylation of such diverse alcohols, phenols and vitamins. Thus the present work has brought out the multifaceted characteristics of these enzymes in the glycosylation of selected phenols and vitamins with structurally diverse carbohydrate molecules employed. Also such glycosylation produces more water-soluble and stable vitamin derivatives with diverse carbohydrate molecules.

References

Adachi S, Kobayashi T (2005) Synthesis of esters by immobilized lipase catalyzed condensation reaction of sugars and fatty acids in water-miscible organic solvent. J Biosci Bioeng 99(2):87–89

Camacho-Paez B, Robles Medina A, Camacho Rubio F, Gonzalez Moreno P, Molina Grim E (2003a) Modeling the effect of free water on enzyme activity in immobilized lipase- catalyzed reactions in organic solvents. Enzyme Microb Technol 33:845–853

Camacho-Paez B, Robles MA, Camacho RF, Esteban-Cerdán L, Molina GE (2003b) Kinetics of lipase-catalysed interesterification of triolein and caprylic acid to produce structured lipids. J Chem Technol Biotechnol 78(4):461–470

Chiba S (1997) Molecular mechanism in α-glucosidase and glucoamylase. Biosci Biotechnol Biochem 61:1233–1239

Dordick JS (1989) Enzymatic catalysis in monophasic organic solvents. Enzyme Microb Technol 11:194–211

Humeau M, Girardin B, Rovel AM (1998) Effect of the thermodynamic water activity and the reaction medium hydrophobicity on the enzymatic synthesis of ascorbyl palmitate. J Biotechnol 63:1–8

Marty A, Chulalaksananukul W, Condoret JS, Willemont RM, Durand G (1992) Comparison of lipase-catalyzed esterification in supercritical carbon dioxide and n-hexane. Biotechnol Lett 12(1):11–16

Pardridge WM (2002) Targeting neurotherapeutic agents through the blood–brain barrier. Arch Neurol 59:35–40

Romero MD, Calvo L, Alba C, Daneshfar A, Ghaziaskar HS (2003) Enzymatic synthesis of isoamyl acetate with immobilized *Candida antarctica* lipase in *n*-hexane. Enzyme Microb Technol 37:42–48

Sivakumar R (2009) Enzymatic synthesis of selected phenolic and vitamin glycosides. PhD thesis, University of Mysore

Valiveti RH, Johnston GA, Suckling CJ, Halling PJ (1991) Solvent effect on biocatalysis in organic systems equilibrium position and rates of lipase catalysed esterification. Biotechnol Bioeng 38:1137–1143

Vic G, Crout DHG (1995) Synthesis of allyl and benzyl β-D-glucopyranosides and allyl β-D-galactopyranoside from D-glucose or D-galactose and the corresponding alcohol using almond β-D-glucosidase. Carbohydr Res 279:315–319

Vic G, Thomas D (1992) Enzyme-catalyzed synthesis of alkyl-β-D-glucosides in organic media. Tetrahedron Lett 33:4567–4570

Vijayakumar GR (2007) Enzymatic synthesis of selected glycosides. PhD thesis, University of Mysore

Vijayakumar GR, Divakar S (2005) Synthesis of guaiacol-α-D-glucoside and curcumin-bis-α- D-glucoside by an amyloglucosidase from *Rhizopus*. Biotechnol Lett 27:1411–1415

Vijayakumar GR, George C, Divakar S (2007) Synthesis of n-alkyl glucosides by amyloglucosidase. Indian J Chem B 46B:314–319

Whitby LG (1971) Glycosides or riboflavin. In: McCormick CDB, Wright LD (eds) Vitamin and coenzymes. Methods in enzymology, 18B. Academic, New York, pp 404–413

10 Kinetics of Some Selected Enzyme-Catalysed Reactions in Organic Solvents

Abstract

Kinetics of few selective enzymatic esterification and glycosylation reactions in organic solvents is discussed. In all the kinetics, initial rates were determined, and from the pattern of the double reciprocal plots of 1/[S] versus 1/*v*, appropriate kinetic models were identified and the equations worked out. Iterative procedures adopted to carry out curve-fitting of the experimental plots resulted in determination of the four kinetic parameters K_i, K_{mA}, K_{mB} and k_{cal} corresponding to the best fit. All the enzymes studied showed Ping-Pong Bi-Bi mechanism. In the enzymatic esterification reaction between L-alanine and D-glucose in dichloromethane, D-glucose was found to be inhibitory to both *Rhizomucor miehei* lipase and *Candida rugosa* lipase. However, both L-phenylalanine and D-glucose in dichloromethane solvent were found to exhibit competitive double substrate inhibition of *Rhizomucor miehei* lipase, leading to dead-end inhibition by RML–D-glucose complex and RML–L-phenylalanyl complexes. On the other hand, esterification of L-phenylalanine with D-glucose using *Candida rugosa* lipase in dichloromethane showed that only D-glucose functions as a competitive inhibitor forming dead-end CRL–D-glucose complex.

Similarly, kinetics of glucosylation was investigated in detail for the glucosylation of curcumin and vanillin using amyloglucosidase in di-isopropyl ether solvent. Both kinetics could be best described by the Ping-Pong Bi-Bi model with a single competitive substrate inhibition by respective curcumin and vanillin of amyloglucosidase leading to dead-end inhibition. Observed kinetic picture was explained in terms of the active site geometry and the geometry of binding of the substrate/s to the enzyme.

10.1 Introduction

In recent years, lipases (EC 3.1.1.3) have been successfully used in organic solvents for the synthesis of esters. Many reports available on lipase-catalysed reactions in organic solvents deal with reaction kinetics of esterification (Yadav and Trivedi 2003; Kiran and Divakar 2002; Janssen et al. 1999), racemisation (Duan et al. 1997) and hydrolysis (Van-Tol et al. 1992). In some esterification

S. Divakar, *Enzymatic Transformation*,
DOI 10.1007/978-81-322-0873-0_10,

reactions, lipases follow Ping-Pong Bi-Bi mechanism (Zhang et al. 2005; Zaidi et al. 2002; Yadav and Lathi 2004). Chulalaksanaukul et al. (1990) first proposed that kinetics of *Rhizomucor miehei* lipase-catalysed esterification of ethyl oleate followed Ping-Pong Bi-Bi mechanism. This mechanism involves binding of acid and alcohol in successive steps releasing water and the product ester again in succession. The kinetic investigation of RML-catalysed synthesis of isoamyl acetate followed Ping-Pong Bi-Bi mechanism with competitive inhibition by the substrates and product forming dead-end inhibitor complexes (Rizzi et al. 1992). Bousquet-Dubouch et al. (2001) reported a competitive inhibition in CRL-catalysed alcoholysis of methylpropionate, in which water was also found to be a competitive inhibitor with a higher inhibition constant than *n*-propanol. Kinetic behaviour of CRL in the esterification of long-chain fatty acids and alcohols (Zaidi et al. 2002) and tetrahydrofurfuryl alcohol with butyric acid (Yadav and Devi 2004) showed that both reactions followed Ping-Pong Bi-Bi mechanism with competitive inhibition by both substrates. In citronellyl laurate synthesis, RML follows the ordered Bi-Bi mechanism wherein β-citronellol binds to the enzyme to yield the β-citronellol–enzyme complex, which again binds to lauric acid to form the ternary enzyme–β-citronellol–lauric acid complex. Finally, it decomposes to give β-citronellyl laurate and water as products in this process (Yadav and Lathi 2004). A series of dead-end RML–lauric acid complexes were also reported in this process.

Kinetic studies on few enzymatic hydrolytic reactions involving glucosidases and amyloglucosidase are known (Hiromi et al. 1983; Tanaka et al. 1983; Ohnishi and Hiromi 1989; Goto et al. 1994). However, report on glycosylation kinetics involving a glycosylating enzyme is practically nil.

In this chapter, kinetics of esterification between L-alanine and D-glucose and L-phenylalanine and D-glucose to form L-alanyl–D-glucose and L-phenylalanyl–D-glucose, respectively, with RML and CRL was carried out (Lohith et al. 2007; Somashekar et al. 2007). Similarly, a detailed kinetic investigation on the glucosylation reaction between curcumin and D-glucose and vanillin and D-glucose involving an amyloglucosidase from *Rhizopus* mould leading to the synthesis of 1,7-*O*-bis-D-glucopyranosyl–curcumin and vanillin–D-glucose, respectively, was also carried out, and the results from these investigations are described below (Sivakumar et al. 2006).

10.2 Methodology

10.2.1 Esterification Procedure

Kinetic experiments were carried out by refluxing 5 mM to 0.1 M of L-alanine/L-phenylalanine and 5 mM to 0.1 M of D-glucose along with 100 mL CH_2Cl_2:DMF (90:10 v/v, 40°C) in the presence of 0.75–0.18 g (50% w/w carbohydrate employed) of *Rhizomucor miehei* lipase (RML) in the presence of 0.1 mM (0.1 mL of 0.1 M) of acetate buffer (pH 4.0) and *Candida rugosa* lipase (CRL) in the presence of 0.1 mM (0.1 mL of 0.1 M) of phosphate buffer (pH 7.0) l for the specified periods of time between 3 and 48 h. An experimental set-up which rendered the reaction mixture dry was employed (Lohith and Divakar 2005; Lohith et al. 2007). The product formation was analysed by HPLC and NMR. NMR data clearly indicated the presence of mono- and *di-O*-esters.

10.2.2 Glycosylation Procedure

Kinetic experiments were carried out by refluxing 5 mM to 0.1 M of vanillin/curcumin and 5 mM to 0.1 M of D-glucose in 100 mL di-isopropyl ether at 68°C in the presence of amyloglucosidase (10–75% w/w of carbohydrate) and 0.5–2.5 mL of 0.01 M pH 4.0–8.0 buffer for an incubation period of 72 h. Here also, the product formation was analysed by HPLC and NMR.

10.3 Esterification Kinetics

10.3.1 Esterification Kinetics of L-Alanine and D-Glucose: Single Substrate Inhibition

Enzymatic esterification reaction between L-alanine and D-glucose was studied in dichloromethane using *Rhizomucor miehei* and *Candida*

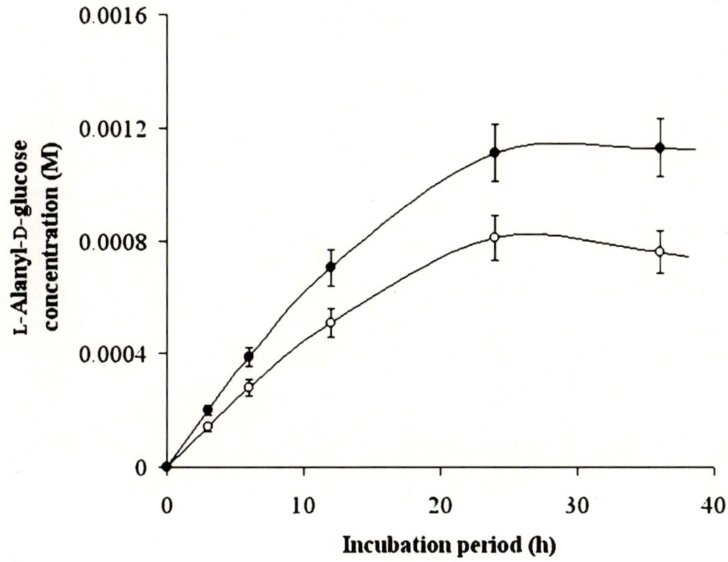

Fig. 10.1 Time courses of esterification reactions – concentrations of L-alanyl–D-glucose versus incubation periods; RML (90 mg) or CRL (90 mg) was allowed to react with 0.005 M L-alanine and 0.020 M D-glucose in 100 mL of mixture of dichloromethane/dimethylformamide (v/v 90:10) mixed with 0.1 mM (0.1 mL of 0.1 M) sodium acetate buffer (pH 4.0) for RML or 0.1 mM (0.1 mL of 0.1 M) sodium phosphate buffer (pH 7.0) for CRL

rugosa lipases (Somashekar et al. 2007). For the concentrations of D-glucose and L-alanine, individual experiments in duplicate (30×2 lipases) were performed for incubation periods of 3, 6, 12, 24 and 36 h. Initial rate (specific reaction rate, v) was determined from the initial slope of the plot of the amount of esters formed (M) versus incubation period (h) and expressed as M h^{-1} (mg protein)$^{-1}$ (Fig. 10.1). R^2 obtained from least-square analysis for the initial rate in each case was found to be within 0.88–0.95. Each plot shown in this work was constructed from all experimentally determined values; a few initial rates were obtained by curve-fitting.

The initial rates (v) for RML were found to be in the range of 15–176×10^{-6} M h^{-1} (mg protein)$^{-1}$. CRL experiments showed the initial rates to be in the range of 20–460 x 10^{-6} M h^{-1} (mg protein)$^{-1}$. The effects of external mass transfer phenomena – internal and external diffusions (Yadav and Devi 2004; Marty et al. 1990) – if any, on the RML and CRL enzymes employed were not tested in this work.

Using initial rates, double reciprocal plots were constructed to graphically evaluate the apparent values of k_{cat}, $K_{m\ \text{L-alanine}}$, $K_{m\ \text{D-glucose}}$ and K_i: RML, Fig. 10.2 ($1/v$ versus 1/[D-glucose]) and Fig. 10.3 ($1/v$ versus 1/[L-alanine]), and CRL, Fig. 10.4 ($1/v$ versus 1/[D-glucose]) and Fig. 10.5 ($1/v$ versus 1/[L-alanine]). Figure 10.6 shows a replot of the slopes from Fig. 10.3 (RML), and Fig. 10.7 shows the replot of the slopes from Fig. 10.5 (CRL). Figure 10.2 from RML reactions and Fig. 10.4 from CRL reactions show a series of curves obtained for different fixed concentrations of L-alanine for varying D-glucose concentration, in which slight increase in initial rates is observed at lower D-glucose concentrations. At higher D-glucose concentrations, the rates markedly decrease. Also, increasing L-alanine concentration increases the initial rates at all D-glucose concentrations. Figure 10.3 from RML reactions and Fig. 10.5 from CRL reactions show a series of parallel lines for different fixed low D-glucose concentrations at varying L-alanine concentration. These change to lines with different slopes at higher D-glucose concentrations.

The plots in Figs. 10.2, 10.3, 10.4, 10.5, and 10.6 show that the kinetics could be best described by Ping-Pong Bi-Bi model in which L-alanine and D-glucose bind in subsequent steps releasing water and L-alanyl–D-glucose also in subsequent

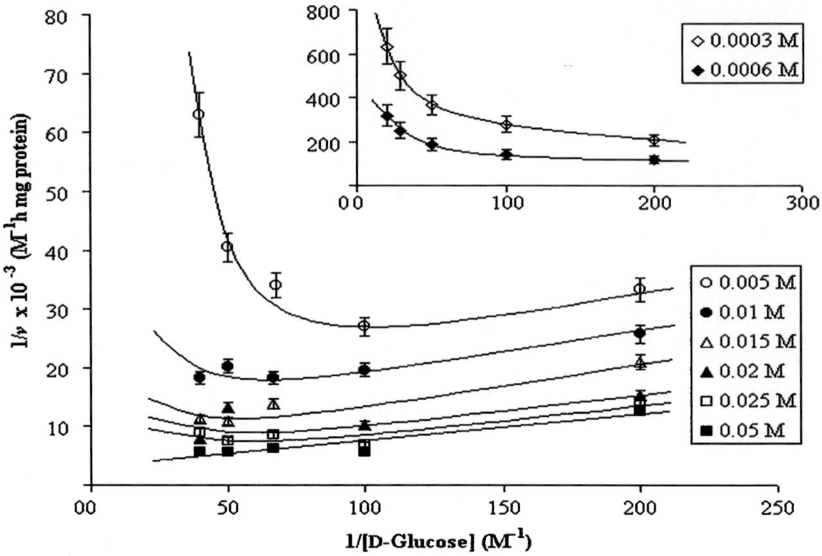

Fig. 10.2 Double reciprocal plots for RML-catalysed L-alanyl–D-glucose reaction: $1/v$ versus 1/[D-glucose] plots; a series of curves show the effect of varying D-glucose concentration at different fixed L-alanine concentrations in the range of 0.005–0.05 M. The inset shows plots obtained by the computer simulation for 0.0003 and 0.0006 M L-alanine concentrations

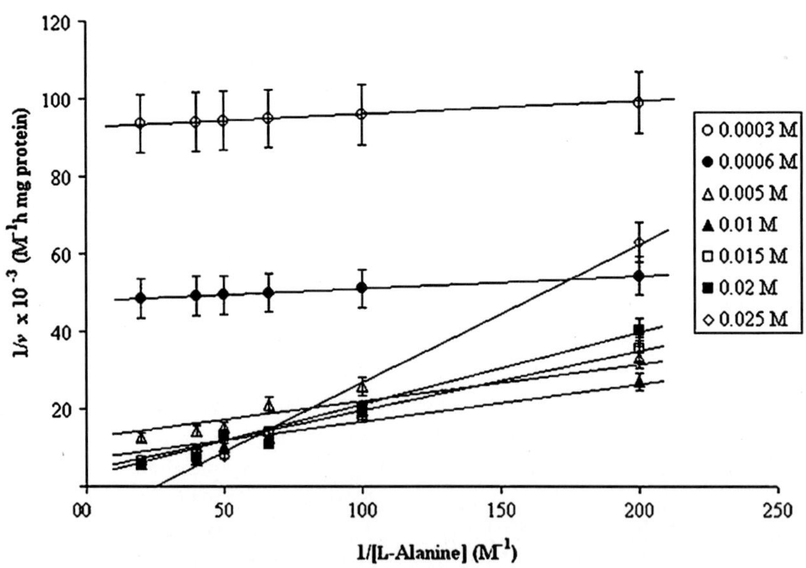

Fig. 10.3 Double reciprocal plots for RML-catalysed L-alanyl–D-glucose reaction: $1/v$ versus 1/[L-alanine] plots; a series of plots shows the effects of varying L-alanine concentration at different fixed D-glucose concentrations in the range of 0.005–0.025 M; plots shown for 0.0003 and 0.0006 M D-glucose concentrations are from computer simulation

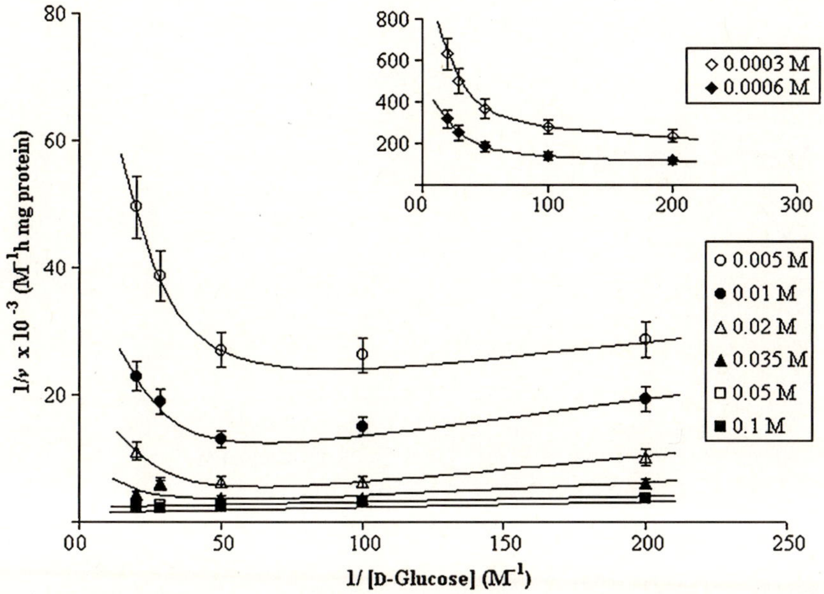

Fig. 10.4 Double reciprocal plots for CRL-catalysed L-alanyl–D-glucose reaction: 1/*v* versus 1/[L-alanine] plots 1/*v* versus 1/[D-glucose] plots; a series of plots show the effect of varying D-glucose concentration at different fixed L-alanine concentrations in the range of 0.005–0.1 M; the plots shown for 0.0003 and 0.0006 M L-alanine concentrations are from computer simulation

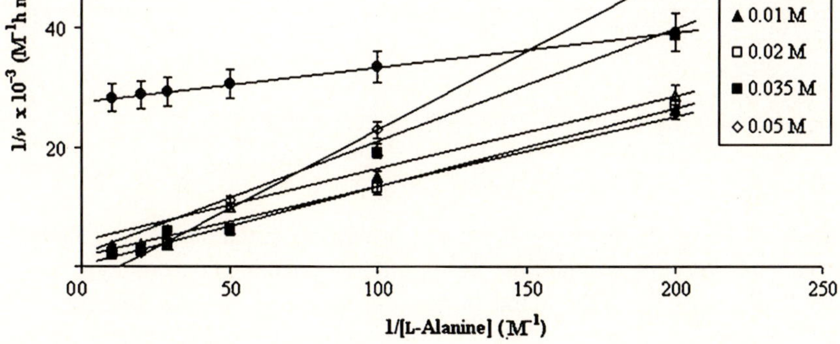

Fig. 10.5 Double reciprocal plots for CRL-catalysed L-alanyl–D-glucose reaction: 1/*v* versus 1/[L-alanine]; a series of plots show the effect of varying L-alanine concentrations at different fixed D-glucose concentrations in the range of 0.005–0.05 M. The plots shown for 0.0003 and 0.0006 M D-glucose concentrations are from the computer simulation

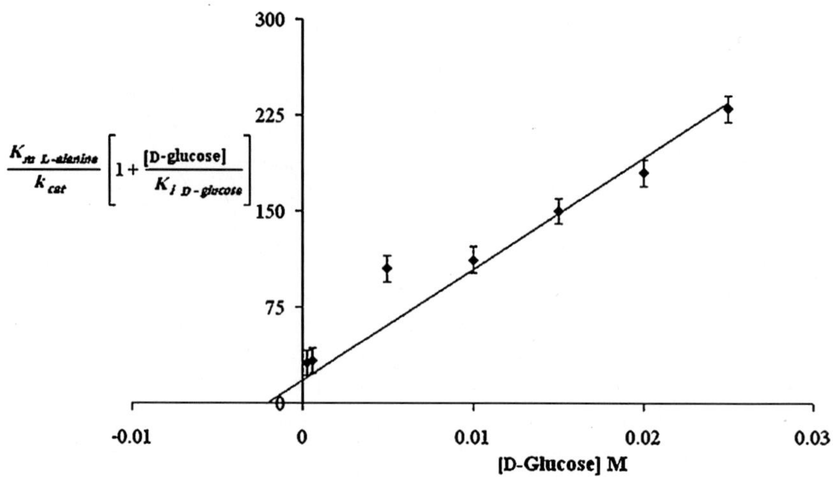

Fig. 10.6 Replot of slopes obtained from Fig. 10.3 versus [D-glucose] (RML)

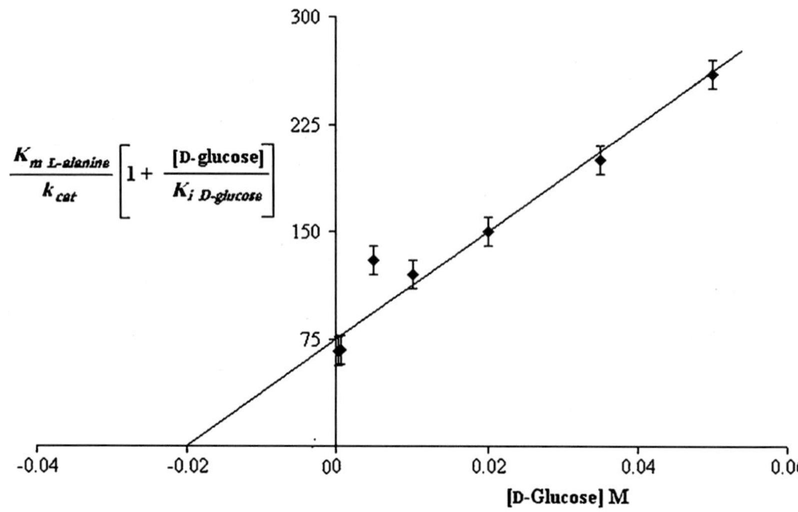

Fig. 10.7 Replot of slopes obtained from Fig. 10.5 versus [D-glucose] (CRL)

steps, (Scheme 10.1) with competitive substrate inhibition that lead to dead-end inhibition (Segel 1993). Both RML and CRL were found to be inhibited by D-glucose.

This model could be described by the following rate equation:

$$\frac{v}{V_{\max}} = \frac{[A][B]}{K_{\mathrm{mA}}[B]\left(1+\frac{[B]}{K_{\mathrm{i}}}\right)+K_{\mathrm{mB}}[A]+[A][B]} \quad (10.1)$$

where v is initial rate, $V_{\max}$ is maximum velocity, $[A]$ is L-alanine concentration, $[B]$ is D-glucose concentration, K_{mA} is Michaelis–Menten constant for the lipase–L-alanine complex, K_{i} is dissociation constant for the lipase–inhibitor (D-glucose) complex and K_{mB} is Michaelis–Menten constant for the lipase–D-glucose complex. Because the initial rates are in M h^{-1} (mg protein)$^{-1}$, $V_{\max}$ is expressed as $k_{\mathrm{cat}} = V_{\max}$/enzyme concentration.

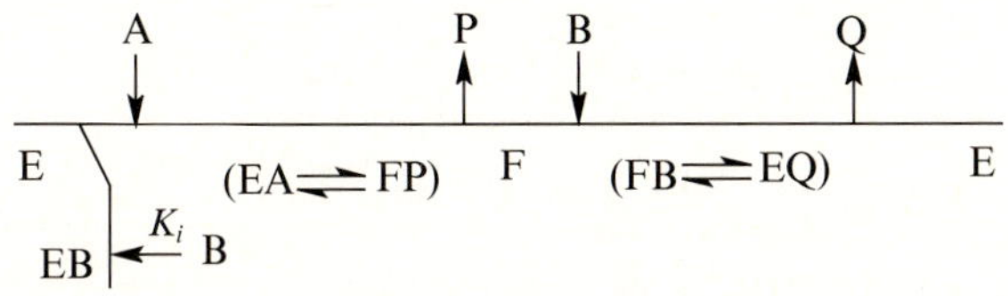

Scheme 10.1 Ping-Pong Bi-Bi mechanism of RML and CRL-catalysed synthesis of L-alanyl D-glucose showing inhibition by D-glucose (Segel 1993)

Table 10.1 Apparent values of kinetic parameters for RML- and CRL-catalysed synthesis of L-alanyl–D-glucose

Name of the lipase		$k_{cat} \times 10^3$ (M h^{-1} mg^{-1})	$K_{mA} \times 10^3$(M)	$K_{mB} \times 10^3$ (M)	$K_i \times 10^3$ (M)
RML	a	0.29±0.028	4.9±0.51	0.21±0.018	1.76±0.19
	b	0.4±0.038	11.2±1.23	0±0.96	5.5±0.59
CRL	a	0.75±0.08	56.2±5.7	16.2±1.8	21.0±1.9
	b	1.0±0.11	56.2±5.4	16.1±1.5	21.0±2.3

A L-alanine, *B* D-glucose, *a* graphical method, *b* curve-fitted values

The apparent values of the four important kinetic parameters $K_{\text{i D-glucose}}$, $K_{\text{m L-alanine}}$, $K_{\text{m D-glucose}}$ and k_{cat} were graphically evaluated. The intercepts of the positive slopes of the curves in Figs. 10.2 and 10.4 on the *Y*-axis, particularly at the highest L-alanine concentration (0.05 M/0.1 M) employed, gave $1/k_{cat}$ for RML and CRL (Table 10.1). Figure 10.5 (RML) and Fig. 10.6 (CRL) show a replot of the slopes from Figs. 10.3 and 10.5 versus [D-glucose], respectively, for which slope = $K_{\text{m L-alanine}}/(k_{cat}K_i)$, Y-intercept = $K_{\text{m L-alanine}}/k_{cat}$ and X-intercept = $-K_i$, where K_i represents the dissociation constant for the lipase–D-glucose complex. $K_{\text{m D-glucose}}$ was obtained using Eq. 10.2 derived by rearranging Eq. 10.1:

$$K_{mB} = \frac{k_{cat}[B]}{v} - \frac{K_{mA}[B]}{[A]} - \frac{K_{mA}[B]^2}{[A]K_i} - [B] \quad (10.2)$$

where K_{mB} = Michaelis–Menten constant for the lipase–D-glucose complex.

To confirm that the kinetics of the RML- and CRL-catalysed syntheses of L-alanyl–D-glucose follows the above-mentioned model, the apparent values of the four important kinetic parameters k_{cat}, K_i, K_{mA} and K_{mB} were also estimated through curve-fitting using Eq. 10.1.

The range of values tested for these parameters and the constraints employed for the iteration procedure are as follows: $k_{cat} < 1$ M h^{-1} mg^{-1}, $K_{\text{i D-glucose}} > K_{\text{m D-glucose}}$, $K_{\text{m D-glucose}} < K_{\text{m L-alanine}}$ and $K_{\text{m L-alanine}} < 10$ M.

The iteration procedure for the curve-fitting method involved non-linear optimisation through minimising the sum of squares of deviations between v_{exptl} and v_{pred}, such that values for the four kinetic parameters mentioned above correspond to the best fit achieved.

Table 10.1 lists graphical as well as the curve-fitted values for comparison. Tables 10.2 and 10.3 show the comparison between experimental and predictive initial rate obtained under different reaction conditions. Although computer-simulated v_{pred} values showed R^2 values of 0.84 for RML and 0.86 for CRL, the discrepancy between v_{exptl} and v_{pred} appeared to be significant at several substrate concentrations. This could be due to (1) the constraints employed in the iteration procedure (curve-fitting method), which limits the flexibility required to examine the real system in solution; (2) the error in the experimental graphical methods based on HPLC measurements, which itself could involve errors of the order of ±10–16%; and (3) the heterogeneous experimental conditions employed involving undissolved carbohydrate and enzyme, on one hand, and dissolved amino acid, on the other, in dichloromethane containing dimethylformamide in a 50:1 proportion.

Table 10.2 Experimental and predicted initial rate values for the synthesis of L-alanyl–D-glucose by RML

L-Alanine (M)	D-Glucose (M)	$v_{experimental} \times 10^6$ (M h^{-1} mg^{-1}) [a]	$v_{predictive} \times 10^6$ (M h^{-1} mg^{-1}) [b]
0.005	0.005	30	51
0.005	0.01	37	45
0.005	0.015	28	37
0.005	0.02	25	31
0.005	0.025	15	27
0.01	0.005	38	73
0.01	0.01	51	72
0.01	0.015	55	64
0.01	0.02	49	56
0.01	0.025	55	49
0.015	0.005	48	84
0.015	0.01	46	91
0.015	0.015	72	84
0.015	0.02	91	75
0.015	0.025	88	67
0.02	0.005	65	92
0.02	0.01	97	104
0.02	0.015	41	99
0.02	0.02	76	91
0.02	0.025	127	83
0.025	0.005	71	97
0.025	0.01	146	114
0.025	0.015	116	112
0.025	0.02	132	104
0.025	0.025	111	96
0.05	0.005	79	109
0.05	0.01	176	142
0.05	0.015	158	149
0.05	0.02	176	147
0.05	0.025	176	141

[a]Graphical method
[b]Curve-fitted values

10.3.2 Esterification Kinetics of L-Phenylalanine and D-Glucose: Double Substrate Inhibition

Comparative kinetic investigations were carried out involving lipases from *Rhizomucor miehei* and *Candida rugosa* on the esterification reaction between L-phenylalanine and D-glucose to form L-phenylalanyl–D-glucose (Lohith et al. 2007). The results from the investigations using RML are described below.

R^2 values obtained from least-square analysis for the initial velocities in both cases were found to be 0.85–0.95. In case of RML-catalysed reactions, initial velocities (v) were found to be in the range 35×10^{-6} to 245×10^{-6} M h^{-1} (mg protein)$^{-1}$, and CRL-catalysed experiments showed initial velocities in the range 40×10^{-6} to 325×10^{-6} M h^{-1} (mg protein)$^{-1}$.

Double reciprocal plots were constructed, and the reciprocal plot patterns for the two substrates in RML-catalysed reactions are almost symmet-

Table 10.3 Experimental and predicted initial rate values for the synthesis of L-alanyl–D-glucose by CRL

L-Alanine (M)	D-Glucose (M)	$v_{experimental} \times 10^6$ (M h^{-1} mg^{-1})[a]	$v_{predictive} \times 10^6$ (M h^{-1} mg^{-1})[b]
0.005	0.005	35	55
0.005	0.01	38	52
0.005	0.02	37	42
0.005	0.035	26	32
0.005	0.05	20	25
0.01	0.005	52	89
0.01	0.01	66	92
0.01	0.02	76	78
0.01	0.035	53	61
0.01	0.05	44	49
0.02	0.005	99	130
0.02	0.01	157	148
0.02	0.02	158	137
0.02	0.035	162	112
0.02	0.05	90	92
0.035	0.005	164	161
0.035	0.01	275	201
0.035	0.02	267	202
0.035	0.035	168	174
0.035	0.05	228	148
0.05	0.005	269	178
0.05	0.01	314	234
0.05	0.02	347	250
0.05	0.035	370	224
0.05	0.05	404	195
0.1	0.005	269	203
0.1	0.01	320	291
0.1	0.02	362	345
0.1	0.035	449	338
0.1	0.05	460	310

[a]Graphical method
[b]Curve-fitted values

rical (Figs. 10.8 and 10.9). The double reciprocal plots shown in the present work were constructed from all the experimentally determined initial rate values and few initial rate values determined from the curve-fitting procedure. Figure 10.8 shows 1/v versus 1/[D-glucose] plots, exhibiting series of curves obtained for different fixed level concentrations of L-phenylalanine at varying D-glucose concentrations. Figure 10.9 shows 1/v versus 1/[L-phenylalanine] plots, exhibiting series of curves obtained for different fixed level concentrations of D-glucose at varying L-phenylalanine concentrations. Figures 10.10 and 10.11 show the replots of slopes from Figs. 10.8 and 10.9, respectively.

The plots in Figs. 10.8 and 10.9 showed that the kinetics could be best described by (Segel 1993) Ping-Pong Bi-Bi model with competitive double substrate inhibition leading to dead-end inhibition by RML–D-glucose complex and L-phenylalanyl–RML–L-phenylalanine complex (Scheme 10.1).

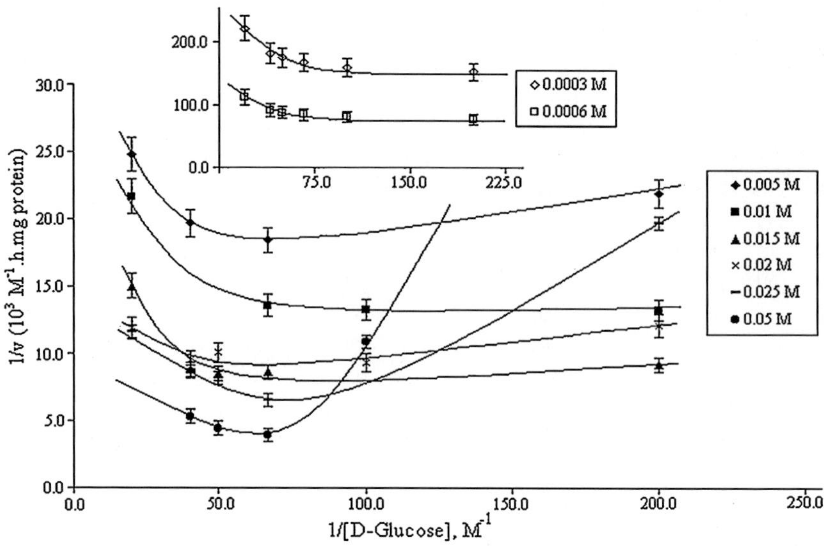

Fig. 10.8 Double reciprocal plots for RML-catalysed L-phenylalanyl–D-glucose reaction: $1/v$ versus 1/[D-glucose], series of curves showing the effect of varying D-glucose concentrations at different fixed concentrations of L-phenylalanine in the range 0.005–0.05 M. Inset shows plots obtained from the computer simulation procedure for L-phenylalanine at 0.0003 and 0.0006 M concentrations

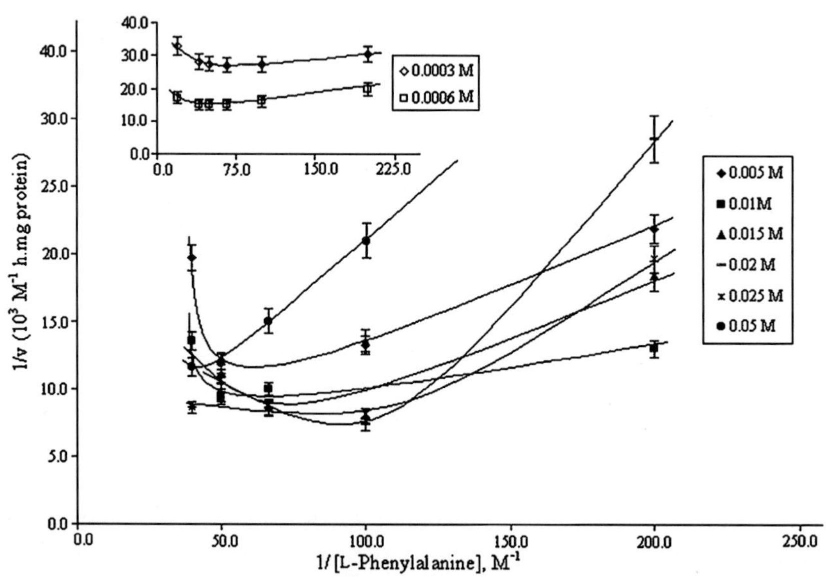

Fig. 10.9 Double reciprocal plots for RML-catalysed L-phenylalanyl–D-glucose reaction: $1/v$ versus 1/[L-phenylalanine], series of curves showing effect of varying L-phenylalanine concentrations at different fixed concentrations of D-glucose in the 0.005–0.05 M range. Inset shown for 0.0003 M and 0.0006 M concentrations of D-glucose is from the computer simulation procedure

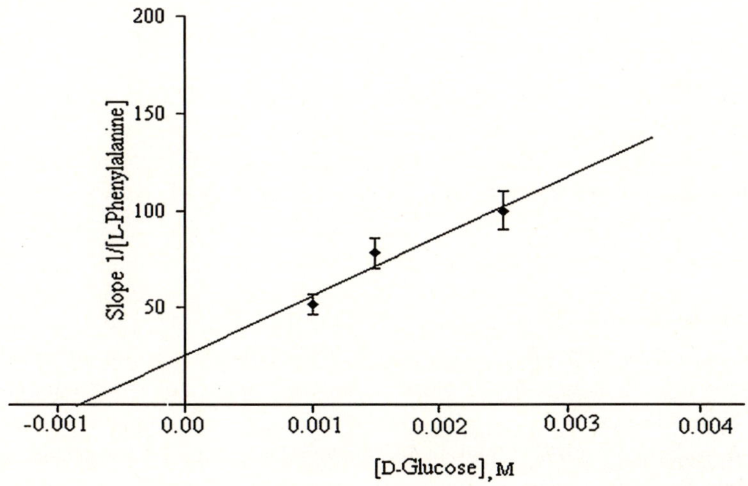

Fig. 10.10 Replot of slopes of 1/[L-phenylalanine] versus [D-glucose] from Fig. 10.8 from RML-catalysed reaction

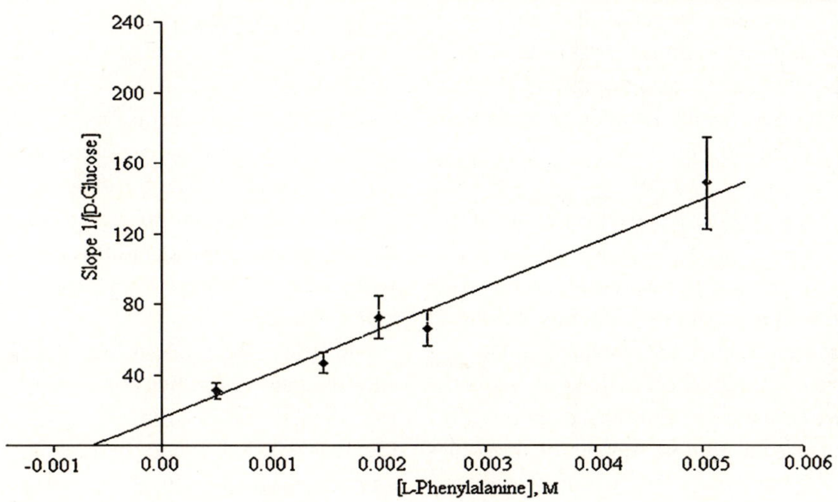

Fig. 10.11 Replot of slopes of 1/[D-glucose] versus [L-phenylalanine] from Fig. 10.9 from RML-catalysed reaction

This model could be described by the following rate equation 10.3:

$$\frac{v}{V_{\max}} = \frac{[A][B]}{K_{\mathrm{mA}}[B]\left(1+\frac{[B]}{K_{\mathrm{iB}}}\right)+K_{\mathrm{mB}}[A]\left(1+\frac{[A]}{K_{\mathrm{iA}}}\right)+[A][B]} \tag{10.3}$$

where v=initial rate, $V_{\max}$=maximum velocity, A=L-phenylalanine concentration, B=D-glucose concentration, K_{mA}=Michaelis–Menten constant for L-phenylalanine, K_{mB}=Michaelis–Menten constant for D-glucose, K_{iB}=dissociation constant of the RML–D-glucose complex and K_{iA}=dissociation constant of the RML–L-phenylalanine complex. Since the initial rates are in M h^{-1} mg^{-1} of the protein, $V_{\max}$ is expressed as $k_{\mathrm{cat}}=V_{\max}$/enzyme concentration.

The five important kinetic parameters $K_{\mathrm{i\ D\text{-}glucose}}$, $K_{\mathrm{i\ L\text{-}phenylalanine}}$, $K_{\mathrm{m\ L\text{-}phenylalanine}}$, $K_{\mathrm{m\ D\text{-}glucose}}$ and k_{cat} were evaluated graphically. Intercept of the positive slopes of Fig. 10.10 on the Y-axis, especially at the higher concentrations of L-phenylalanine employed,

Table 10.4 Apparent values of kinetic parameters for RML-catalysed synthesis of L-phenylalanyl–D-glucose

Name of the compound		k_{cat} (mM h^{-1} (mg protein)$^{-1}$)	K_{mA} (mM)	K_{mB} (mM)	K_{iA} (mM)	K_{iB} (mM)
L-Phenylalanyl–D-Glucose	a	0.13±0.002	3.13±0.33	2.13±0.2	6.0±0.58	9.0±0.86
	b	2.24±0.23	95.6±9.7	80.0±8.5	90.0±9.2	13.6±1.42

A L-phenylalanine, *B* D-glucose, *a* graphical method, *b* computer-simulated values

gave $1/k_{cat}$ (Table 10.4). From Fig. 10.10, slope$=K_{m\,\text{L-phenylalanine}}/(k_{cat}K_{iB})$, Y-intercept$=K_{m\,\text{L-phenylalanine}}/k_{cat}$ and X-intercept$=-K_{iB.}$ From Fig. 10.11, slope$=K_{m\,\text{D-glucose}}/(k_{cat}\,K_{iA})$, Y-intercept$=K_{m\,\text{D-glucose}}/k_{cat}$ and X-intercept$=-K_{iA}$, where K_{iA} and K_{iB} represent dissociation constant for RML–L-phenylalanine and RML–D-glucose complexes, respectively.

As a confirmatory approach, the values of the five important kinetic parameters K_{mA}, K_{mB}, k_{cat}, K_{iA} and K_{iB} were also estimated through a curve-fitting procedure by using Eq. 10.3. The range of values tested for these parameters and the constraints employed for the iteration procedure are $k_{cat\,\text{D-glucose}}<10$ mM h^{-1}mg^{-1}, $K_{i\,\text{D-glucose}}>K_{m\,\text{D-glucose}}$, $K_{m\,\text{D-glucose}}<K_{m\,\text{L-phenylalanine}}$ and $K_{m\,\text{L-phenylalanine}}<90$ mM. The iteration procedure involved determination of initial velocities (v_{pred}) by incrementing the above-mentioned five kinetic parameters in Eq. 10.3 from their lowest approximations (bound by the above-mentioned constraints) and subjecting the v_{pred} (obtained for all the concentrations of D-glucose and L-phenylalanine) to non-linear optimisation, by minimising the sum of squares of deviations between v_{pred} and $v_{exptl.}$ The set of five kinetic parameters which resulted in minimum sum of squares of deviation between v_{pred} and v_{exptl} were considered the best set (Table 10.4). Table 10.5 shows the comparison between experimental and predictive initial rate values obtained under different reaction conditions. Computer-simulated v_{pred} values showed an R^2 value of 0.65.

10.3.3 Esterification Kinetics of L-Phenylalanine and D-Glucose: Single Substrate Inhibition

The results from the above kinetic studies using Candida rugosa lipase are described below. Double reciprocal plots constructed for the CRL-catalysed reaction showed that Figs. 10.12 and 10.13 are not similar unlike those of the RML-catalysed reaction. Figure 10.12 represents $1/v$ versus 1/[D-glucose], and Fig. 10.13 represents $1/v$ versus 1/[L-phenylalanine]. Figure 10.14 shows the replot of slopes from Fig. 10.12. Figure 10.12 shows series of curves obtained for different fixed levels concentrations of L-phenylalanine at varying D-glucose concentrations where slight increase in rates is observed at lower D-glucose concentrations and, at higher concentrations of D-glucose, the rates reduce drastically. Figure 10.13 shows series of parallel lines for different fixed lower D-glucose concentrations at various L-phenylalanine concentrations, and this changes to lines with different slopes at higher D-glucose concentrations.

Here also, the plots in Figs. 10.12 and 10.13 showed that the kinetics could be best described by (Segel 1993) Ping-Pong Bi-Bi model (Scheme 10.1), however, with a difference that only one substrate (D-glucose) functions as a competitive inhibitor forming dead-end CRL–D-glucose complex.

This model could be described by the following rate equation 10.1.

The four important kinetic parameters $K_{i\,\text{D-glucose}}$, $K_{m\,\text{L-phenylalanine}}$, $K_{m\,\text{D-glucose}}$ and k_{cat} were evaluated graphically. Intercept of the positive slopes of Fig. 10.12 on the Y-axis, especially at the higher concentration of L-phenylalanine, gave $1/k_{cat}$ (Table 10.6). From Fig. 10.14, slope$=K_{m\,\text{L-phenylalanine}}/(k_{cat}\;K_i)$, Y-intercept$=K_{m\,\text{L-phenylalanine}}/k_{cat}$ and X-intercept$=-K_i$ where K_i represents dissociation constant for the CRL–D-glucose complex. $K_{m\,\text{D-glucose}}$ was obtained from Eq. 10.2 obtained by rearranging Eq. 10.1.

Table 10.5 Experimental and predicted initial rate values for the synthesis of L-phenylalanyl–D-glucose by RML

L-Phenylalanine conc. (M)	D-Glucose conc. (M)	$K_{cat\text{-}experimental}$ [a] (10^{-6} M h^{-1} (mg protein)$^{-1}$)	$K_{cat\text{-}predictive}$ [b] (10^{-6} M h^{-1} (mg protein)$^{-1}$)
0.005	0.005	46	51
0.005	0.01	79	53
0.005	0.015	54	48
0.005	0.02	35	43
0.005	0.025	50	38
0.005	0.050	40	24
0.01	0.005	75	70
0.01	0.01	75	84
0.01	0.015	74	83
0.01	0.02	135	77
0.01	0.025	123	71
0.01	0.050	46	47
0.015	0.005	109	79
0.015	0.01	100	705
0.015	0.015	116	709
0.015	0.02	117	105
0.015	0.025	113	98
0.015	0.050	67	69
0.02	0.005	83	83
0.02	0.01	107	117
0.02	0.015	89	127
0.02	0.02	96	126
0.02	0.025	105	121
0.02	0.050	84	89
0.025	0.005	51	84
0.025	0.01	74	125
0.025	0.015	153	141
0.025	0.02	123	144
0.025	0.025	116	140
0.025	0.050	86	107
0.05	0.01	92	133
0.05	0.015	245	168
0.05	0.02	224	187
0.05	0.025	186	196

[a]Graphical method
[b]Curve-fitted values
Average absolute deviation in initial velocities – 21.7%

Here also, a curve-fitting procedure was carried out using Eq. 10.1 to estimate $k_{cat,}$ $K_{i,}$ K_{mA} and $K_{mB,}$ to confirm that the kinetics of CRL-catalysed L-phenylalanyl–D-glucose reactions followed the above-mentioned model. The range of values tested and the constraints employed are $k_{cat\ D\text{-}glucose} <$ 1 mM h^{-1} mg^{-1}, $K_{i\ D\text{-}glucose} > K_{m\ D\text{-}glucose}$, $K_{m\ D\text{-}glucose} < K_{m\ L\text{-}phenylalanine}$ and $K_{m\ L\text{-}phenylalanine} < 10$ mM. Here also, the iteration procedure was the same as that employed for the RML-catalysed reaction. Table 10.6 lists graphical as well as the computer-simulated values for comparison. Table 10.7 shows the comparison between experimental and predictive initial rate values obtained under

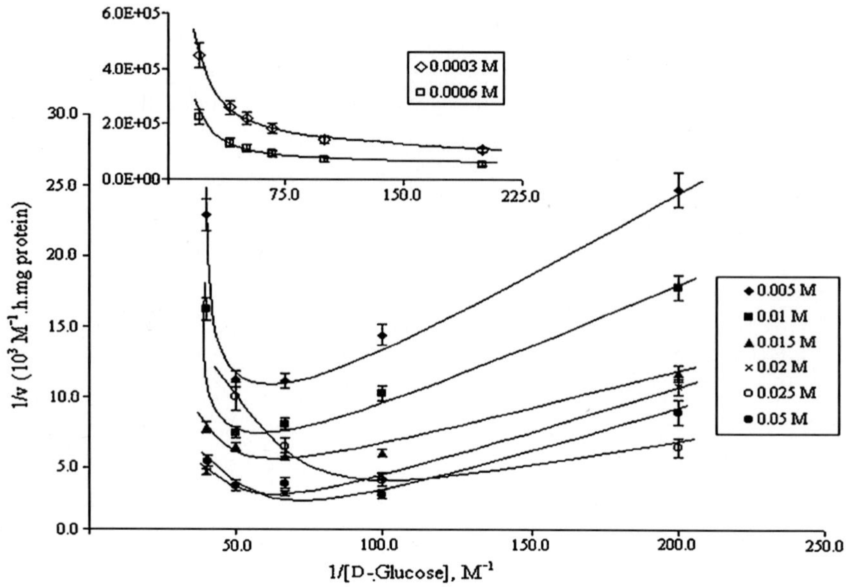

Fig. 10.12 Double reciprocal plots for CRL-catalysed L-phenylalanyl–D-glucose reaction: $1/v$ versus 1/[D-glucose], series of curves showing the effect of varying D-glucose concentrations at different fixed concentrations of L-phenylalanine in the 0.005–0.05 M concentration range. Inset shown for 0.0003 and 0.0006 M concentrations of L-phenylalanine is from the computer simulation procedure

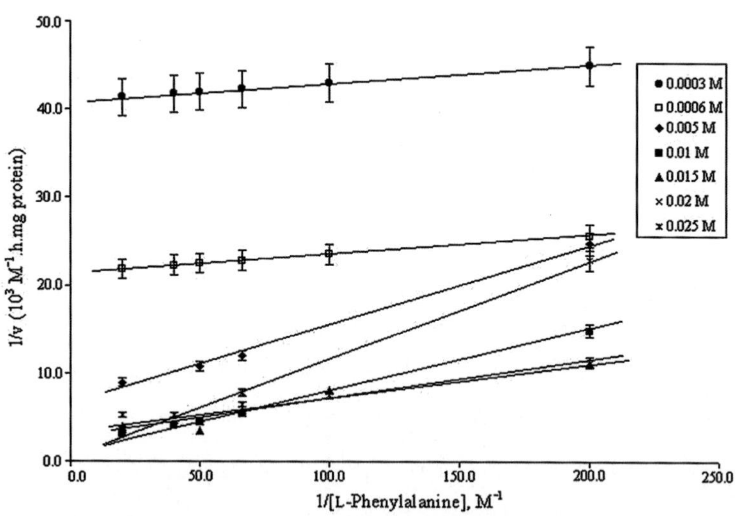

Fig. 10.13 Double reciprocal plots for CRL-catalysed L-phenylalanyl–D-glucose reaction: $1/v$ versus 1/[L-phenylalanine], series of plots showing effect of varying L-phenylalanine concentrations at different fixed concentrations of D-glucose in the 0.005–0.025 M concentration range, and the plots for 0.0003 and 0.0006 M concentrations of D-glucose are from the computer simulation procedure

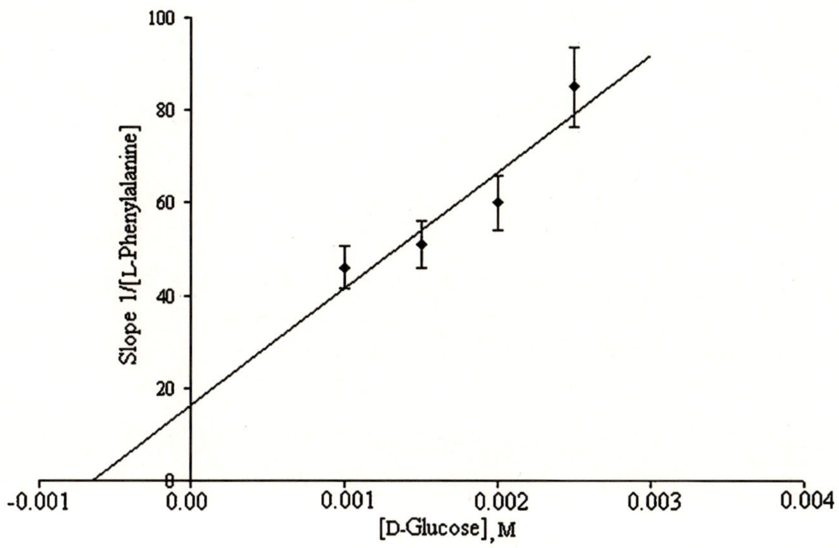

Fig. 10.14 Replot of slopes of 1/[L-phenylalanine] versus [D-glucose] from Fig. 10.12 from CRL-catalysed reaction

Table 10.6 Apparent values of kinetic parameters for CRL-catalysed synthesis of L-phenylalanyl–D-glucose

Name of the compound		k_{cat} (mM h^{-1} (mg protein)$^{-1}$)	K_{mA} (mM)	K_{mB} (mM)	K_{iB} (mM)
L-Phenylalanyl–D-glucose	a	0.40±0.005	5.6±0.58	6.48±0.69	7.0±0.73
	b	0.51±0.06	0±0.98	6.0±0.64	8.5±0.81

A L-phenylalanine, *B* D-glucose, *a* graphical method, *b* computer-simulated values

different reaction conditions. Computer-simulated v_{pred} values showed an R^2 of value 0.63.

In case of CRL, at increasing fixed concentrations of L-phenylalanine (Fig. 10.12), the rate increases at lower concentrations of D-glucose. At higher concentrations of D-glucose corresponding to minimum $1/v$, the rate decreases; the plots tend to become closer to $1/v$ axis. Figure 10.13 (CRL) also reflects the same behaviour, where at lower concentrations of D-glucose, the plots appear parallel probably so for as $K_i > K_{mB}$. However, at higher fixed concentrations of D-glucose, the slopes vary drastically. Thus the kinetic data clearly shows the inhibitory nature of D-glucose in this reaction.

10.4 Glycosylation Kinetics

10.4.1 Glycosylation Kinetics of Curcumin and D-Glucose: Single Substrate Inhibition

Typical rate plot for the curcumin glucosidic reaction is shown in Fig. 10.15 (Sivakumar et al. 2006). Initial velocities (v) were found to be in the range 0.08 to 5.2×10^{-5} M h^{-1} (mg protein)$^{-1}$.

Double reciprocal plots were constructed by plotting $1/v$ versus 1/[curcumin] and $1/v$ versus 1/[D-glucose]. The plots are shown in Figs. 10.16 and 10.17. The plots in Figs. 10.16 and 10.17 showed that the kinetics could be best described

Table 10.7 Experimental and predicted initial rate values for the synthesis of L-phenylalanyl–D-glucose by CRL

L-Phenylalanine conc. (M)	D-Glucose conc. (M)	$K_{cat\text{-}experimental}$ (10^{-6} M h^{-1} (mg protein)$^{-1}$)[a]	$K_{cat\text{-}predictive}$ (10^{-6} M h^{-1} (mg protein)$^{-1}$)[b]
0.005	0.005	40	96
0.005	0.01	67	86
0.005	0.015	90	74
0.005	0.02	89	64
0.005	0.025	44	56
0.01	0.005	56	136
0.01	0.01	92	136
0.01	0.015	123	123
0.01	0.02	133	110
0.01	0.025	62	99
0.015	0.005	85	158
0.015	0.01	187	168
0.015	0.015	173	158
0.015	0.02	155	145
0.015	0.025	128	133
0.02	0.005	93	171
0.02	0.01	224	191
0.02	0.015	317	185
0.02	0.02	100	173
0.02	0.025	213	160
0.025	0.005	155	181
0.025	0.01	243	208
0.025	0.015	155	205
0.025	0.02	100	195
0.025	0.025	191	182
0.05	0.005	256	204
0.05	0.01	325	252
0.05	0.015	258	263
0.05	0.02	269	261
0.05	0.025	291	253

[a]Graphical method
[b]Curve-fitted values
Average absolute deviation in initial velocities – 26.3%

by (Segel 1993) Ping-Pong Bi-Bi model (Scheme 10.1) with competitive substrate inhibition leading to dead-end inhibition.

The rate equation describing the model is shown in Eq. 10.1.

The four important kinetic parameters $K_{curcumin}$, $K_{m\ D\text{-}glucose}$, $K_{m\ curcumin}$ and $k_{cat\ curcumin}$ were evaluated graphically. Intercept of the positive slopes of Fig. 10.16 on the *Y*-axis, especially at the highest concentration of D-glucose (0.1 M) employed, gave $1/k_{cat}$ for curcumin (Table 10.8). Figure 10.18 shows the replot of slopes (Fig. 10.17) of 1/[D-glucose] versus [curcumin], from which slope$=K_{m\ D\text{-}glucose}/(k_{cat}\ K_i)$, *Y*-intercept$=K_{m\ D\text{-}glucose}/k_{cat}$ and *X*-intercept$=-K_i$, where K_i represents dissociation constant for the amyloglucosidase–curcumin complex. $K_{m\ curcumin}$ was obtained by Eq. 10.2 generated by rearranging Eq. 10.1.

The values of the four important kinetic parameters, $k_{cat,}$ K_i, K_{mA} and K_{mB}, were also

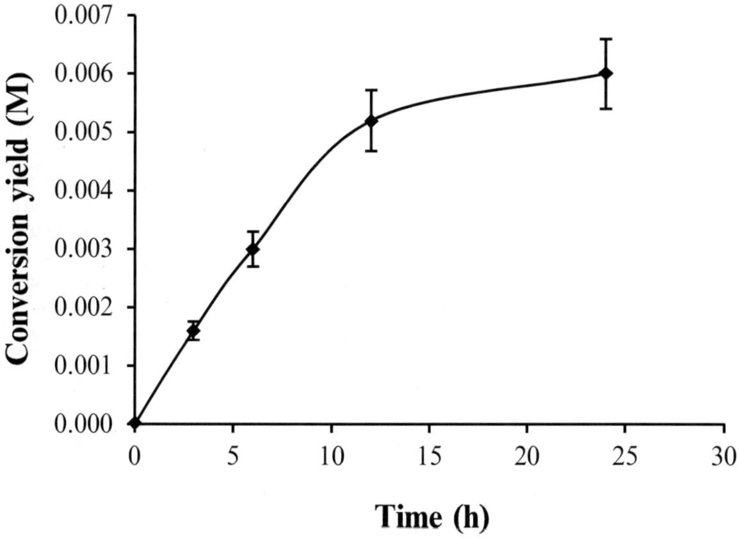

Fig. 10.15 Initial rate (v) plot – conversion yields versus incubation period – curcuminyl-bis-α-D-glucoside: D-glucose 20 mM, curcumin 5 mM, amyloglucosidase 90 mg and 0.6 mL (0.06 mM) of 0.01 M, pH 6.0 phosphate buffer

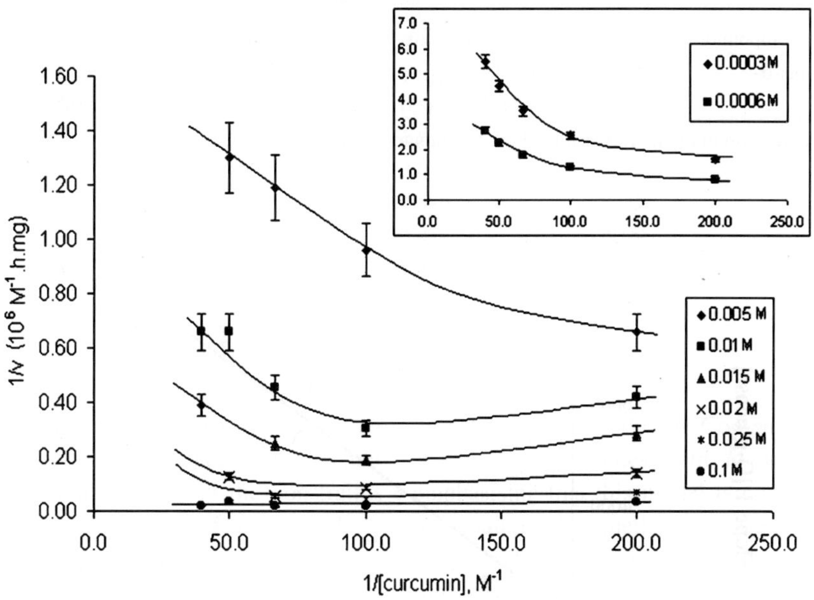

Fig. 10.16 Double reciprocal plot: $1/v$ versus 1/[curcumin], series of plots from experimentally measured initial rate values showing the effect of varying curcumin concentrations at different fixed concentrations of D-glucose in the range 5 mM to 0.1 M. Insets show plots obtained from the computer simulation procedure for 0.3 and 0.6 mM concentrations of D-glucose

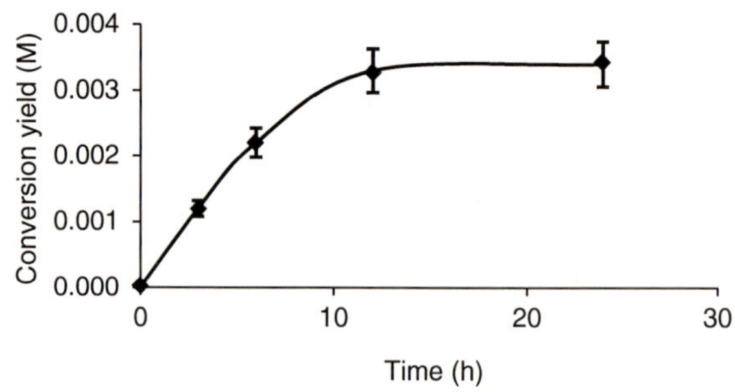

Fig. 10.19 Initial rate (v) plot: D-glucose 10 mM, vanillin 5 mM, amyloglucosidase 90 mg and 0.1 mM (1 mL) of 0.01 M, pH 4 acetate buffer

of curcumin, the slopes vary drastically where $K_i < K_{mB}$. Thus the kinetic data clearly shows the inhibitory nature of curcumin in this reaction. Competition between D-glucose and curcumin for the active site (binding site) of amyloglucosidase could result in predominant curcumin binding at higher concentrations, displacing D-glucose, leading to the formation of the dead-end amyloglucosidase–curcumin complex.

10.4.2 Glucosylation Kinetics of Vanillin and D-Glucose: Single Substrate Inhibition

Here also amyloglucosidase was employed (Sivakumar et al. 2006). To graphically evaluate the apparent values of the kinetic parameters $V_{max,}$ K_i, $K_{m\ vanillin}$ and $K_{m\ D\text{-}glucose}$, initial rates (specific reaction rate) were evaluated by measuring of 4-*O*-(D-glucopyranosyl)vanillin formation at different incubation periods. For each concentration of vanillin (5 mM–0.1 M) and D-glucose (5 mM–0.1 M), individual experiments were performed for incubation periods of 3, 6, 12 and 24 h (30×4 for each system). R^2 values obtained from least-square analysis for the initial velocities in both cases were found to be around 0.95. The plots shown in this work were constructed from all the experimentally determined and few computer-generated initial rate values. A typical rate plot for vanillin glucosidic reactions is shown in Fig. 10.19, and the initial velocities (v) were found to be in the range 0.17 to 5×10^{-5} M h^{-1} (mg protein)$^{-1}$. The enzyme lost only 10% of its activity after incubation for 24 h.

Double reciprocal plot was constructed by plotting $1/v$ versus 1/[vanillin]. The plot is shown in Fig. 10.20, which shows a series of curves obtained for different fixed concentrations of D-glucose at varying vanillin concentrations, where slight increase in initial rates at lower vanillin concentrations is observed and at higher concentrations of vanillin the rates reduce drastically. Figure 10.21 shows a series of lines obtained for different fixed concentrations of vanillin at varying D-glucose concentrations where at fixed lower vanillin concentrations, the lines were parallel and at fixed higher vanillin concentrations, lines with different slopes were observed. The plots in Figs. 10.20 and 10.21 showed that the kinetics could be best described by (Segel 1993) Ping-Pong Bi-Bi model (Scheme 10.1) with competitive substrate inhibition leading to dead-end inhibition (Eq. 10.1).

The four important kinetic parameters $K_{i\ vanillin}$, $K_{m\ D\text{-}glucose}$, $K_{m\ vanillin}$ and $k_{cat\ vanillin}$ were evaluated graphically. Intercept of the positive slope of Fig. 10.20 on the *Y*-axis, especially at the highest concentration of D-glucose (0.1 M) employed, gave $1/k_{cat}$ for vanillin (Table 10.10).

Figure 10.22 shows the replot of slope of Fig. 10.21 (1/[D-glucose] versus [vanillin] plot) from which slope $= K_{m\ D\text{-}glucose}/(k_{cat}\ K_i)$, Y-intercept $= K_{m\ D\text{-}glucose}/k_{cat}$ and X-intercept $= -K_i$, where K_i represents dissociation constant for the amyloglucosidase–vanillin complex. $K_{m\ vanillin}$ was obtained from Eq. 10.2 generated by rearranging Eq. 10.1.

The values of the four important kinetic parameters, $k_{cat,}$ K_i, K_{mA} and K_{mB}, were also estimated mathematically through computer simula-

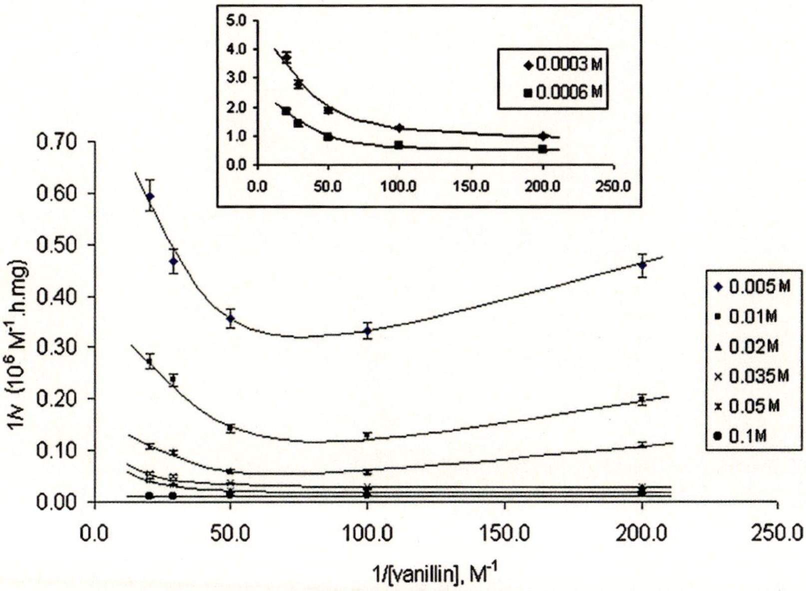

Fig. 10.20 Double reciprocal plot: 1/*v* versus 1/[vanillin]. Series of plots from experimentally measured initial rate values showing the effect of varying vanillin concentrations at different fixed concentrations of D-glucose in the 5 mM to 0.1 M range. Insets show plots obtained from the computer simulation procedure for 0.3 and 0.6 mM concentrations of D-glucose

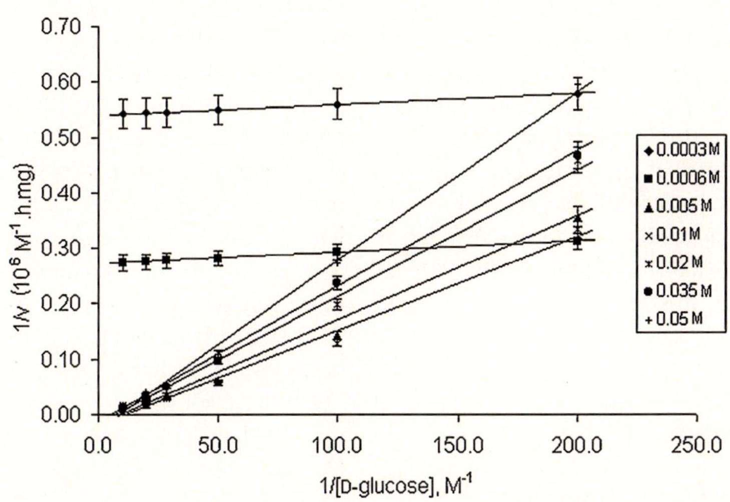

Fig. 10.21 Double reciprocal plot: 1/*v* versus 1/[D-glucose]. Series of plots from experimentally measured initial rate values showing the effect of varying D-glucose concentrations at different fixed concentrations of vanillin in the 5 mM to 0.05 M range. The plots shown for 0.3 and 0.6 mM concentrations of vanillin are from the computer simulation procedure

Table 10.10 Kinetic parameters for the synthesis of 4-*O*-(D-glucopyranosyl)vanillin

	k_{cat} 10^{-5} M $h^{-1}mg^{-1}$	$K_{m\ D\text{-}glucose}$ (mM)	$K_{m\ vanillin}$ (mM)	K_i (mM)
Graphical method	0±1	65.0±6.7	45.6±4.4	12.5±1.3
Computer-simulated values	35.0±3.2	60.0±6.2	50.0±4.8	5±1.1

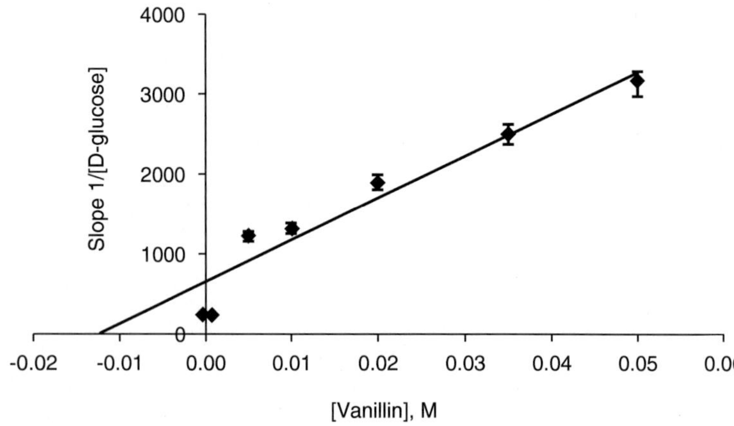

Fig. 10.22 Replot of slope (from Fig. 10.21): 1/[D-glucose] versus [vanillin]

tion. The range of values tested for these parameters and the constraints employed for the iteration procedure are as follows: $k_{cat\ vanillin} < 0.01\ M\,h^{-1}mg^{-1}$, $K_{i\ vanillin} < K_{m\ vanillin}$, $K_{m\ vanillin} < K_{m\ D\text{-}glucose}$ and $K_{m\ D\text{-}glucose} < 0.1$ M. The set of four kinetic parameters which resulted from minimum sum of squares of deviation between v_{pred} and v_{exptl} were considered to be the best set, and they are shown in Table 10.10 which lists graphical as well as the computer-simulated values for comparison. Table 10.11 shows the comparison between experimental and predictive initial rate values obtained under different reaction conditions. Computer simulation showed v_{pred} values with R^2 values of 0.85 for vanillin reaction emphasising that this model is reasonably good in explaining the kinetics of this reaction.

10.5 Discussion

10.5.1 Esterification Kinetics

In case of L-alanine and D-glucose reaction, the kinetic data clearly shows the inhibitory nature of D-glucose towards both RML and CRL. With increasing L-alanine concentration (Fig. 10.2 for RML and Fig. 10.4 for CRL), the initial rate increases with decreasing D-glucose concentration. With increasing D-glucose concentration up to the minimum $1/v$, the initial rate decreases, and the plots tend to become closer to the $1/v$ axis (Y-axis).

Figure 10.3 (RML) and Fig. 10.5 (CRL) also show the same behaviour, in which at low D-glucose concentrations, the plots appear parallel probably as long as $K_i > K_{mB}$ is concerned. However, at high fixed D-glucose concentrations, the slopes of the plots drastically vary. Thus, in these reactions, the kinetic data clearly shows the inhibitory nature of D-glucose. The competition between L-alanine and D-glucose for the active site (binding site) of lipases (RML/CRL) could result in a predominant binding of D-glucose at high concentrations, displacing L-alanine and thus leading to the formation of the dead-end lipase–D-glucose complex.

For the RML reaction, K_{mA} ($4.9 \pm 0.51 \times 10^{-3}$M) is always higher than K_{mB} ($0.21 \pm 0.018 \times 10^{-3}$ M, Table 10.1), which shows that L-alanine is bound to RML less firmly than D-glucose ($K_{mA}/K_{mB} = 23.3$). A similar behaviour is also observed with CRL (Table 10.1) K_{mA} ($56.2 \pm 5.7 \times 10^{-3}$ M), K_{mB} ($16.2 \pm 1.8 \times 10^{-3}$ M) and $K_{mA}/K_{mB} = 3.5$. However, the respective values are very much higher for CRL than for RML, indicating that CRL can yield better conversions than RML. Between RML and CRL, the K_i for D-glucose is lower for RML ($5.5 \pm 0.59 \times 10^{-3}$ M) than for CRL ($21.0 \pm 2.3 \times 10^{-3}$ M), indicating that the RML is inhibited by D-glucose far more efficiently than

Table 10.11 Experimental and predicted initial rate values for the synthesis of 4-OD-glucopyranosyl)vanillin

D-Glucose (M)	Vanillin (M)	$v_{experimental}$ (10^{-5} M h^{-1} mg^{-1})	$v_{predictive}$ (10^{-5} M h^{-1} mg^{-1})
0.005	0.005	0.218	1.219
0.005	0.01	0.300	1.189
0.005	0.02	0.280	0.913
0.005	0.035	0.214	0.643
0.005	0.05	0.168	0.492
0.01	0.005	0.506	1.763
0.01	0.01	0.774	1.976
0.01	0.02	0.702	1.672
0.01	0.035	0.422	1.231
0.01	0.05	0.366	0.957
0.02	0.005	1.003	2.269
0.02	0.01	1.746	2.952
0.02	0.02	1.666	2.866
0.02	0.035	1.033	2.269
0.02	0.05	0.917	1.815
0.035	0.005	2.069	2.587
0.035	0.01	4.128	3.745
0.035	0.02	2.757	4.128
0.035	0.035	1.987	3.551
0.035	0.05	1.774	2.947
0.05	0.005	3.475	2.741
0.05	0.01	1.618	4.195
0.05	0.02	4.968	5.010
0.05	0.035	3.027	4.588
0.05	0.05	2.401	3.926
0.1	0.005	6.110	2.945
0.1	0.01	7.340	4.881
0.1	0.02	7.510	6.676
0.1	0.035	9.170	6.961
0.1	0.05	480	6.414

CRL. This could also explain the better conversion observed with CRL than with RML.

The kinetic data of L-phenylalanine and D-glucose reaction clearly shows the inhibitory nature of both D-glucose and L-phenylalanine towards RML. Competition between L-phenylalanine and D-glucose for the active site (binding site) of RML could result in predominant binding of either substrate at their higher concentrations leading to the formation of the dead-end complexes (RML–D-glucose/L-phenylalanine). K_{mA} (95.6±9.7 mM) is slightly higher than K_{mB} (80.0±8.5 mM, Table 10.4), which shows that both substrates possess almost equal propensity for the reaction (K_{mA}/K_{mB}=1.19). This could also inferred from the dissociation constant values, $K_{i\ \text{L-phenylalanine}}$ (90.0±9.2 mM)>$K_{i\ \text{D-glucose}}$ (13.6±1.42 mM).

In case of CRL, competition between L-phenylalanine and D-glucose for the active site of CRL could result in predominant binding of D-glucose at higher concentrations, displacing L-phenylalanine, leading to the formation of the dead-end lipase–D-glucose complex. Here also, K_{mA} (0±0.98 mM) is slightly higher (Table 10.6) than K_{mB} (6.0±0.64 mM, K_{mA}/K_{mB}=1.67). The respective kinetic parameter values (K_{mA} and K_{mB})

are higher for RML compared to CRL indicating that RML can give better conversions than CRL.

Lipase is a single-domain molecule belonging to the family of α/β-hydrolase proteins (Derewenda et al. 1992; Grochulski et al. 1994). Most of the lipases reported contain Ser-His-Asp/Glu catalytic triads in their active site (Grochulski et al. 1994) with exceptions like esterases from *Streptomyces scabies* which contain only Ser 14 and His 283 (Wei et al. 1995).

Both RML and CRL contain amino acids in their active sites capable of forming hydrogen bonds with suitable donor molecules. The catalytic triad in RML consists of Ser-144, His-257 and Asp-203 (Brady et al. 1990). CRL contains Ser-209, Glu-341 and His-449 (Grochulski et al. 1994). Brzozowski et al. (1991) showed in an atomic model of the inhibitor *n*-hexylchlorophosphonate ethyl ester–RML complex that in the oxyanion hole, which is directly responsible for the substrate binding, a direct covalent bond formation between the nucleophilic O_γ of Ser-144 and the phosphorous atom of *n*-hexylchlorophosphonate ethyl ester is possible. In CRL, the oxyanion hole O_γ (Ser-209) is formed by the amide backbones of Gly-123, Gly-124 and Ala-210 through the hydrogen bonding between the amide –CO–NH– and the hydroxyl of Ser-209, which is stabilised by the helix dipole (Grochulski et al. 1994). D-Glucose possesses five hydroxyl groups, and L-alanine possesses carboxyl and amino groups capable of forming hydrogen bonds with polar side chains of amino acids. Ser-144 hydroxyl and Asp-203 carboxyl groups of RML and Ser-209 and Glu-341 of CRL (Grochulski et al. 1993) residues are very good candidate molecules for exhibiting hydrogen-bonding interactions.

Between D-glucose and L-alanine/L-phenylalanine, the former possesses more hydrogen-bonding functional groups than the amino or carboxyl groups of L-alanine/L-phenylalanine. Ser-144 in RML and Ser-209 in CRL can form hydrogen bonds with the amino *N* atom of L-alanine as well as the O atom of D-glucose. Because the $K_{m\ \text{L-alanine}}$ value is higher than the $K_{m\ \text{D-glucose}}$ values for both enzymes, D-glucose could strongly bind to these enzymes than L-alanine/L-phenylalanine.

Zaidi et al. (2002) reported that the interaction between nylon-immobilised CRL and alcohol through hydrogen bonding could block the nucleophilic site of the enzyme engaged in acylation, leading to inhibition. A similar behaviour can also be envisaged between D-glucose hydroxyl groups and the above-mentioned oxygen of serine and the carboxylate groups of glutamic acids. Hence, this kinetic study could clearly explain the inhibition of both RML and CRL by D-glucose. Also, for the first time, it has been found that D-glucose could be inhibitor to both lipases at higher concentrations.

10.5.2 Glycosylation Kinetics

In the kinetic reaction between D-glucose and curcumin, $K_{m\ \text{D-glucose}}$ (0±0.9 mM, Table 10.8) is always higher than K_{mB} (4.6±0.5 mM) which shows that while glucose binding could lead to product formation, curcumin binding to the active site could result in inhibition of the amyloglucosidase activity.

The inhibitory nature of vanillin towards amyloglucosidase from *Rhizopus* mould can be deduced clearly from the kinetic data of D-glucose and vanillin reaction. With increasing concentrations of D-glucose (Fig. 10.20), the rate increases at lower concentrations of vanillin. At higher concentrations of vanillin corresponding to minimum $1/v$, the rate decreases; the plots tend to become closer to $1/v$ axis. Figure 10.21 also reflects the same behaviour, where at lower concentrations of vanillin, the lines appear parallel probably so for as $K_i > K_{mB}$. However, at higher fixed concentrations of vanillin, the slopes vary drastically where $K_i < K_{mB}$. Thus the kinetic data clearly shows the inhibitory nature of vanillin in this reaction. Competition between D-glucose and vanillin for the active site (binding site) of amyloglucosidase could result in predominant vanillin binding at higher concentrations, displacing D-glucose, leading to the formation of the dead-end amyloglucosidase–vanillin complex. In this reaction, $K_{m\ \text{D-glucose}}$ (60.0±6.2 mM, Table 10.4) is always higher than K_{mB} (50.0±4.8 mM) which shows that while glucose binding could lead to product formation, vanillin binding to the active site could result in inhibition of the amyloglucosidase activity.

Glucoamylases possess $(\alpha/\alpha)_6$-barrel fold structure which is different from the $(\beta/\alpha)_8$-barrel fold structure of α-amylase, β-amylase and α-glucosidase (Svensson et al. 1990; Aleshin et al. 1992, 1994; Chiba 1997). In the catalytic domain, two glutamic acids Glu314 and Glu544 in *Rhizopus oryzae* (Ashikari et al. 1986; Aleshin et al. 1992) are reported to be the catalytic amino acid residues directly involved as acid–base catalysts in the hydrolytic reaction (Sierks et al. 1990; Chiba 1997). It has also been shown that oxocarbenium ion mechanism is the most suitable for the hydrolytic reaction in both 'retaining' and 'inverting' enzymes (Chiba 1995). Although no decisive mechanism has been proposed so far, for the glycosylation reactions, it is generally believed that the oxocarbenium ion mechanism could be the most probable one.

Catalysis occurs mainly between subsites 1 and 2 of glucoamylase, and the active site of *Aspergillus niger* glucoamylase is identical to that of *Rhizopus oryzae* (Stoffer et al. 1995). Sugar OH groups are held firmly in the active site subsites 1 and 2 of *Rhizopus oryzae* through hydrogen bonds with Arg191, Asp192, Leu312, Trp313, Glu314, Glu315 and Arg443 (Ashikari et al. 1986; Aleshin et al. 1994). The above-mentioned residues can also stabilise planar curcumin/vanillin bound to the active site through hydrogen bonds. Curcumin/vanillin could form effective hydrogen bonds between the enolic, phenolic OH and the carbonyl group of curcumin/vanillin and the Arg191, Asp192, Trp313, Glu315 and Arg443 residues. Hence, higher concentrations of curcumin/vanillin are capable of displacing the glucose–oxocarbenium ion from the active site and occupy its position instead, leading thereby to dead-end inhibition. This may not happen at lower concentrations of curcumin/vanillin.

Several lipase-catalysed esterification reactions have been described to follow Ping-Pong Bi-Bi mechanism, which deals with two substrates (acid and alcohol) and two products (water and ester). So far, enzyme-mediated glycosylation, especially the one involving a carbohydrate molecule and an aglycon molecule, has not been reported to follow Ping-Pong Bi-Bi model. This could be the first report of its kind.

References

Aleshin AE, Golubev A, Firsov LM, Honzatko RB (1992) Crystal structure of glucoamylase from *Aspergillus awamori* var X100 to 2.2-Å resolution. J Biol Chem 267:19291–19298

Aleshin AE, Firsov LM, Honzatko RB (1994) Refined structure for the complex of acarbose with glucoamylases from *Aspergillus awamori* var. X100 to 2.4 A° resolution. J Biol Chem 269:15631–15639

Ashikari T, Nakamura N, Tanaka Y, Kiuchi N, Shibano Y, Tanaka T, Amachi T, Yoshizumi H (1986) *Rhizopus* raw-starch-degrading glucoamylase: ts cloning and expression in yeast. Agric Biol Chem 50:957–964

Bousquet-Dubouch MP, Graber M, Sousa N, Lamare S, Legoy MD (2001) Alcoholysis catalysed by *Candida rugosa* lipase B in a gas/solid system obeys a Ping-Pong Bi-Bi mechanism with competitive inhibition by the alcohol substrate and water. Biochem Biophys Acta 1550(1):90–99

Brady L, Brzozowski AM, Derewenda U, Derewenda ZS, Dodson GG, Tolley S, Turkenburg JP, Christiansen L, Huge-Jensen B, Nashkov L, Thim L, Menge U (1990) A serine protease triad forms the catalytic center of triglycerol lipase. Nature 343:767–770

Brzozowski AM, Derewenda U, Derewenda ZS, Dodson GG, Lawson DM, Turkenburg JP, Bjorkling F, Huge-Jensen B, Patkar SA, Thim L (1991) A model for interfacial activation in lipases from the structure of a fungal lipase-inhibitor complex. Nature 351:491–494

Chiba S (1995) Enzyme chemistry and molecular biology of amylase and related enzymes. The amylase research society of Japan (ed). CRC Press, Boca raton/Ann arbor/London/Tokyo, pp 68–82

Chiba S (1997) Molecular mechanism in α-glucosidase and glucoamylase. Biosci Biotech Biochem 61:1233–1239

Chulalaksanaukul W, Condort JS, Delorme P, Willemot RM (1990) Kinetic study of esterification by immobilized lipase in *n*-hexane. FEBS Lett 276:181–184

Derewenda U, Brzozwski AM, Lawson DM, Derewenda ZS (1992) Catalysis at the interface the anatomy of a conformational change in a triglyceride lipase. Biochemistry 31:1532–1541

Duan G, Ching CB, Lim E, Ang CH (1997) Kinetic study of enantioselective esterification of ketoprofen with n-propanol catalysed by an lipase in an organic medium. Biotechnol Lett 19:1051–1055

Goto M, Tanigava K, Kanlayakrit W, Hayashida S (1994) The molecular mechanism of binding of glucoamylases I from A*spergillus awamori* var. kawachi to cyclodextrin and raw starch. Biosci Biotech Biochem 58:49–54

Grochulski P, Li Y, Schrag JD, Bouthillier F, Smith P, Harrison D, Rubin B, Cygler M (1993) Insight into interfacial activation from an open structure of *Candida rugosa* lipase. J Biol Chem 268:12843–12847

Grochulski P, Bouthillier F, Kazlauskas RJ, Serreqi AN, Schrag JD, Ziomek E, Cygler M (1994) Analogs of reaction intermediates identify a unique substrate binding site in *Candida rugosa* lipase. Biochemistry 33:3494–3500

Hiromi K, Ohnishi M, Tanaka A (1983) Subsite structure and ligand binding mechanism of glucoamylase. Mol Cell Biochem 51:79–95

Janssen AEM, Sjursnes BJ, Vakurov AV, Halling PJ (1999) Kinetics of lipase catalyzed esterification in organic media correct model and solvent effects on parameters. Enzyme Microb Technol 24:463–470

Kiran KR, Divakar S (2002) Enzyme inhibition by *p*-cresol and lactic acid in lipase mediated syntheses of *p*-cresyl acetate and stearoyl lactic acid: a kinetic study. World J Microbiol Biotechnol 18:707–712

Lohith K, Divakar S (2005) Lipase catalysed synthesis of L-phenylalanine esters of D-glucose. J Biotechnol 117:49–56

Lohith K, Manohar B, Divakar S (2007) Competitive inhibition by substrates in *Rhizomucor miehei* and *Candida rugosa* lipases catalysed esterification reaction between L-phenylalanine and D-glucose. World J Microbiol Biotechnol 23:955–964

Marty A, Chulalaksananukul W, Condoret JS, Willemont RM, Durand G (1990) Comparison of lipase-catalyzed esterification in supercritical carbon dioxide and n-hexane. Biotechnol Lett 12(1):11–16

Ohnishi M, Hiromi K (1989) Binding of maltose to *Rhizopus niveus* glucoamylases in the pH range where the catalytic carboxyl groups are ionized. Carbohyd Res 195:138–144

Rizzi M, Stylos P, Riek A, Reuss M (1992) A kinetic study of immobilized lipase catalyzing the synthesis of isoamyl acetate by transesterification in n-hexane. Enzyme Microb Technol 14:709–714

Segel IH (1993) Enzyme kinetics, 2nd edn. Wiley, New York, pp 826–882

Sierks MR, Ford C, Reilly PJ, Svensson B (1990) Catalytic mechanism of fungal glucoamylases as defined by mutagenesis of Asp 176, Glu179, and Glu180 in the enzyme from *Aspergillus awamori*. Protein Eng 3:193–198

Sivakumar R, Vijayakumar GR, Manohar B, Divakar S (2006) Competitive substrate inhibition of amyloglucosidase from *Rhizopus mold* by vanillin and curcumin in respective glucosylation reactions. Biocatal Biotrans 24:299–305

Somashekar BR, Lohith K, Manohar B, Divakar S (2007) Inhibition of *Rhizomucor miehei* and *Candida rugosa* lipases by D-glucose in the esterification reaction between L-alanine and D-glucose. J Biosci Bioeng 103(2):122–128

Stoffer B, Aleshin AE, Firsov LM, Svensson B, Honzatko RB (1995) Refined structure for the complex of D-gluco-dihydroacarbose with glucoamylases from *Aspergillus awamori* var. X100 to 2.2 Å resolution dual conformation for extended inhibitors bound to the active site of glucoamylases. FEBS Lett 358:57–61

Svensson B, Clarke AJ, Svendsen I, Moller H (1990) Identification of carboxylic acid residues in glucoamylase G2 from *Aspergillus niger* that participate in the catalysis and substrate binding. Eur J Biochem 18:29–38

Tanaka A, Yamashita T, Ohnishi M, Hiromi K (1983) Steady-state and transient kinetic studies on the binding of maltooligosaccharides to glucoamylases. J Biochem 93:1037–1043

Van-Tol JBA, Odenthal JB, Jongejan JA, Duine JA (1992) Relation of enzyme reaction rate and hydrophobicity of the solvent. In: Tramper J, Vermue MH, Beetink HH, Von-Stocker U (eds) Biocatalysis in non-conventional media. Elsevier, Amsterdam, pp 229–235

Wei Y, Schottel JL, Derewenda U, Swenson L, Patkar S, Derewenda ZS (1995) A novel variant of the catalytic triad in the *Streptomyces scabies* esterase. Nat Struct Biol 2:218–223

Yadav GD, Lathi PS (2004) Synthesis of citronellol laurate in organic media catalyzed by immobilized lipases kinetic studies. J Mol Cat B Enzyme 27:113–119

Yadav GD, Devi KM (2004) Immobilized lipase-catalysed esterification and transesterification reactions in non-aqueous media for the synthesis of tetrahydrofurfuryl butyrate comparison and kinetic modeling. Chem Eng Sci 59:373–383

Yadav GD, Trivedi AH (2003) Kinetic modeling of immobilized-lipase catalysed transesterification of *n*-octanol with vinyl acetate in non-aqueous media. Enzyme Microb Technol 32:783–789

Zaidi A, Gainer JL, Carta G, Mrani A, Kadiri T, Belarbi Y, Mir A (2002) Esterification of fatty acids using nylon-immobilized lipase in *n*-hexane kinetic parameters and chain length effects. J Biotechnol 93:209–216

Zhang T, Yang L, Zhu Z (2005) Determination of internal diffusion limitation and its macroscopic kinetics of the transesterification of CPB alcohol catalyzed by immobilized lipase in organic media. Enzyme Microb Technol 36:203–209

11 ACE Inhibition and Antioxidant Activities of Enzymatically Synthesised Aminoacyl Esters and Glycosides

Abstract

Since most of the ACE inhibitory drugs are peptides, this chapter deals with exploration of ACE inhibitory activities for some enzymatically synthesised L-alanyl, L-valyl, L-leucyl, L-isoleucyl, L-proline, L-phenylalanine, L-tryptophan and L-histidine esters of carbohydrates and glycosides of *n*-octanol, phenolic glycosides of guaiacol, eugenol, curcumin, vanillin, N-vanillyl-nonanamide, DL-dopa, dopamine and vitamin glycosides of retinol, thiamin, riboflavin, pyridoxine, ergocalciferol, cholecalciferol and α-tocopherol. A few glycosides were also tested for antioxidant activities. These results exhibiting both the phenolic and vitamin glycosides holding promising potential as antioxidants and ACE inhibitors are discussed in terms of structure–function relationship between these compounds and those of the commercially known inhibitors.

11.1 ACE Inhibition by Aminoacyl Esters of Carbohydrates

Angiotensin-converting enzyme (dipeptidyl carboxypeptidase, EC 3.4.15.1) is a zinc-containing nonspecific dipeptidyl carboxypeptidase widely distributed in mammalian tissues (Li et al. 2004). Angiotensin-converting enzyme (ACE) regulates the blood pressure by modulating renin–angiotensin system as shown in Scheme 11.1 (Vermeirssen et al. 2002). This enzyme increases the blood pressure by converting the decapeptide angiotensin I into the potent vasoconstricting octapeptide, angiotensin II. Angiotensin II brings about several central effects, all leading to a further increase in blood pressure. ACE is a multifunctional enzyme that also catalyses the degradation of bradykinin (blood pressure-lowering nanopeptide), and therefore inhibition of ACE results in an overall anti-hypertensive effect (Li et al. 2004; Johnston 1992).

Several synthetic drugs and biomolecules are available for ACE inhibition. Captopril is a successful synthetic anti-hypertensive drug, and similar such molecules like enalapril, perindopril, ceranopril, ramipril, quinapril and fosinopril also show ACE inhibitory activities (Hyuncheol et al. 2003; Dae-Gill et al. 2003). The mechanism of ACE inhibition by captopril is shown in Scheme 11.2 (De Lima 1999). The hypothetical representation of inhibitors (hydrolysed products of peptides) binding to the ACE is shown in Scheme 11.3 and also reported that glycine, valine and leucine at the carboxyl terminus of the peptide inhibitor are the potent inhibitors (De

S. Divakar, *Enzymatic Transformation*,
DOI 10.1007/978-81-322-0873-0_11,

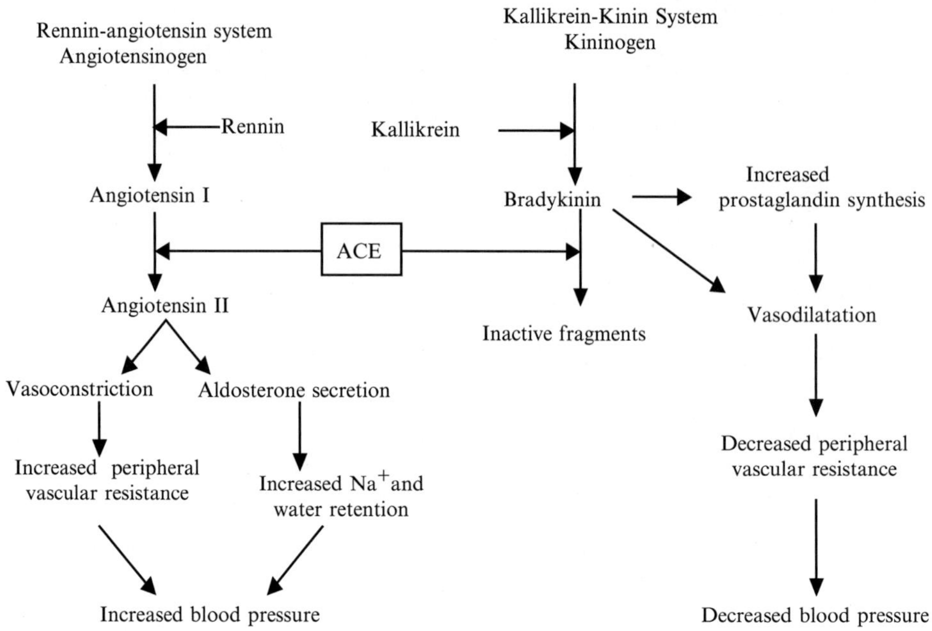

Scheme 11.1 Role of angiotensin-converting enzyme (ACE) in regulating blood pressure (Li et al. 2004)

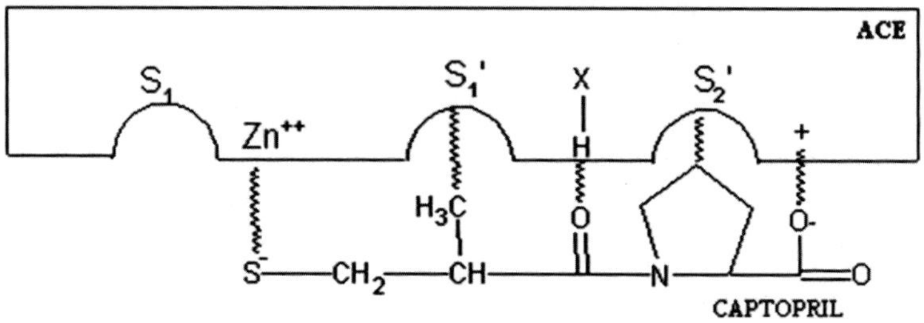

Scheme 11.2 Hypothetical representation of ACE inhibition by captopril binding to the active sites (De Lima 1999)

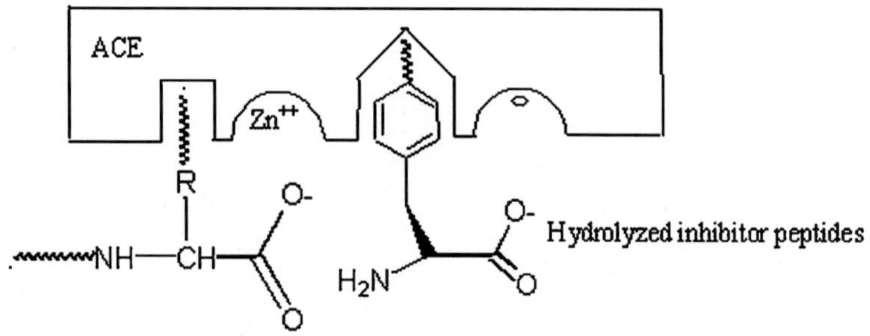

Scheme 11.3 Hypothetical representation of ACE active sites and binding of inhibitors (De Lima 1999)

Table 11.1 Protease inhibition assay for D-glucose ester[a]

System	Protease activity (min^{-1} mg^{-1} enzyme protein[b])	Percentage of protease activity with respect to ACE activity[c]
Control		
ACE – 0.5 mL + 0.5 mL of 0.6% haemoglobin + 0.5 mL Buffer	0.0436	13.3
L-Isoleucyl-D-glucose – 0.5 mL ester + ACE – 0.5 mL + 0.5 mL of 0.6% haemoglobin	0.0267	8.9

[a]Conditions: ACE – 0.5 mL (0.5 mg). All the solutions were prepared in 0.1 M tris-HCl (pH 7.5): incubation period, 30 min; temperature, 37°C; 0.5 mL of 10% trichloroacetic acid added to arrest the reaction; Blank performed without enzyme and ester; absorbance measured at 440 nm; ester – 0.5 mL of 0.8 mM

[b]Average absorbance values from three individual experiments

[c]Percentage protease activity with respect to an ACE activity of 0.327 μmol (min mg protein)$^{-1}$

Lima 1999; Wu and Liu 2002; Kim et al. 2001, 2003).

Some naturally occurring 'biologically active peptides' also act as ACE inhibitors. Deloffre et al. (2004) reported that a neuropeptide from leach brain showed ACE inhibition with an IC_{50} value of 19.8 μM. The N-terminal dipeptide (Tyr-Leu) of β-lactorphin was found to be the most potent inhibitor (Mullally et al. 1996). Many peptide inhibitors are derived from different food proteins like Asp-Leu-Pro and Asp-Gly from soy protein hydrolysis (Wu and Liu 2002) and Gly-Pro-Leu and Gly-Pro-Val from bovine skin gelatin hydrolysis (Kim et al. 2001). Cooke et al. (2003) prepared 4-substituted phenylalanyl esters of alkyl or benzyl derivatives, which exhibited ACE inhibitory activity.

Aminoacyl esters of carbohydrates find wide variety of applications in food and pharmaceutical industries. Aminoacyl esters have not been shown so far to exhibit ACE inhibition activity. Since most of the ACE inhibitory drugs are peptides, it was envisaged that the aminoacyl esters of carbohydrates also could possess ACE inhibition activities as they contain aminoacyl groups as part of their structure. Hence, this chapter deals with exploration of ACE inhibition activities for some enzymatically synthesised L-alanyl, L-valyl, L-leucyl, L-isoleucyl, L-proline, L-phenylalanine, L-tryptophan and L-histidine esters of carbohydrates using lipases in organic media. ACE inhibition activity of the above-mentioned aminoacyl esters of carbohydrates was carried out by the Cushman and Cheung method (1969, 1971). Since hippuryl-L-histidyl-L-leucine (HHL) mimics the carboxyl dipeptide of angiotensin I, it has been routinely used as the substrate for screening ACE inhibitors.

Underivatised L-amino acids and carbohydrates were also tested for ACE inhibition as controls, and they did not show any ACE inhibitory activities (Vasudeva Kamath et al. 2006; Lohith et al. 2006). Only esters showed activities. Isolated ACE inhibitor tested for lipase and protease activity (Table 11.1) showed a small extent of protease activity (13.3%) compared to ACE activity but no lipase activity. In presence of aminoacyl esters prepared, the isolated ACE showed 8.9% protease activity (Table 11.1) compared to the ACE activity. This confirmed that the ACE inhibition observed in the presence of aminoacyl esters prepared is more due to ACE inhibition rather than protease inhibition.

Figure 11.1 shows a typical ACE inhibition plot for captopril which showed an IC_{50} value of 0.060 ± 0.006 mM. ACE inhibition plots for all the tested esters, such as carbohydrate esters of L-alanine (Fig. 11.2), L-valine (Fig. 11.3), L-leucine (Fig. 11.4) and L-isoleucine (Fig. 11.5), L-prolyl esters (Fig. 11.6), L-phenylalanyl esters (Fig. 11.7), L-tryptophanyl esters (Fig. 11.8) and L-histidyl esters (Fig. 11.9), are shown. Tables 11.2 and 11.3 show the compounds tested, their conversion yields from the respective enzy-

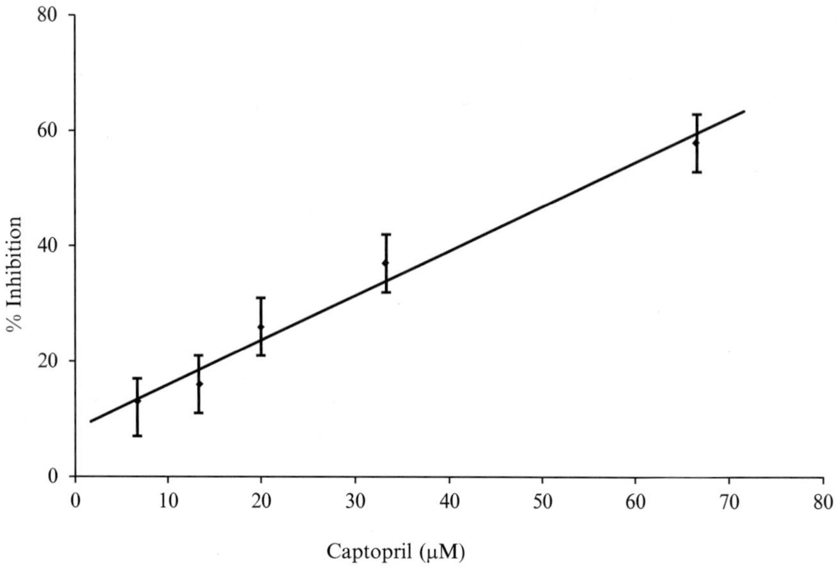

Fig. 11.1 A typical ACE inhibition plot for captopril. Concentration range, 6.7–66.7 μM; substrate, 0.1 mL hippuryl-histidyl-leucine (5 mM); buffer, 100 mM phosphate buffer (pH 8.3) containing 300 mM NaCl; incubation period, 30 min; temperature, 37°C. IC_{50} value – 0.060±0.006 mM

matic reactions, proportions and nature of the esters formed and ACE inhibitory activities for these compounds.

The compounds were characterised by two-dimensional Heteronuclear Single Quantum Coherence Transfer (2D-HSQCT) NMR spectra recorded for the samples. From NMR, it was confirmed that mono- and diesters in different proportions were detected (Tables 11.2 and 11.3). In some cases like L-valyl-maltose, L-valyl-D-mannitol, L-leucyl-D-fructose, L-leucyl-D-mannitol and L-isoleucyl-D-glucose, L-prolyl-D-glucose, L-prolyl-D-ribose, L-prolyl-lactose, L-phenylalanyl-D-fructose, L-tryptophanyl-D-fructose, L-histidyl-D-fructose and L-histidyl-D-mannitol, only monoesters were found to be formed. A 1-*O*-monoester was formed in case of L-valyl-D-fructose, L-valyl-D-mannitol, L-leucyl-D-sorbitol, L-isoleucyl-D-fructose, L-prolyl-D-fructose, L-phenylalanyl-D-fructose, L-phenylalanyl-D-mannitol, L-tryptophanyl-D-fructose and L-histidyl-D-mannitol. A 2-*O*-monoester was found to be formed in case of L-alanyl-β-D-glucose, L-valyl-D-glucose, L-leucyl-D-glucose and L-isoleucyl-maltose, L-prolyl-D-glucose, L-prolyl-D-ribose, L-phenylalanyl-D-glucose, L-phenylalanyl-D-galactose, L-tryptophanyl-D-glucose and L-histidyl-D-glucose. A 3-*O*-monoester was found to be formed in case of L-alanyl-β-D-glucose, L-valyl-D-glucose, L-leucyl-D-glucose and L-isoleucyl-D-glucose. A 6′-*O*-monoester was found to be formed in case of L-alanyl-lactose, L-valyl-maltose and L-isoleucyl-maltose. All the esters invariable showed formation of 6-*O*-monoester except L-alanyl-D-ribose, L-leucyl-D-ribose, L-isoleucyl-D-ribose and L-prolyl-D-ribose where the primary C-5 hydroxyl group reacted to form 5-*O*-ester. Diesters such as 1,6-di-*O*-, 2,6-di-*O*-, 3,6-di-*O*-, 3,5-di-*O*- and 6,6′-di-*O*- were found to be formed in case of L-alanyl-β-D-glucose, L-alanyl-D-ribose, L-alanyl-lactose, L-valyl-D-glucose, L-valyl-D-fructose, L-leucyl-D-glucose, L-leucyl-D-ribose, L-isoleucyl-D-fructose and L-isoleucyl-D-ribose. Similarly *1,6-di-O-, 2,6-di-O-, 3,5-di-O-* and *6,6′-di-O-* were found to be formed in case of L-prolyl-D-fructose, L-phenylalanyl-D-glucose, L-phenylalanyl-D-galactose, L-phenylalanyl-lactose, L-phenylalanyl-D-mannitol, L-tryptophanyl-D-glucose and L-histidyl-D-glucose. It was not

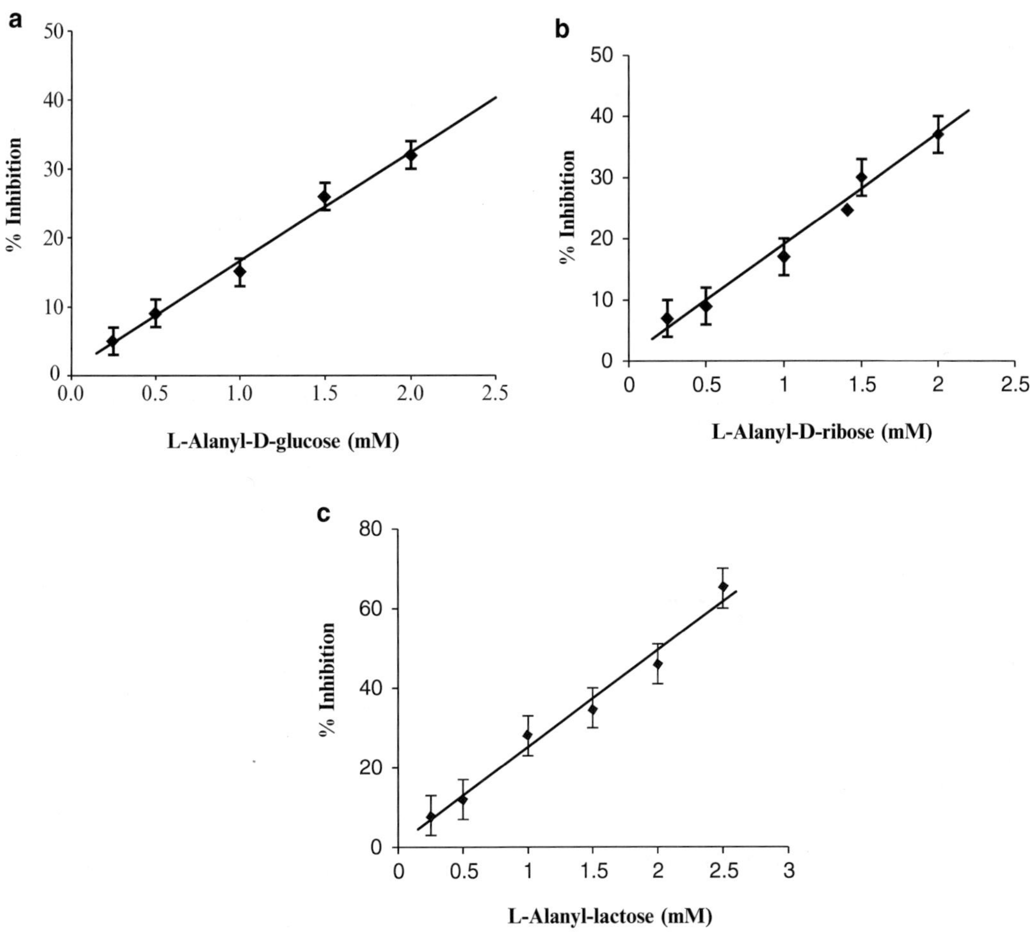

Fig. 11.2 ACE inhibition plots for L-alanyl esters of carbohydrates. (**a**) L-alanyl-D-glucose – concentration range, 0.25–2.0 mM; substrate, 0.1 mL hippuryl-histidyl-leucine (5 mM); buffer, 100 mM phosphate buffer (pH 8.3) containing 300 mM NaCl; incubation period, 30 min; temperature, 37°C. (**b**) L-alanyl-D-ribose – concentration range, 0.25–2.0 mM. (**c**) L-alanyl-D-lactose – concentration range, 0.25–2.5 mM

possible to separate the individual esters from their reaction mixtures even through chromatography on Sephadex G-10 or Bio Gel P2. Thus, the activities described are for the mixtures of these mono- and diesters. From NMR, it was confirmed that mono- and diesters in different proportions were detected (Table 11.3).

Among the esters, L-isoleucyl-D-glucose (0.7 ± 0.07 mM) was found to exhibit the best inhibitory activity. With increase in alkyl side chain branching, D-glucose esters of L-alanine (3.1 ± 0.30 mM), L-valine (6.0 ± 0.59 mM), L-leucine (2.8 ± 0.27 mM) and L-isoleucine (0.7 ± 0.07 mM) showed better inhibition (lesser IC_{50} values) than the other esters, which could be directly correlated to increase in hydrophobicity (Table 11.2). IC_{50} values ≤1.0 mM were detected for L-valyl-D-mannitol (1.0 ± 0.09 mM), L-isoleucyl-D-glucose (0.7 ± 0.07 mM), L-isoleucyl-D-fructose (0.9 ± 0.09 mM), L-isoleucyl-maltose (0.9 ± 0.09 mM) and L-leucyl-D-fructose (0.9 ± 0.08 mM). Similarly, the best IC_{50} values ≤1.0 mM were obtained for L-phenylalanyl-D-glucose (1.0 ± 0.09 mM), L-tryptophanyl-D-fructose (0.9 ± 0.09 mM) and L-histidyl-D-fructose (0.9 ± 0.09 mM). Among aminoacyl esters tested for ACE inhibition activity, L-phenylalanyl-D-glucose (1.0 ± 0.09 mM), L-tryptophanyl-D-fruc-

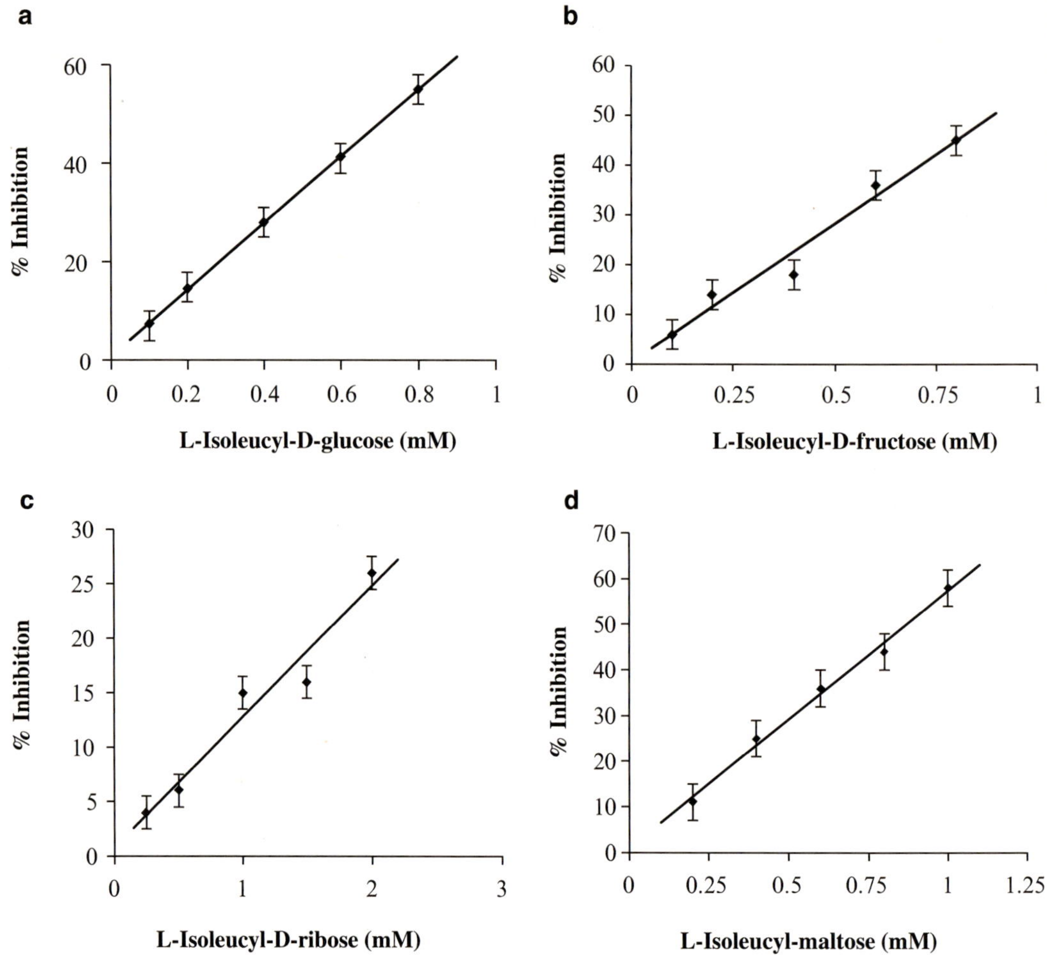

Fig. 11.5 ACE inhibition plots for L-isoleucyl esters of carbohydrates. (**a**) L-isoleucyl-D-glucose – concentration range, 0.2–0.8 mM; substrate, 0.1 mL HHL (5 mM); buffer, 100 mM phosphate buffer (pH 8.3) containing 300 mM NaCl; incubation period, 30 min; temperature, 37°C. (**b**) L-isoleucyl-D-fructose – concentration range, 0.2–0.8 mM; (**c**) L-isoleucyl-D-ribose – concentration range, 0.25–2.0 mM. (**d**) L-isoleucyl-maltose – concentration range, 0.2–1.0 mM

11.2 ACE Inhibition of Glycosides

Glycosides from the leaves of *Abeliophyllum distichum* like acteoside, isoacteoside, rutin and hirsutin moderately inhibited the angiotensin I-converting enzyme activity (Hyuncheol et al. 2003). Glycosides like 3-*O*-methyl crenatoside from *Microtoena prainiana* also showed more than 30% ACE inhibitory activity. Phenylpropanoid glycosides from *Clerodendron trichotomum* such as acteoside, leucosceptoside A, martynoside, acteoside isomer and isomartynoside also showed ACE inhibitory effect (Dae-Gill et al. 2003).

Phenolic glycosides found in a variety of fruits, vegetables and other food materials have been studied extensively for their antioxidant properties (Moon and Terao 1998; Moon et al. 2007). Antioxidative action is reported to protect living organisms from oxidative damages, resulting in the prevention of various diseases such as cancer, cardiovascular diseases, diabetes and aging (Azuma et al. 1999). Flavonols and their glycosides protect red blood cells against free

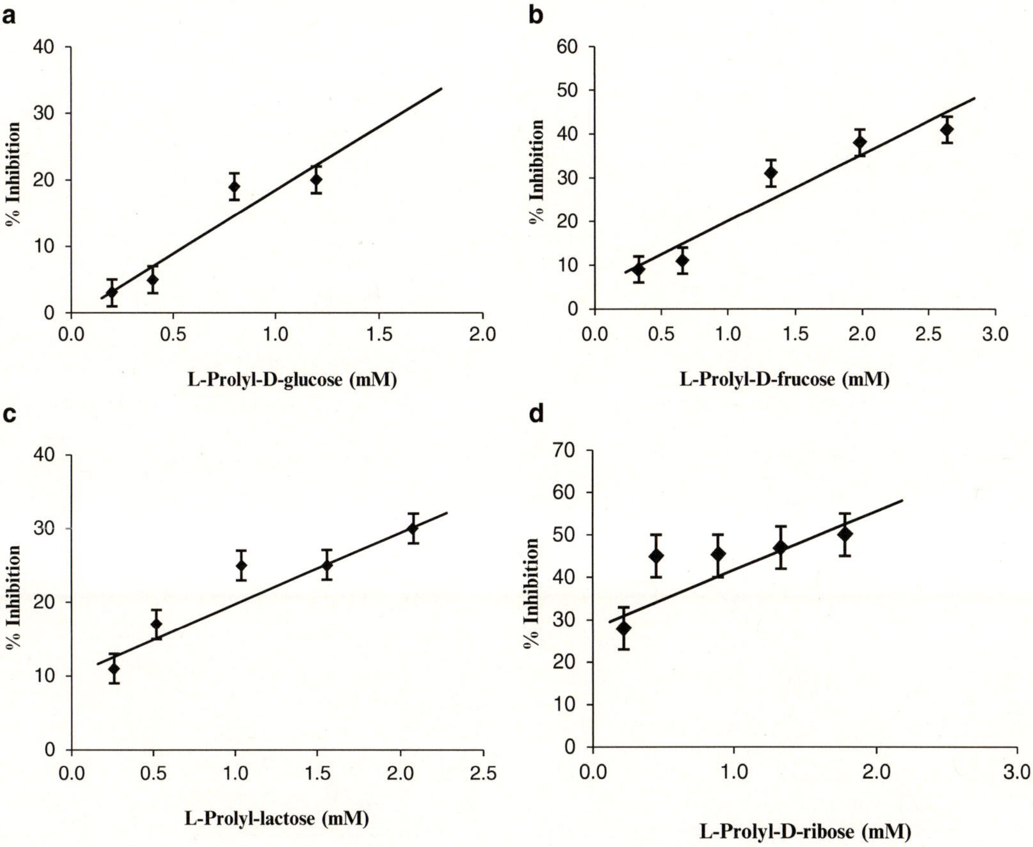

Fig. 11.6 ACE inhibition plots for L-prolyl esters of carbohydrates. (**a**) L-prolyl-D-glucose – concentration range, 0.2–1.6 mM; substrate, 0.1 mL HHL (5 mM); buffer, 100 mM phosphate buffer pH 8.3 containing 300 mM NaCl; incubation period, 30 min; temperature, 37°C. (**b**) L-prolyl-D-fructose – concentration range, 0.33–2.64 mM. (**c**) L-prolyl-lactose – concentration range, 0.2–1.6 mM. (**d**) L-prolyl-D-ribose – concentration range, 0.2–1.8 mM

radical-induced oxidative haemolysis (Dai et al. 2006). The key role of phenols as antioxidants stems from the presence of hydroxyl groups attached to their aromatic rings, which enable them to scavenge free radicals (Kefalas et al. 2003; Villano et al. 2007). The present work describes the ACE inhibition and antioxidant activities of the synthesised glycosides.

Several glycosides were tested for the ACE inhibitory activities. ACE was isolated from pig lung. The enzymatic reactions were carried out under optimised conditions worked out for these reactions. The enzymatic procedure employed unprotected and unactivated alcohols, phenols and carbohydrates. ACE inhibition activity of the above-mentioned glycosides of carbohydrates was determined by the Cushman and Cheung method (1971). Since hippuryl-L-histidyl-L-leucine (HHL) mimics the carboxyl dipeptide of angiotensin I, it has been used as the substrate for screening ACE inhibitors. Here also, eugenyl-D-glycoside showed inhibition of protease in ACE (Table 11.4).

Glycosides of *n*-octanol, phenolic glycosides of guaiacol, eugenol, curcumin, vanillin, N-vanillyl-nonanamide, DL-dopa, dopamine and vitamin glycosides of retinol, thiamin, riboflavin, pyridoxine, ergocalciferol, cholecalciferol and α-tocopherol were tested for ACE inhibition (Tables 11.5 and 11.6 and 11.7, Vasudeva, Kamath et al. 2006; Lohith et al. 2006; Einstein Charles et al. 2009). A few glycosides were also tested for antioxidant activities. Antioxidant

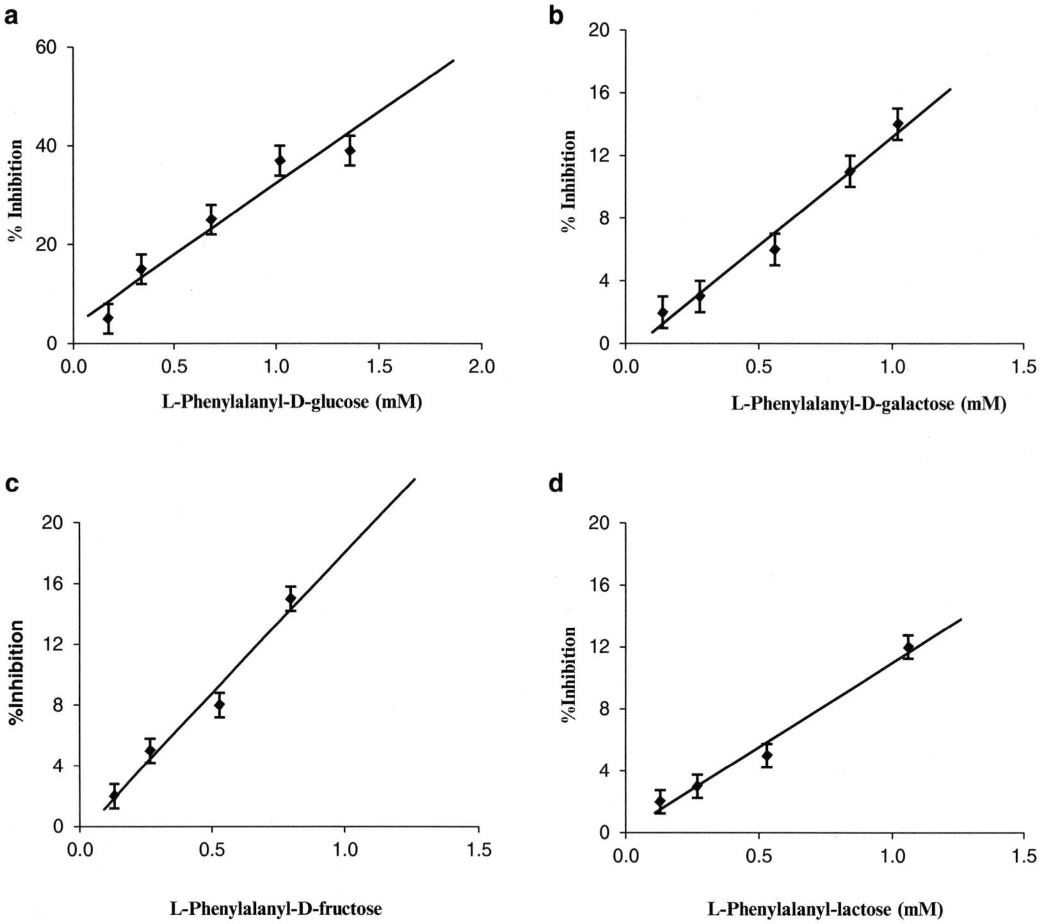

Fig. 11.7 ACE inhibition plots for L-phenylalanyl esters of carbohydrates (**a**) L-phenylalanyl-D-glucose – concentration range, 0.17–1.36 mM; substrate, 0.1 mL hippuryl-histidyl-leucine (5 mM); buffer, 100 mM phosphate buffer pH 8.3 containing 300 mM NaCl; incubation period, 30 min; temperature, 37°C. (**b**) L-phenylalanyl-D-galactose – concentration range, 0.14–1.02 mM. (**c**) L-phenylalanyl-D-fructose – concentration range, 0.13–1.36 mM. (**d**) L-phenylalanyl-lactose – concentration range, 0.13–1.06 mM

activities of few glycosides prepared are also shown in Table 11.8.

Typical ACE inhibition plot for captopril, which showed an IC_{50} value of 0.060 ± 0.006 mM, is shown in Fig. 11.10. Typical ACE inhibition plots for all the tested glycosides such as *n*-octyl glycosides (Fig. 11.11), guaiacyl glycosides (Fig. 11.12), eugenyl glycosides (Fig. 11.13), curcuminyl and α-tocopheryl glycosides (Fig. 11.14) are shown 4-*O*-(D-galactopyranosyl) vanillin (Fig. 11.15), 4-*O*-(D-sorbitol)vanillin (Fig. 11.15), 4-*O*-(α-D-ribofuranosyl)N-vanillyl-nonanamide (Fig. 11.15), 4-*O*-(α-D-gluco pyranosyl-(1′→4)β-D-glucopyranosyl) N-vanillyl-nonanamide (Fig. 11.15), 1,7-*O*-(bis-D-galactopyranosyl)curcumin (Fig. 11.15), 1,7-*O*-(bis-β-D-galactopyranosyl-(1′→4)D-glucopyranosyl) curcumin (Fig. 11.15), DL-3-hydroxy-4-*O*-(D-gluco pyranosyl)phenylalanine (Fig. 11.16), DL-3-hydroxy-4-*O*-(6-D-sorbitol) phenyl alanine (Fig. 11.16), 3-hydroxy-4-*O*-(D-glucopyranosyl)phenylethylamine (Fig. 11.16),

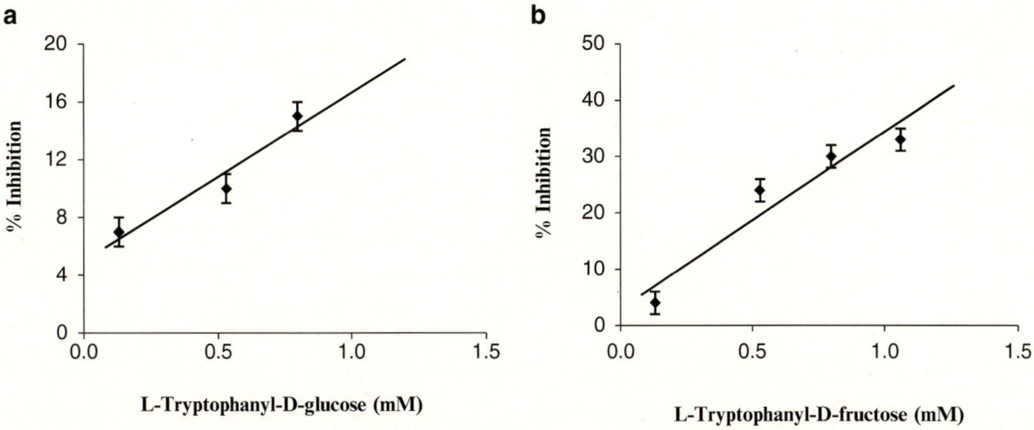

Fig. 11.8 ACE inhibition plots for L-tryptophanyl esters of carbohydrates (**a**) L-tryptophanyl-D-glucose – concentration range, 0.13–1.06 mM; substrate, 0.1 mL hippuryl-histidyl-leucine (5 mM); buffer, 100 mM phosphate buffer pH 8.3 containing 300 mM NaCl; incubation period, 30 min; temperature, 37°C. (**b**) L-tryptophanyl-D-fructose – concentration range, 0.13–1.06 mM, concentration range – 0.2–1.2 mM

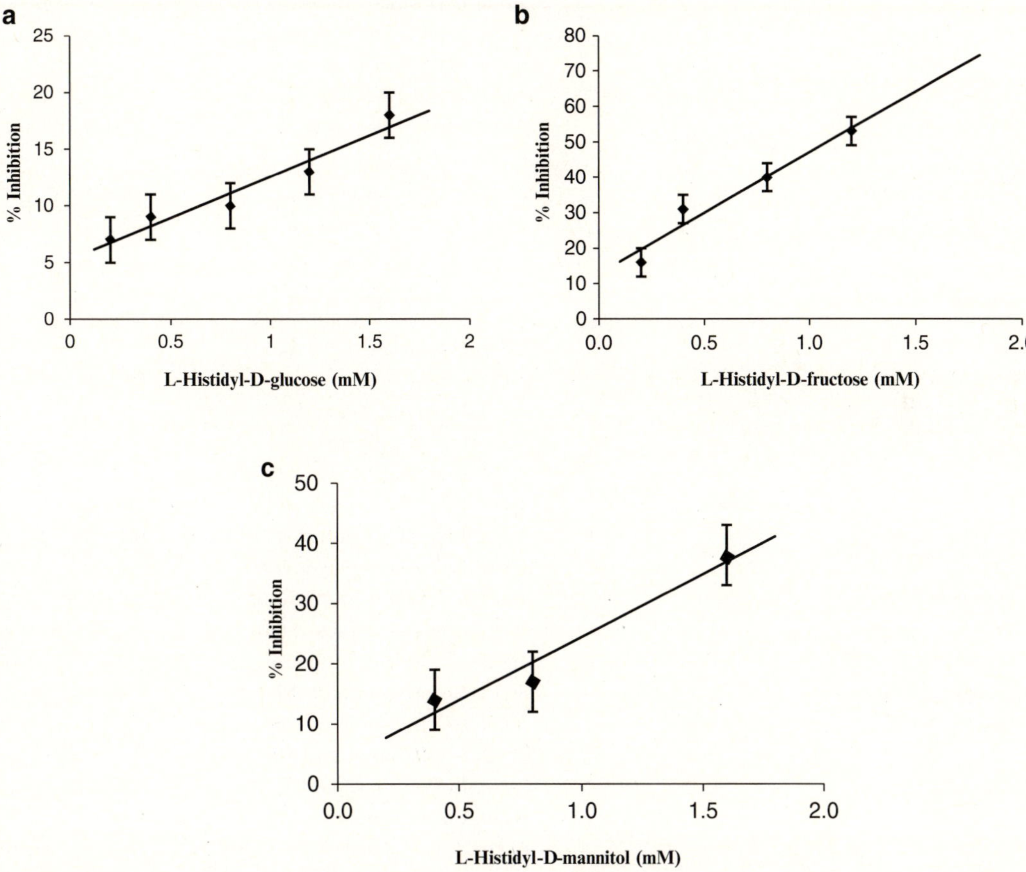

Fig. 11.9 ACE inhibition plots for L-histidyl esters of carbohydrates (**a**) L-histidyl-D-glucose – concentration range, 0.2–1.6 mM; substrate, 0.1 mL hippuryl-histidyl-leucine (5 mM); buffer, 100 mM phosphate buffer pH 8.3 containing 300 mM NaCl; incubation period, 30 min; temperature, 37°C. (**b**) L-histidyl-D-fructose – concentration range, 0.2–1.6 mM. (**c**) L-histidyl-D-mannitol, concentration range – 0.2–1.6 mM

Table 11.2 IC_{50} values for ACE inhibition by alkyl side chain containing aminoacyl esters of carbohydrates[a]

Aminoacyl ester of carbohydrates	Conversion yield (%)[b]	Products (% proportion)[c]	IC_{50} value (mM)[d]
L-Alanyl-β-D-glucose	30	2-*O*-L-Alanyl-β-D-glucose (47)	3.1 ± 0.30
		3-*O*-L-Alanyl-β-D-glucose (12)	
		6-*O*-L-Alanyl-β-D-glucose (20)	
		2,6-di-*O*-L-Alanyl-β-D-glucose (15)	
		3,6-di-*O*-L-Alanyl-β-D-glucose (6)	
L-Alanyl-D-ribose	48	3-*O*-L-Alanyl-D-ribose (16)	2.7 ± 0.25
		5-*O*-L-Alanyl-D-ribose (32)	
		3,5-di-*O*-L-Alanyl-D-ribose (52)	
L-Alanyl-lactose	20	6-*O*-L-Alanyl-lactose (34)	2.0 ± 0.20
		6′-*O*-L-Alanyl-lactose (34)	
		6,6′-di-*O*-L-Alanyl-lactose (32)	
L-Valyl-D-glucose	68	2-*O*-L-Valyl-D-glucose (10)	6.0 ± 0.59
		3-*O*-L-Valyl-D-glucose (12)	
		6-*O*-L-Valyl-D-glucose (31)	
		2,6-di-*O*-L-Valyl-D-glucose (23)	
		3,6-di-*O*-L-Valyl-D-glucose (24)	
L-Valyl-D-fructose	34	1-*O*-L-Valyl-D-fructose (29)	2.8 ± 0.28
		6-*O*-L-Valyl-D-fructose (34)	
		1,6-di-*O*-L-Valyl-D-fructose (37)	
L-Valyl-maltose	42	6-*O*-L-Valyl-maltose (49)	3.1 ± 0.30
		6′-*O*-L-Valyl-maltose (51)	
L-Valyl-D-mannitol	56	1-*O*-L-Valyl-D-mannitol	1.0 ± 0.09
L-Leucyl-D-glucose	43	2-*O*-L-Leucyl-D-glucose (17)	2.8 ± 0.27
		3-*O*-L-Leucyl-D-glucose (20)	
		6-*O*-L-Leucyl-D-glucose (48)	
		2,6-di-*O*-L-Leucyl-D-glucose (8)	
		3,6-di-*O*-L-Leucyl-D-glucose (7)	
L-Leucyl-D-fructose	48	6-*O*-L-Leucyl-D-fructose	0.9 ± 0.08
L-Leucyl-D-ribose	38	3-O-L-Leucyl-D-ribose (16)	1.5 ± 0.14
		5-O-L-Leucyl-D-ribose (32)	
		3,5-di-*O*-L-Leucyl-D-ribose (52)	
L-Leucyl-D-sorbitol	60	1-*O*-L-Leucyl-D-sorbitol	2.7 ± 0.25
L-Isoleucyl-D-glucose	47	3-*O*-L-Isoleucyl-D-glucose (42)	0.7 ± 0.07
		6-*O*-L-Isoleucyl-D-glucose (58)	
L-Isoleucyl-D-fructose	42	1-O-L-Isoleucyl-D-fructose (36)	0.9 ± 0.09
		6-O-L-Isoleucyl-D-fructose (30)	
		1,6-di-*O*-L-Isoleucyl-D-fructose (34)	
L-Isoleucyl-D-ribose	53	3-*O*-L-Isoleucyl-D-ribose (52)	3.8 ± 0.37
		5-*O*-L-Isoleucyl-D-ribose (20)	
		3,5-di-*O*-L-Isoleucyl-D-ribose (28)	
L-Isoleucyl-maltose	54	2-*O*-L-Isoleucyl-maltose (38)	0.9 ± 0.09
		6-*O*-L-Isoleucyl-maltose (40)	
		6′-*O*-L-Isoleucyl-maltose (22)	

[a]Respective amino acids and carbohydrates as controls showed no ACE inhibition activity

[b]Conversion yields were from HPLC within ±10% errors in HPLC yield measurements

[c]Product proportions determined from ^{13}C, 2D HSQCT NMR C6 peak areas (C5 cross peaks in case of ribose) or their cross peaks

[d]IC_{50} values compared to that of captopril 0.060 ± 0.006 mM determined by Cushman and Cheung method

Table 11.3 IC_{50} values for ACE inhibition by aminoacyl esters of carbohydrates[a]

Aminoacyl ester of carbohydrates	Conversion yield (%)[b]	Products (% proportion)[c]	IC_{50} value (mM)[d]
L-Prolyl-D-glucose	62	*2-O*-L-Prolyl-D-glucose (26)	1.7±0.17
		3-O-L-Prolyl-D-glucose (26)	
		6-O-L-Prolyl-D-glucose (48)	
L-Prolyl-D-fructose	61	*1-O*-L-Prolyl-D-fructose (31)	4.4±0.43
		6-O-L-Prolyl-D-fructose (42)	
		1,6-di-O-L-Prolyl-D-fructose (27)	
L-Prolyl-D-ribose	41	*3-O*-L-Prolyl-D-ribose (35)	2.0±0.19
		5-O-L-Prolyl-D-ribose (65)	
L-Prolyl-lactose	68	6-*O*-L-Prolyl-lactose (58)	1.6±0.15
		6′-*O*-L-Prolyl-lactose (42)	
L-Phenylalanyl-D-glucose	79	*2-O*-L-Phenylalanyl-D-glucose (19)	1.0±0.09
		3-O-L-Phenylalanyl-D-glucose (23)	
		6-O-L-Phenylalanyl-D-glucose (25)	
		2,6-di-*O*-L-Phenylalanyl-D-glucose (17)	
		3,6-di-O-L-Phenylalanyl-D-glucose (16)	
L-Phenylalanyl-D-galactose	45	*2-O*-L-Phenylalanyl-D-galactose (32)	4.6±0.45
		3-O-L-Phenylalanyl-D-galactose (20)	
		6-O-L-Phenylalanyl-D-galactose (19)	
		2,6-di-O-L-Phenylalanyl-D-galactose (16)	
		3,6-di-O-L-Phenylalanyl-D-galactose (13)	
L-Phenylalanyl-D-fructose	50	*1-O*-L-Phenylalanyl-D-fructose (72)	13.6±1.35
		6-O-L-Phenylalanyl-D-fructose (28)	
L-Phenylalanyl-lactose	61	*6-O*-L-Phenylalanyl-lactose (42)	7.8±0.77
		6′-O-L-Phenylalanyl-lactose (31)	
		6,6′-di-O-L-Phenylalanyl-lactose (27)	
L-Phenylalanyl-D-mannitol	43	*1-O*-L-Phenylalanyl-D-mannitol (62)	2.6±0.25
		1,6-di-O-L-Phenylalanyl-D-mannitol (38)	
L-Tryptophanyl-D-glucose	42	*2-O*-L-Tryptophanyl-D-glucose (22)	7.4±0.73
		3-O-L-Tryptophanyl-D-glucose (21)	
		6-O-L-Tryptophanyl-D-glucose (38)	
		2,6-di-O-L-Tryptophanyl-D-glucose (10)	
		3,6-di-O-L-Tryptophanyl-D-glucose (9)	
L-Tryptophanyl-D-fructose	18	*1-O*-L-Tryptophanyl-D-fructose (45)	0.9±0.09
		6-O-L-Tryptophanyl-D-fructose (55)	

(continued)

Table 11.3 (continued)

Aminoacyl ester of carbohydrates	Conversion yield (%)[b]	Products (% proportion)[c]	IC_{50} value (mM)[d]
L-Histidyl-D-glucose	42	*2-O*-L-Histidyl-D-glucose (25)	3.5±0.34
		3-O-L-Histidyl-D-glucose (24)	
		6-O-L-Histidyl-D-glucose (28)	
		2,6-di-O-L-Histidyl-D-glucose (12)	
		3,6-di-O-L-Histidyl-D-glucose (11)	
L-Histidyl-D-fructose	58	*6-O*-L-Histidyl-D-fructose	0.9±0.09
L-Histidyl-D-mannitol	8	*1-O*-L-Histidyl-D-mannitol	1.7±0.16

[a]Respective amino acids and carbohydrates as controls showed no ACE inhibition activity
[b]Conversion yields were from HPLC with ±10–15% errors in HPLC yield measurements
[c]Product proportions determined from ^{13}C, 2D HSQCT NMR C6 peak areas (C5 cross peaks in case of ribose) or their cross peaks
[d]IC_{50} values compared to that of captopril 0.060±0.006 mM determined by Cushman and Cheung (1969) method

Table 11.4 Inhibition of protease in ACE by eugenyl-α-D-glucoside[a]

System	Protease activity (min^{-1} mg^{-1} enzyme protein)[b]	Percentage of protease activity with respect to ACE activity[c]
Control: ACE – 0.5 mL + 0.5 mL of 0.6% haemoglobin + 0.5 mL buffer	0.0436	13.3
Eugenyl-α-D-glucoside: 0.5 mL glycoside + ACE – 0.5 mL + 0.5 mL of 0.6% haemoglobin	0.0292	8.2

[a]Conditions: ACE – 0.5 mL (0.5 mg). All the solutions were prepared in 0.1 M pH 7.5 tris-HCl: incubation period, 30 min; temperature, 37°C; 0.5 mL of 10% trichloroacetic acid added to arrest the reaction; Blank performed without enzyme and glycoside; absorbance measured at 440 nm; eugenyl-α-D-glucoside – 0.5 mL of 0.8 mM
[b]Average absorbance values from three individual experiments
[c]Percentage protease activity with respect to an ACE activity of 0.327 μmol (min mg protein)$^{-1}$

Table 11.5 IC_{50} values for ACE inhibition by glycosides[a]

Glycoside	Yield[b] (%)	Products formed (% proportions)[c]	IC_{50} value (mM)[d]
n-Octyl-D-glucoside	46	C1α-glucoside (63), C1β-glucoside (25), C6-alkylated (12)	1.0±0.09
n-Octyl maltoside	22	C1α-maltoside	1.5±0.13
n-Octyl sucrose,	13	C1-*O*-alkylated (44), C6-*O*-alkylated (56)	1.7±0.15
Guaiacyl-α-D-glucoside	52	C1α glucoside (52), C6-*O*-arylated (48)	3.7±0.36
Guaiacyl-α-D-galactoside	17	C1α-galactoside (95), C6-*O*-arylated (5)	2.3±0.22
Eugenyl-α-D-glucoside	32	C1α glucoside (53), C6-*O*-arylated (47)	0.5±0.04
Eugenyl-α-D-mannoside	8	C1α-mannoside	5.3±0.51
Eugenyl maltoside	17	C1α-maltoside (52), 6-*O*-arylated (28), C6″-*O*-arylated (20)	0.7±0.06
Eugenyl sucrose	7	C1-*O*-arylated (45), C6-*O*-arylated (35), C6″-*O*-arylated (20)	1.7±0.15

(continued)

Table 11.5 (continued)

Glycoside	Yield[b] (%)	Products formed (% proportions)[c]	IC_{50} value (mM)[d]
Eugenyl-D-mannitol	7	C1-*O*-arylated	2.1±0.21
Curcuminyl-bis-α-D-glucoside	48	C1α-glucoside (62), C6-*O*-arylated (38)	1.5±0.13
Curcuminyl-bis-α-D-mannoside	9	C1α-mannoside	1.0±0.09
Curcuminyl-bis-maltoside	19	C1α-maltoside (37), C6-*O*-arylated (36), C6″-*O*-arylated (27)	1.2±0.11
Curcuminyl-bis-sucrose	19	C1-*O*-arylated (12), C6-*O*-arylated (70), C6″-*O*-arylated (18)	1.8±0.17
Curcuminyl-bis-D-mannitol	14	C1-*O*-arylated	1.8±0.17
α-Tocopheryl-α-D-glucoside	52	C1α-glucoside	1.2±0.11

[a]Respective alcohols, phenols and carbohydrates as controls did not show any ACE inhibition activities; nonreducing sugar unit carbons of disaccharide are double primed

[b]Conversion yields were from HPLC

[c]Product proportions determined from 2D-HSQCT NMR C1/C6 cross-peak areas

[d]IC_{50} values compared to that of captopril 0.060±0.005 mM determined by Cushman and Cheung method

Table 11.6 Antioxidant and angiotensin-converting enzyme inhibitory activities of various phenolic and vitamin glycosides[a]

Compounds	Antioxidant activity IC_{50} value (mM)[b]	ACE inhibition IC_{50} value (mM)[c]
Butylated hydroxy anisole (BHA)	0.046±0.002	–
Enalapril	–	0.071±0.004
Vanillin 1	1.65±0.08	1.87±0.09
4-*O*-(D-Glucopyranosyl)vanillin	2.66±0.13	1.11±0.06
4-*O*-(β-D-Glucopyranosyl)vanillin	0.9±0.45	0.61±0.03
4-*O*-(α-D-Galactopyranosyl)vanillin	1.62±0.08	1.12±0.06
4-*O*-(D-Galactopyranosyl)vanillin	1.18±0.06	0.61±0.03
4-*O*-(α-D-Mannopyranosyl)vanillin	1.55±0.08	1.02±0.05
4-*O*-(D-Mannopyranosyl)vanillin	1.08±0.05	2.3±0.1
4-*O*-(α-D-Glucopyranosyl-(1′→4)D-glucopyranosyl)vanillin	1.17±0.06	1.63±0.08
4-*O*-(α-D-Glucopyranosyl-(1′→4)β-D-glucopyranosyl)vanillin	2.64±0.13	1.89±0.09
4-*O*-(D-Fructofuranosyl-(2→1′)α-D-glucopyranosyl)vanillin	1.23±0.06	15.7±0.79
4-*O*-(β-D-Galactopyranosyl-(1′→4)β-D-glucopyranosyl)vanillin	0.8±0.04	0.92±0.05
4-*O*-(D-Sorbitol)vanillin	1.24±0.06	0.81±0.04
N-Vanillyl-nonanamide	0.054±0.003	1.53±0.08
4-*O*-(D-Glucopyranosyl)N-vanillyl-nonanamide	1.18±0.06	1.33±0.07
4-*O*-(β-D-Glucopyranosyl)N-vanillyl-nonanamide	1.4±0.07	3.33±0.17
4-*O*-(D-Galactopyranosyl)N-vanillyl-nonanamide	2.9±0.15	2.05±0.1
4-*O*-(β-D-Galactopyranosyl)N-vanillyl-nonanamide	0.94±0.05	2±0.1
4-*O*-(β-D-Mannopyranosyl)N-vanillyl-nonanamide	1.14±0.06	2.57±0.13
4-*O*-(α-D-Ribofuranosyl)N-vanillyl-nonanamide	0.98±0.05	1±0.05
4-*O*-(α-D-Glucopyranosyl-(1′→4)D-glucopyranosyl)N-vanillyl-nonanamide	0.8±0.04	2.41±0.12

(continued)

Table 11.6 (continued)

Compounds	Antioxidant activity IC_{50} value (mM)[b]	ACE inhibition IC_{50} value (mM)[c]
4-*O*-(α-D-Glucopyranosyl-(1′→4)β-D-glucopyranosyl)N-vanillyl-nonanamide	0.75±0.04	0.8±0.04
4-*O*-(β-D-Galactopyranosyl-(1′→4)β-D-glucopyranosyl)N-vanillyl-nonanamide	1.04±0.05	1.82±0.09
Curcumin	0.053±0.003	0.83±0.04
1,7-*O*-(Bis-β-D-glucopyranosyl)curcumin	0.8±0.04	1.09±0.05
1,7-*O*-(Bis-D-galactopyranosyl)curcumin	0.92±0.05	0.88±0.04
1,7-*O*-(Bis-D-mannopyranosyl)curcumin	0.75±0.04	1.9±0.1
1,7-*O*-(Bis-β-D-galactopyranosyl-(1′→4)D-glucopyranosyl)curcumin	0.95±0.05	0.67±0.03
DL-Dopa	0.045±0.002	0.6±0.03
DL-3-Hydroxy-4-*O*-(D-glucopyranosyl)phenylalanine	1.11±0.06	1.2±0.06
DL-3-Hydroxy-4-*O*-(D-glucopyranosyl)phenylalanine	0.98±0.05	1.26±0.06
DL-3-Hydroxy-4-*O*-(D-galactopyranosyl)phenylalanine	2.26±0.11	1.71±0.08
DL-3-Hydroxy-4-*O*-(β-D-mannopyranosyl)phenylalanine	1.13±0.06	1.87±0.09
DL-3-Hydroxy-4-*O*-(β-D-galactopyranosyl-(1′→4)β-D-glucopyranosyl)phenylalanine	0.9±0.05	3.33±0.17
DL-3-Hydroxy-4-*O*-(6-D-sorbitol)phenylalanine	1.86±0.09	0.56±0.03
DL-Dopa-D-mannitol	1.9±0.09	1.58±0.08
Dopamine	0.04±0.002	1.93±0.1
3-Hydroxy-4-*O*-(D-glucopyranosyl)phenylethylamine	1.45±0.07	1.27±0.06
3-Hydroxy-4-*O*-(β-D-glucopyranosyl)phenylethylamine	0.98±0.05	2.38±0.12
3-Hydroxy-4-*O*-(D-galactopyranosyl)phenylethylamine	0.93±0.05	2.38±0.12
3-Hydroxy-4-*O*-(D-mannopyranosyl)phenylethylamine	1.8±0.09	1.93±0.1
Riboflavin	–	1.08±0.05
5-*O*-(D-Glucopyranosyl)riboflavin	–	1.27±0.06
5-*O*-(β-D-Glucopyranosyl)riboflavin	–	1.75±0.09
5-*O*-(D-Galactopyranosyl)riboflavin	–	0.83±0.04
5-*O*-(α-D-Mannopyranosyl)riboflavin	–	2.08±0.1
5-*O*-(D-Mannopyranosyl)riboflavin	–	1.92±0.1
5-*O*-(D-Ribofuranosyl)riboflavin	–	1.11±0.06
5-*O*-(α-D-Glucopyranosyl-(1′→4)D-glucopyranosyl)riboflavin	–	0.8±0.04
5-*O*-(1-D-Fructofuranosyl-(2→1′)α-D-glucopyranosyl)riboflavin	–	1.03±0.05
5-*O*-(β-D-Galactopyranosyl-(1′→4)β-D-glucopyranosyl)riboflavin	–	1.09±0.05
Ergocalciferol	–	1.2±0.06
20-*O*-(D-Glucopyranosyl)ergocalciferol	0.9±0.05	1.17±0.06
α-Tocopherol	0.054±0.003	1.07±0.05
6-*O*-(β-D-Glucopyranosyl)α-tocopherol	1.2±0.06	1.33±0.07
6-*O*-(D-Galactopyranosyl)α-tocopherol	0.72±0.04	2.59±0.13
6-*O*-(D-Mannopyranosyl)α-tocopherol	0.5±0.03	1.8±0.09

[a]Glucosidases catalysed synthesis of phenolic glycosides of vanillin (Table 8.12), N-vanillyl-nonanamide (Table 8.9), curcumin (Table 8.4), DL-dopa (Table 8.18), dopamine (Table 8.23) and vitamin glycosides of riboflavin (Table 8.29), ergocalciferol (Table 8.34) and α-tocopherol (Table 8.38) where the conversion yields and product proportions are shown. Carbohydrates did not show any antioxidant and ACE inhibition activity. Error in measurements is ±5%

[b]Antioxidant activity values determined by DPPH radical scavenging method (Moon and Terao 1998)

[c]ACE activity determined by Cushman and Cheung method (1971)

Table 11.7 IC_{50} values for ACE inhibition by glycosides[a]

Compounds	Product (% proportion)	Amyloglucosidase catalysis (% yield)	β-glucosidase catalysis (% yield)	IC_{50} values (mM)[b]
Enalapril	–	–	–	0.071 ± 0.004
Vanillin	–	–	–	1.87 ± 0.09
Vanillyl-D-glucoside 4-*O*-(β-D-Glucopyranosyl)vanillin		–	10	0.61 ± 0.03
Vanillyl-D-galactoside 4-*O*-(α-D-Galactopyranosyl)vanillin (23) 4-*O*-(β-D-Galactopyranosyl)vanillin (77)		–	6	0.61 ± 0.03
DL-Dopa	–	–	–	0.6 ± 0.03
DL-Dopa-D-sorbitol 3-Hydroxy-4-*O*-(6-D-sorbitol) phenylalanine		12	–	0.56 ± 0.03
Curcumin	–	–	–	0.83 ± 0.04
Curcuminyl-bis-D-glucoside 1,7-*O*-(Bis-β-D-glucopyranosyl) curcumin		–	44	1.09 ± 0.05
Pyridoxine		–	–	1.05 ± 0.06
Pyridoxine-D-glucoside 7-*O*-(α-D-Glucopyranosyl) pyridoxine (39) 7-*O*-(β-D-Glucopyranosyl) pyridoxine (41) 6-*O*-(α-D-Glucopyranosyl) pyridoxine (20)		–	35	0.84 ± 0.04
Thiamin	–	–	–	3.33 ± 0.17
Thiaminyl-D-fructoside 11-*O*-(1-D-Fructofuranosyl) thiamin (54) 11-*O*-(6-D-Fructofuranosyl) thiamin (46)		–	54	0.52 ± 0.03
Riboflavin	–	–	–	1.08 ± 0.05
Riboflavinyl-maltose 5-*O*-(α-D-Glucopyranosyl-(1′→4)α-D-glucopyranosyl)riboflavin (35) 5-*O*-(α-D-Glucopyranosyl-(1′→4)6-D-glucopyranosyl)riboflavin (48) 5-*O*-(α-D-Glucopyranosyl-(1′→4)6′-D-glucopyranosyl)riboflavin (17)		25	–	0.8 ± 0.04

[a]No activity for carbohydrates
[b]Values are an average from two measurements

3-hydroxy-4-*O*-(D-mannopyranosyl)phenylethylamine (Fig. 11.16), 5-*O*-(D-galactopyranosyl)riboflavin (Fig. 11.16), 5-*O*-(α-D-glucopyranosyl-(1′ 4)D-glucopyranosyl)riboflavin (Fig. 11.16), 20-*O*-(D-glucopyranosyl) ergocalciferol (Fig. 11.17), 6-*O*-(β-D-glucopyranosyl) α-tocopherol (Fig. 11.17) and 6-*O*-(D-mannopyranosyl)α-tocopherol (Fig. 11.17) are shown.

The compounds were characterised by two-dimensional Heteronuclear Single Quantum Coherence Transfer (2D-HSQCT) NMR spectra. In case of glycoside syntheses, the major product was the glycosylated product, and relatively lesser amounts of C6-*O*-alkylated or C6-*O*-arylated products were also detected. Thus, the activities described are for the mixtures of these compounds.

Table 11.8 IC_{50} values for antioxidant activities of glycosides[a]

Compound	IC_{50} value (mM)[b]
Butylated hydroxyanisole	0.046 ± 0.002
Vanillin	1.65 ± 0.08
4-*O*-(β-D-Glucopyranosyl)vanillin	0.9 ± 0.05
N-vanillyl-nonanamide	0.054 ± 0.003
4-*O*-(α-D-glucopyranosyl)N-vanillyl-nonanamide	1.18 ± 0.06
DL-Dopa	0.045 ± 0.002
DL-Dopa-D-glucoside	0.98 ± 0.05
3-Hydroxy-4-*O*-(β-D-galactopyranosyl-(1′→4)β-D-glucopyranosyl)phenylalanine	0.9 ± 0.05
Dopamine	0.045 ± 0.002
3-Hydroxy-4-*O*-(β-D-glucopyranosyl)phenylethylamine	0.98 ± 0.05
Dopamine-D-galactoside	0.93 ± 0.05
Curcumin	0.79 ± 0.03
1,7-*O*-(Bis-β-D-glucopyranosyl)curcumin	0.8 ± 0.04
α-Tocopherol	0.054 ± 0.003
6-*O*-(α-D-Glucopyranosyl)α-tocopherol	1.04 ± 0.05
20-*O*-(D-Glucopyranosyl)ergocalciferol	0.9 ± 0.05

[a]No activity for carbohydrates
[b]Values are an average from two measurements

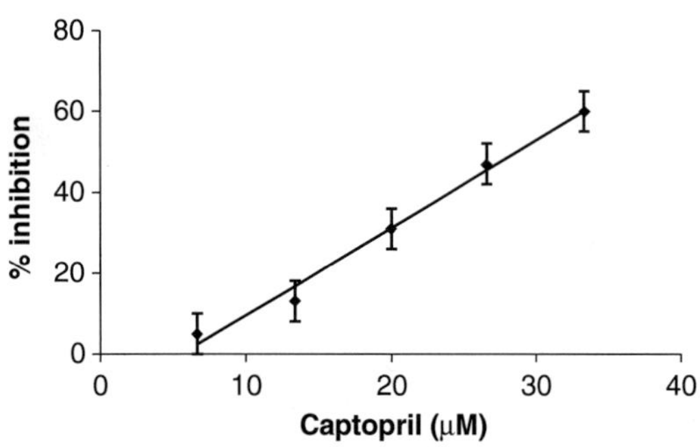

Fig. 11.10 A typical ACE inhibition plot for captopril: concentration range, 6.7–33.3 μM; substrate, 0.1 mL hippuryl-histidyl-leucine (5 mM); buffer, 100 mM phosphate buffer pH 8.3 containing 300 mM NaCl; incubation period, 30 min; temperature, 37°C. IC_{50} value – 0.060 ± 0.006 mM

11.3 Antioxidant Activity of Glycosides

DPPH (2,2-diphenyl-1-picrylhydrazyl) is a highly coloured commercially available radical source, widely used for rough estimation of the ability of antioxidants to trap potentially damaging one-electron oxidants, i.e. the number of DPPH molecules reduced by one molecule of an antioxidant (Potier et al. 1999). Many methods to evaluate the antioxidative activity of specific compounds have been described, but the most widely documented one deals with DPPH radical (Portes et al. 2007; Roche et al. 2005). The radical scavenging efficiency of phenolic and vitamin glycosides tested in this investigation is listed in Table 11.8.

Butylated hydroxy anisole (BHA) was used as a control. The plot obtained is shown in Fig. 11.18. Plots for the antioxidant activity of a few selected glycosides, 4-*O*-(β-D-glucopyranosyl)vanillin (Fig. 11.19), 4-*O*-(β-D-galactopyranosyl-(1′→4)β-D-gluco pyranosyl) vanillin (Fig. 11.19), 4-*O*-(β-D-galactopyranosyl) N-vanillyl-nonanamide (Fig. 11.19), 4-*O*-(α-D-glucopyranosyl-(1′→4)β-D-glucopyranosyl) N-vanillyl-nonanamide (Fig. 11.19), 1,7-*O*-(bis-β-D-glucopyranosyl)curcumin (Fig. 11.19), 1,7-*O*-(bis-D-manno-pyranosyl) curcumin (Fig. 11.19), DL-3-hydroxy-4-*O*-(D-gluco pyranosyl)phenylalanine (Fig. 11.20), DL-3-hydroxy-4-*O*-(β-D-galactopyranosyl-(1′→4)β-D-glucopyranosyl)phenylalanine (Fig. 11.20), 3-hydroxy-4-*O*-(β-D-gluco pyranosyl) phenylethylamine (Fig. 11.20), 3-hydroxy-4-*O*-(D-galactopyranosyl) phenylethylamine (Fig. 11.20),

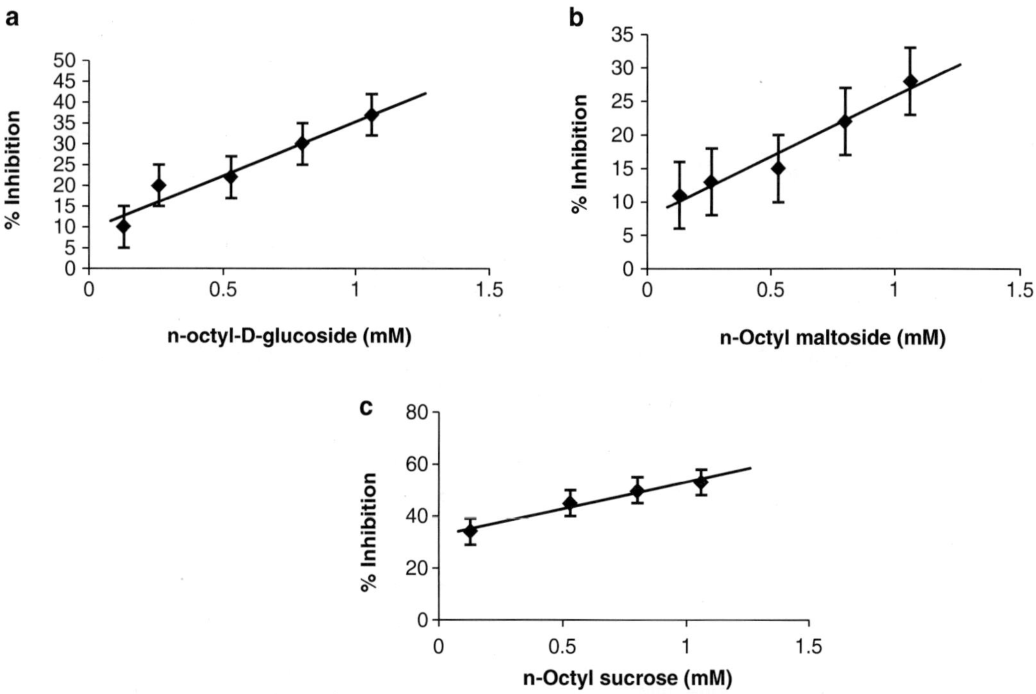

Fig. 11.11 ACE inhibition plots for *n*-octyl glycosides (**a**) *n*-octyl-D-glucoside, (**b**) *n*-octyl maltoside, (**c**) *n*-octyl sucrose. Concentration range, 0.13–1.06 mM; substrate, 0.1 mL hippuryl-histidyl-leucine (5 mM); buffer, 100 mM phosphate buffer pH 8.3 containing 300 mM NaCl; incubation period, 30 min; temperature, 37°C

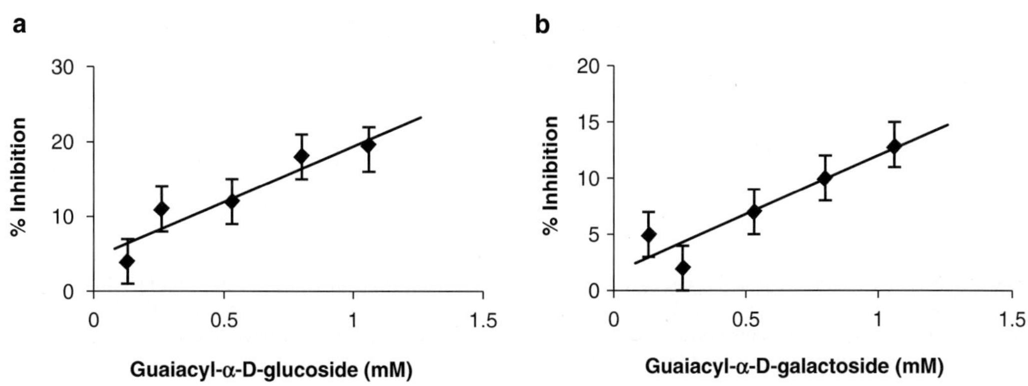

Fig. 11.12 ACE inhibition plots for guaiacyl glycosides (**a**) guaiacyl-α-D-glucoside, (**b**) guaiacyl-α-D-galactoside. Concentration range, 0.13–1.06 mM; substrate, 0.1 mL hippuryl-histidyl-leucine (5 mM); buffer, 100 mM phosphate buffer pH 8.3 containing 300 mM NaCl; incubation period, 30 min; temperature, 37°C

20-*O*-(D-glucopyranosyl) ergocalciferol (Fig. 11.20) and 6-*O*-(D-mannopyranosyl) α-tocopherol (Fig. 11.20), are shown.

Totally, 99 glycosides were tested for antioxidant activity and 49 glycosides tested for angiotensin-converting enzyme (ACE) inhibitory activities. Enzymatic glycosylation produced only monoglycosides, and no diglycosides were detected except curcumin, which showed bis glycosylation. In spite of possessing OH groups at third and fourth positions, DL-dopa and dopamine gave a mixture of 4-OH, 3-OH and 4-*O*-C6-

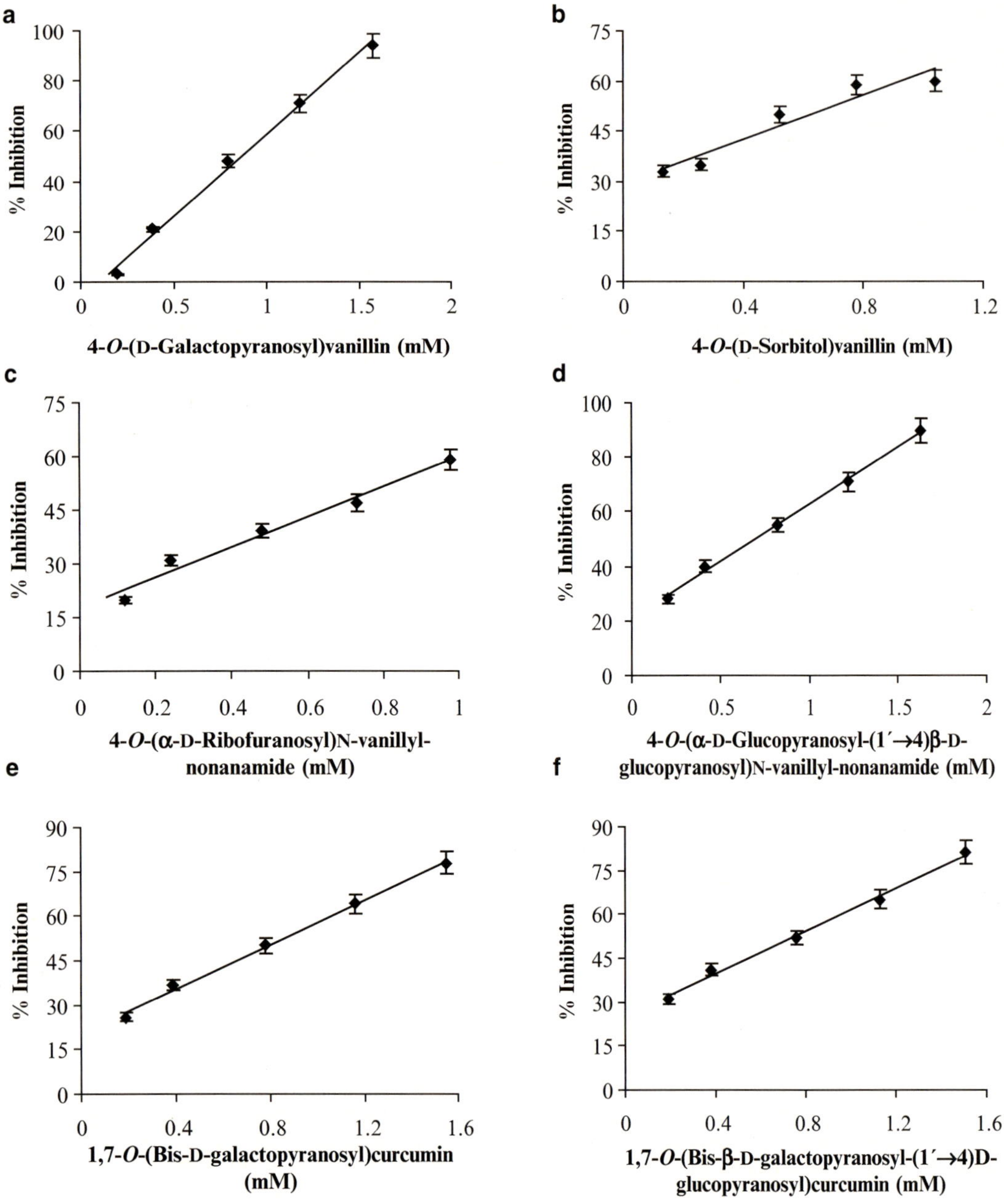

Fig. 11.15 ACE inhibition plots for phenolic glycosides: ACE, 0.1 mL (10 mg in 25-mL stock solution); glycoside concentration range, 0.2–1.8 mM; substrate, 0.1 mL hippuryl-L-histidyl-L-leucine (5 mM); buffer, 100 mM phosphate buffer pH 8.3 containing 0.3 M sodium chloride; incubation period, 30 min; and temperature, 37°C. (**a**) 4-*O*-(D-Galactopyranosyl)vanillin. (**b**) 4-*O*-(D-Sorbitol) vanillin. (**c**) 4-*O*-(α-D-Ribofuranosyl)N-vanillyl-nonanamide. (**d**) 4-*O*-(α-D-Glucopyranosyl-(1′→4)β-D-glucopyranosyl) N-vanillyl-nonanamide. (**e**) 1,7-*O*-(bis-D-Galactopyranosyl) curcumin; and (**f**) 1,7-*O*-(bis-β-D-Galactopyranosyl-(1′→4)D-gluco pyranosyl)curcumin

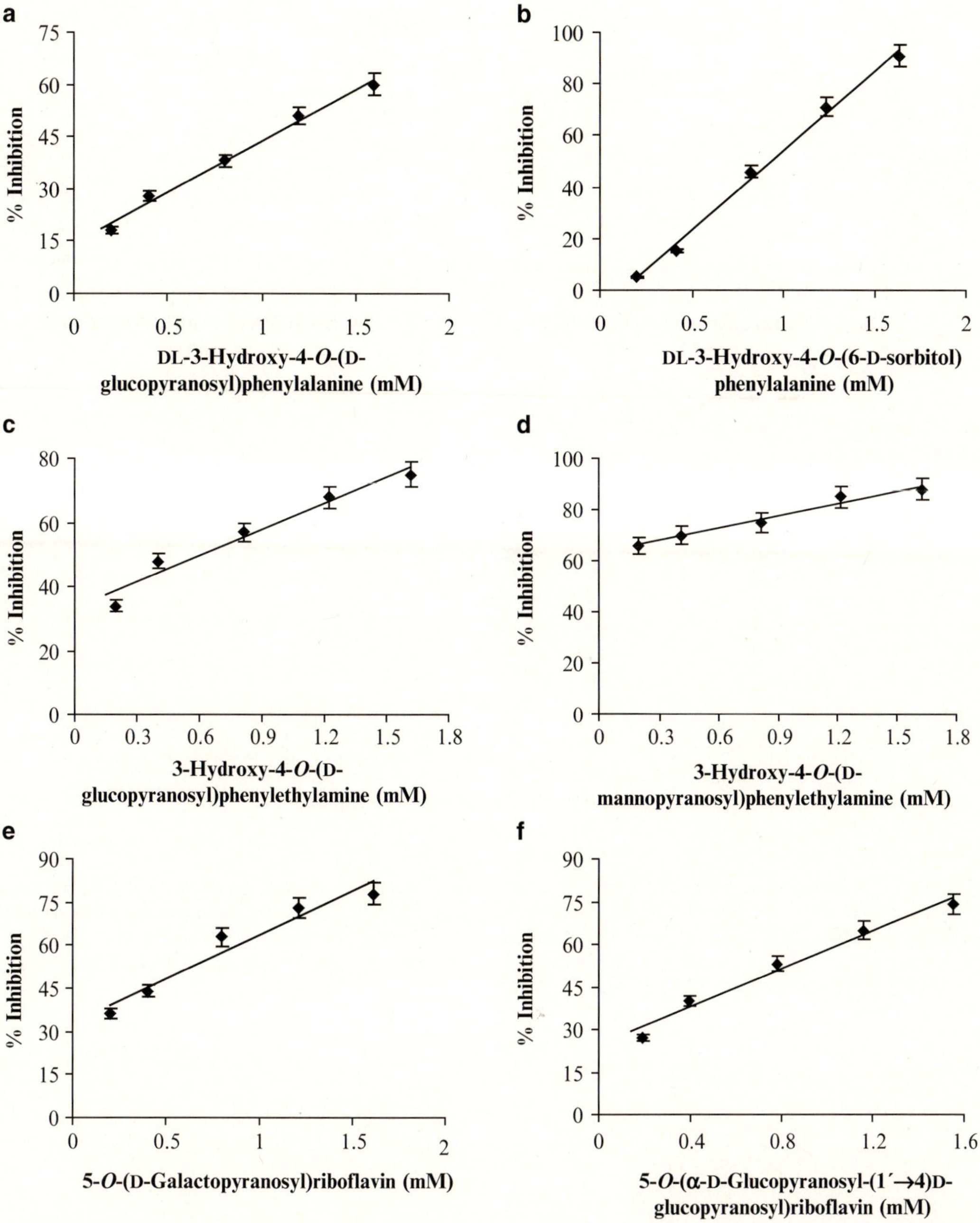

Fig. 11.16 ACE inhibition plots for phenolic and vitamin glycosides: ACE, 0.1 mL (10 mg in 25-mL stock solution); glycoside concentration range, 0.2–1.8 mM; substrate, 0.1 mL hippuryl-L-histidyl-L-leucine (5 mM); buffer, 100 mM phosphate buffer pH 8.3 containing 0.3 M sodium chloride; incubation period, 30 min; and temperature, 37°C. (**a**) DL-3-hydroxy-4-*O*-(D-Glucopyranosyl) phenylalanine. (**b**) DL-3-Hydroxy-4-*O*-(6-D-sorbitol)phenylalanine. (**c**) 3-Hydroxy-4-*O*-(D-glucopyranosyl)phenyl ethylamine. (**d**) 3-Hydroxy-4-*O*-(D-mannopyranosyl)phenylethylamine. (**e**) 5-*O*-(D-Galactopyranosyl)riboflavin. (**f**) 5-*O*-(α-D-Glucopyranosyl-(1′→4)D-glucopyranosyl) riboflavin

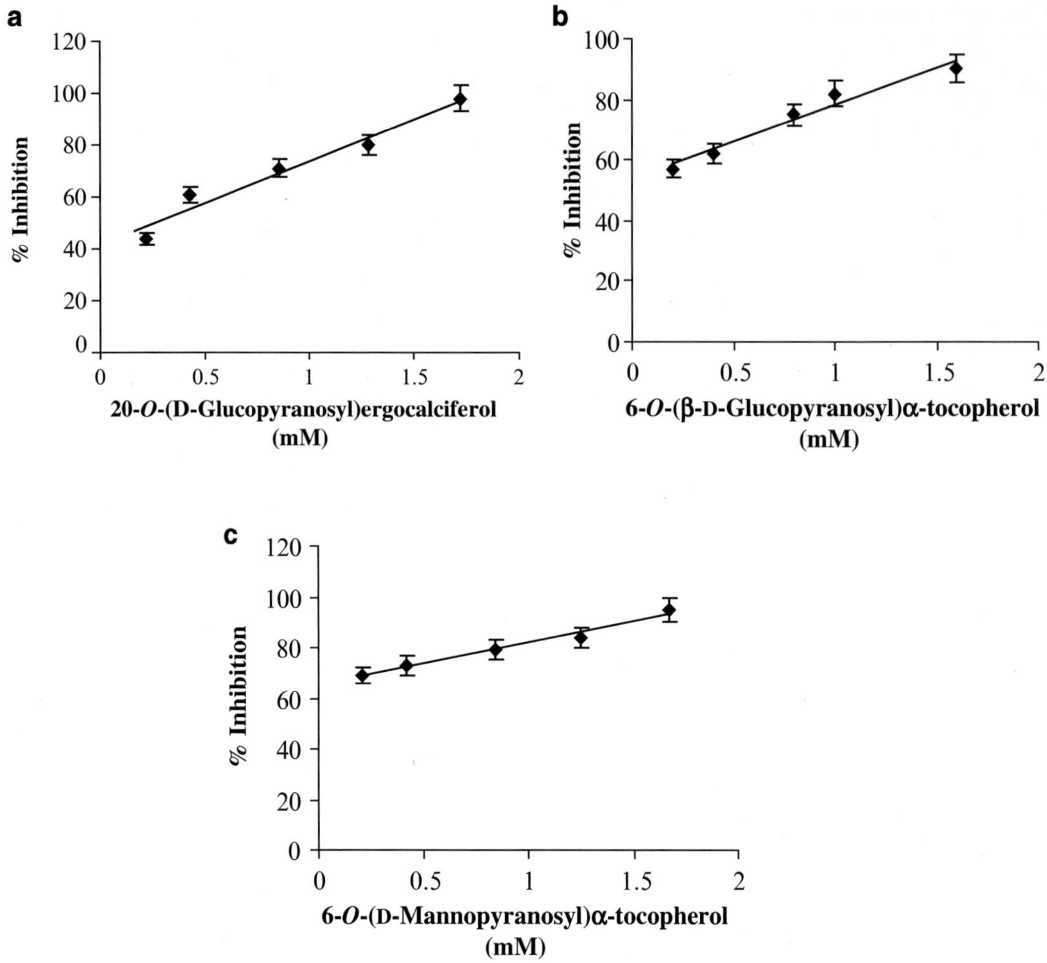

Fig. 11.17 ACE inhibition plots for phenolic and vitamin glycosides: ACE, 0.1 mL (10 mg in 25-mL stock solution); glycoside concentration range, 0.2–1.8 mM; substrate, 0.1 mL hippuryl-L-histidyl-L-leucine (5 mM); buffer, 100 mM phosphate buffer pH 8.3 containing 0.3 M sodium chloride; incubation period, 30 min; and temperature, 37°C. (**a**) 20-*O*-(D-Glucopyranosyl)ergocalciferol. (**b**) 6-*O*-(β-D-Glucopyranosyl)α-tocopherol. (**c**) 6-*O*-(D-Mannopyranosyl)α-tocopherol

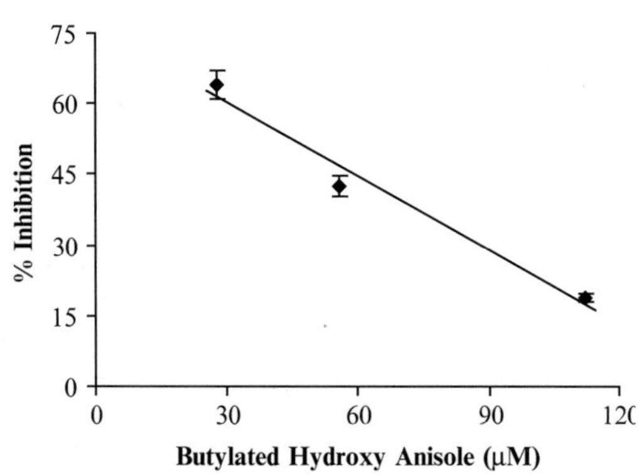

Fig. 11.18 Antioxidant inhibition plot for butylated hydroxyanisole (BHA). Concentration range, 0–120 μM; DPPH, 1 mL (3.6 mM); buffer, 0.1-M tris-HCl (pH 7.4); incubation period, 20 min; and temperature, 37°C. IC_{50} value – 0.046 ± 0.002 mM

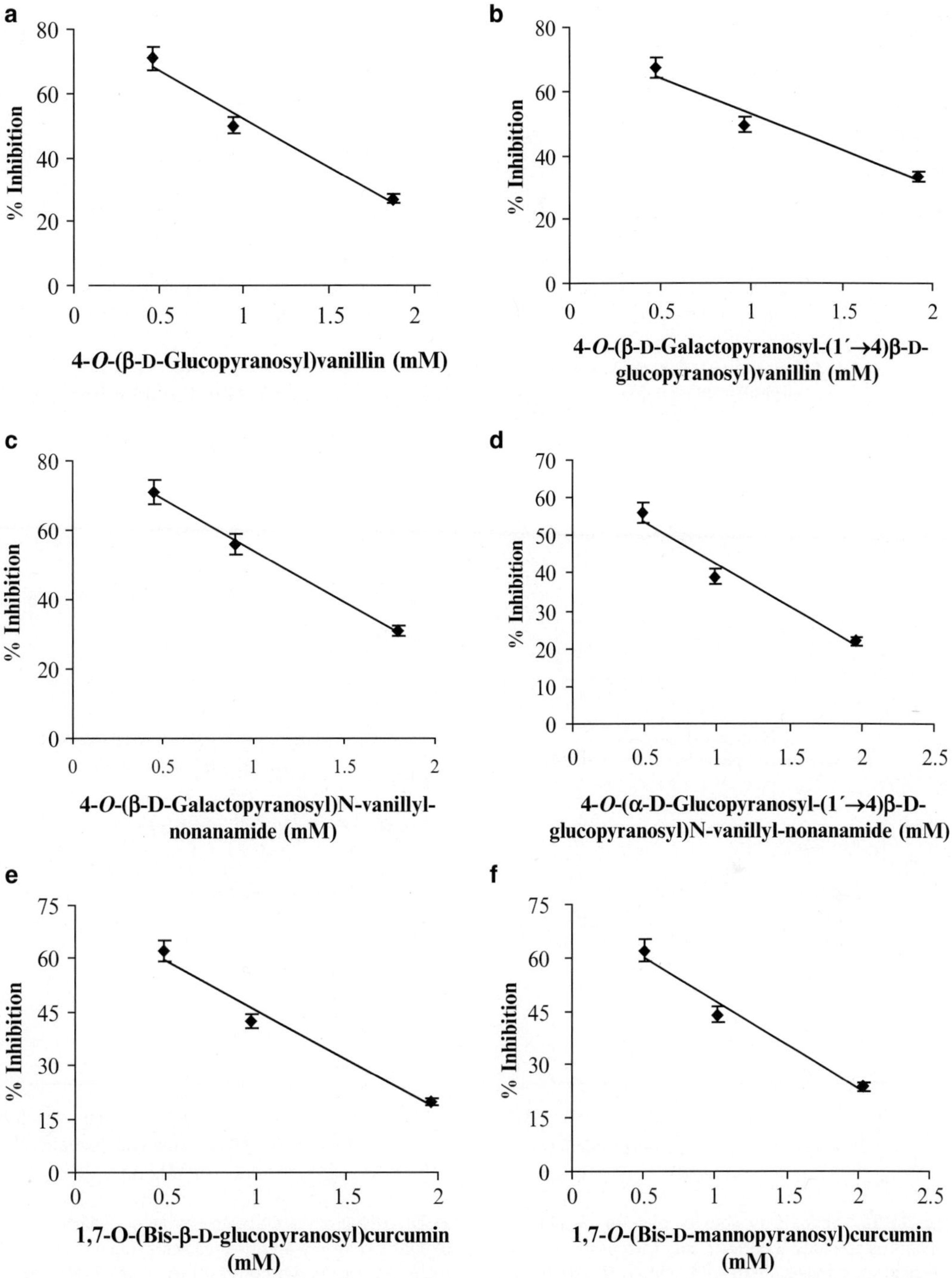

Fig. 11.19 Antioxidant activity plot for phenolic glycosides. DPPH, 1 mL (3.6 mM); glycoside concentration range, 5–10 mM; buffer, 0.1-M tris-HCl (pH 7.4); incubation period, 20 min; and temperature, 37°C. IC_{50} value for the antioxidant activity was obtained as the concentration of the glycoside corresponding to 50% decrease in DPPH absorbance from these plots. (**a**) 4-*O*-(β-D-Glucopyranosyl) vanillin. (**b**) 4-*O*-(β-D-Galactopyranosyl-(1′→4)β-D-glucopyranosyl)vanillin. (**c**) 4-*O*-(β-D-Galactopyranosyl) N-vanillyl-nonanamide. (**d**) 4-*O*-(α-D-Glucopyranosyl-(1′ 4)β-D-glucopyranosyl)N-vanillyl-nonanamide. (**e**) 1,7-*O*-(bis-β-D-Glucopyranosyl)curcumin. (**f**) 1,7-*O*-(bis-D-Mannopyranosyl)curcumin

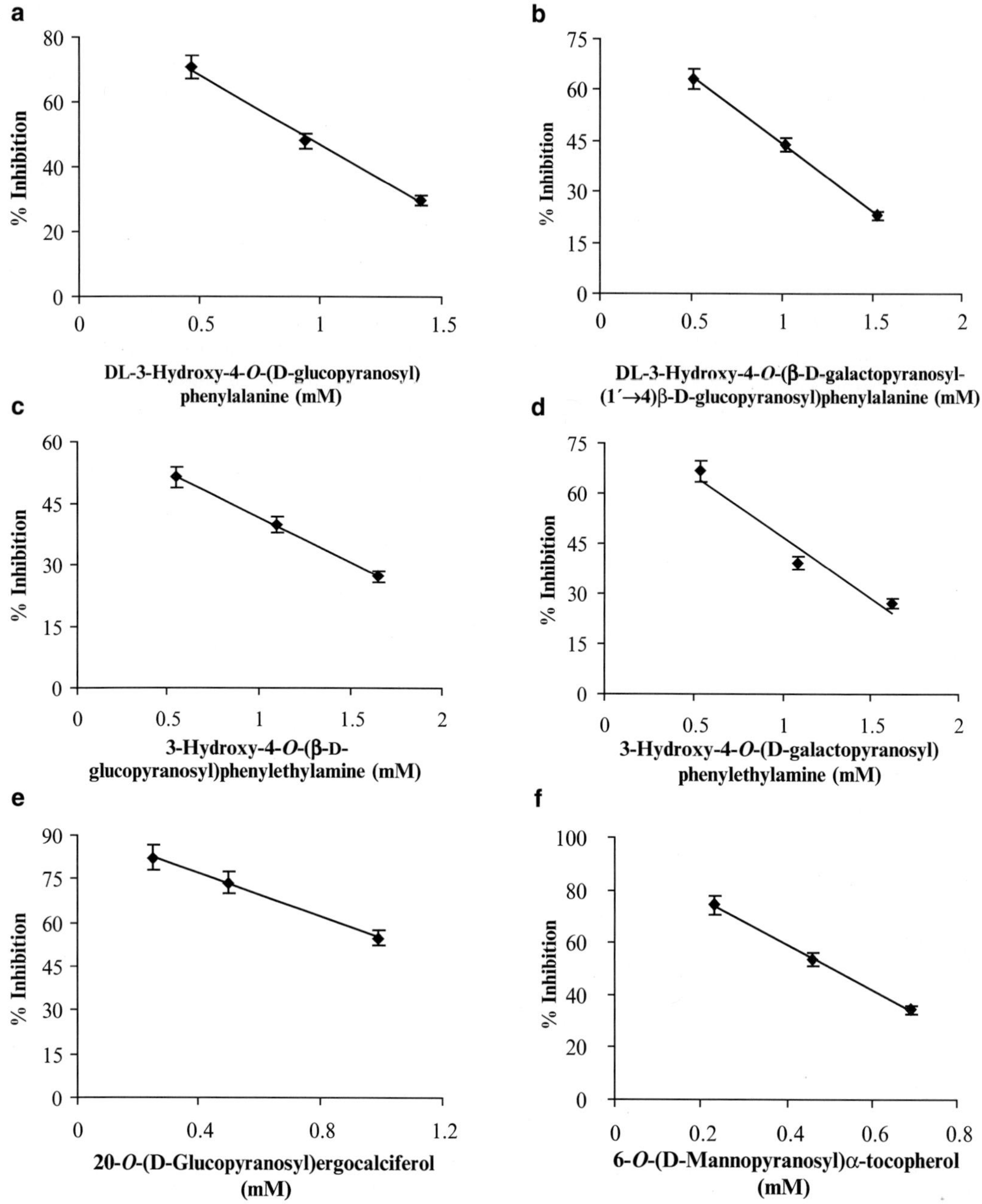

Fig. 11.20 Antioxidant activity plot for phenolic and vitamin glycosides. DPPH, 1 mL (3.6 mM); glycoside concentration range, 5–10 mM; buffer, 0.1 M tris-HCl (pH 7.4); incubation period, 20 min; and temperature, 37°C. IC_{50} value for the antioxidant activity was obtained as the concentration of the glycoside corresponding to 50% decrease in DPPH absorbance from these plots. (**a**) DL-3-Hydroxy-4-*O*-(D-glucopyranosyl)phenylalanine. (**b**) DL-3-Hydroxy-4-*O*-(β-D-galactopyranosyl-(1′→4)β-D-glucopyranosyl)phenylalanine. (**c**) 3-Hydroxy-4-*O*-(β-D-glucopyranosyl)phenylethylamine. (**d**) 3-Hydroxy-4-*O*-(D-galactopyranosyl) phenylethylamine. (**e**) 20-*O*-(D-Glucopyranosyl)ergocalciferol. (**f**) 6-*O*-(D-Mannopyranosyl)α-tocopherol

The aglycon phenols and free vitamins were also subjected to measurement of antioxidant activity and ACE inhibition as controls. Antioxidant activities were determined (Table 11.8) for phenols and also vitamins possessing phenolic OH group and alicyclic OH group like ergocalciferol (vitamin D2). Antioxidant activities of the phenolic and vitamin glycosides were in the 0.5±0.03 mM to 2.66±0.13 mM range when compared to the free aglycons whose values were quite high (0.053±0.003 mM to 1.65±0.08 mM). Butylated hydroxyanisole (BHA) showed the lowest IC_{50} value at 0.046±0.002 mM, and no other glycosides could come closer to this value. Many of the glycosides showed lesser IC_{50} values of <1 mM (Table 11.6). The best IC_{50} values (≤0.75 mM) observed are 4-*O*-(α-D-glucopyranosyl-(1′→4)β-D-glucopyranosyl) N-vanillyl-nonanamide, 0.75±0.04 mM; 1,7-*O*-(bis-D-mannopyranosyl)curcumin, 0.75±0.04 mM; 6-*O*-(D-galactopyranosyl)α-tocopherol, 0.72±0.04 mM; and 6-*O*-(D-mannopyranosyl)α-tocopherol, 0.5±0.03 mM.

IC_{50} values for ACE inhibition of glycosides range from 0.56±0.03 mM to 3.33±0.17 mM (Tables 11.6 and 11.7). Enalapril, a known synthetic drug, showed an IC_{50} value of 0.071±0.004 mM, but no other phenolic or vitamin glycosides could exhibit such a low value. The present work for the first time has shown the ACE inhibitory potency of the above-mentioned glycosides prepared enzymatically. Among the glycosides tested, eugenyl -α- D-glucoside (0.5±0.04 mM), eugenyl maltoside (0.7±0.06 mM), octyl-D-glucoside (1.0±0.09 mM) and curcuminyl-bis-α-D-mannoside (1.0±0.09 mM) exhibited the best ACE inhibitory activities (IC_{50}≤1.0 mM; Table 11.4). Among the carbohydrates employed, both glucosides and maltosides showed the best ACE inhibitory activities. Alkyl glycosides showed better inhibitory activities than the phenolic glycosides. However, the best ACE inhibitory activities for the glycosides (<0.75mM) detected were 4-*O*-(β-D-glucopyranosyl) vanillin, 0.61±0.03 mM; 4-*O*-(D-galacto pyranosyl)vanillin, 0.61±0.03mM; 1,7-*O*-(bis-β-D-galactopyranosyl-(1′→4)D-glucopyranosyl) curcumin, 0.67±0.03 mM; and DL-3-hydroxy-4-*O*-(6-D-sorbitol) phenylalanine, 0.56±0.03 mM. Among the glycosides tested, phenolic glycosides showed better ACE activities than the vitamin glycosides. Both phenolic and vitamin glycosides showed comparable IC_{50} values to aglycon units. Derivatisation by the introduction of carbohydrate molecule to the phenolic/alcoholic OH had very little effect on the IC_{50} values compared to the control. Hence, it can be concluded that the presence of free phenolic or derivated OH groups is not essential for ACE inhibition. Many commercial inhibitors like enalapril are peptides containing the essential prolyl units. Among the carbohydrates employed, aglycons modified with D-glucose, D-galactose and D-sorbitol showed less IC_{50} values (from 0.56±0.03 mM to 0.75±0.04 mM) than those modified with the other carbohydrate molecules.

These results suggest that both the phenolic and vitamin glycosides hold promising potential as antioxidants and ACE inhibitors although the IC_{50} values are slightly on the higher side, a drawback which could be corrected through suitable modifications.

References

Azuma K, Nakayama M, Koshioka M, Ippoushi K, Yamaguchi Y, Kohata K, Yamauchi Y, Ito H, Higashio HJ (1999) Phenolic antioxidants from the leaves of *Corchorus olitorius* L. Agric Food Chem 47:3963–3966

Cooke JWB, Hayes D, Henson RA, Hermitage SA, Ward RA, Whitehead AJ (2003) Process for the preparation of 4-substituted phenylalanyl ester derivatives. PCT Int Appl WO 2003068725:A2 (CA 138 180345)

Cushman DW, Cheung HS (1969) A simple substrate for assay for dog lung angiotensin converting enzyme. Fed Proc 28:3019

Cushman DW, Cheung HS (1971) Spectrophotometric assay and properties of the angiotensin-converting enzyme of rabbit lung. Chem Pharmacol 20:1637–1638

Dae-Gill K, Yong-Sup L, Hyoung-Ja K, Yun-Mi L, Ho-Sub L (2003) Angiotensin converting enzyme inhibitory phenylpropanoid glycosides from *Clerodendron trichotomum*. J Ethnopharmacol 89:151–154

Dai F, Miao Q, Zhou B, Yang L, Liu ZL (2006) Protective effects of flavanols and their glycosides against free radical induced oxidative hemolysis of red blood cells. Life Sci 78:2488–2493

De Lima DP (1999) Synthesis of angiotensin converting enzyme (ACE) inhibitors an important class of antihypertensive drugs. Quim Nova 22:375–381

Deloffre L, Sautiere PE, Huybrechts R, Hens K, Vieau D, Salzet M (2004) Angiotensin converting enzyme inhibition studies by natural leech inhibitors by capillary electrophoresis and competition assay. Eur J Biochem 271:2101–2106

Einstein Charles R, Ponrasu T, Sivakumar R, Divakar S (2009) Angiotensin converting enzyme inhibitory and antioxidant activities of enzymatically synthesized phenolic and vitamin glycosides. Biotech Appl Biochem 52:177–184

Hyuncheol O, Dae-Gill KC, Hun-Taeg L, Ho-Sub L (2003) Four glycosides from the leaves of *Abeliophyllum distichum* with inhibitory effects on angiotensin converting enzyme. Phytother Res 17:811–813

Johnston CI (1992) Renin-angiotensin system: a dual tissue and hormonal system for cardiovascular control. J Hypertens 10:13–26

Kamath V, Rajini PS, Lohith K, Somashekar BR, Divakar S (2006) Angiotensin Converting enzyme inhibitory activity of amino acid esters of carbohydrates. Int J Biol Macromol 38:89–93

Kefalas P, Kallithraka S, Parejo I, Makris DP (2003) A comparative study on the in vitro antiradical activity and hydroxyl free radical scavenging activity in aged red wines. Food Sci Technol Int 9:383–387

Kim SK, Byun HG, Park PJ, Shahidi F (2001) Angiotensin I converting enzyme inhibitory peptides purified from bovine skin gelatin hydrolysate. J Agric Food Chem 49:2992–2997

Kim WD, Kaneko S, Park GG, Tanaka H, Kusakabe I, Kobayashi H (2003) Purification and characterization of α-galactosidase from sunflower seeds. Biotechnol Lett 25:353–358

Li GH, Le GW, Yong-Hui S, Shrestha S (2004) Angiotensin I–converting enzyme inhibitory peptides derived from food proteins and their physiological and pharmacological effects. Nutr Res 24:469–486

Lohith K, Vijayakumar GR, Somashekar BR, Sivakumar R, Divakar S (2006) Glycosides and amino acyl esters of carbohydrates as potent inhibitors of Angiotensin Converting Enzyme. Eur J Med Chem 41:1059–1072

Michaud A, Williams TA, Chauvet MT, Corvol P (1997) Substrate dependence of angiotensin I-converting enzyme inhibition captopril displays a partial selectivity for inhibition of N-acetyl-seryl-aspartyl-lysyl-proline hydrolysis compared with that of angiotensin I. Mol Pharmacol 51:1070–1076

Moon JH, Terao J (1998) Antioxidant activity of caffeic acid and dihydrocaffeic acid in Lard and human low-density lipoprotein. J Agr Food Chem 46:5062–5065

Moon YH, Lee JH, Jhon DY, Jun WJ, Kang SS, Sim J, Choi H, Moon JH, Kim D (2007) Synthesis and characterization of novel quercetin-β-D-glucopyranosides using glucansucrase from *Leuconostoc mesenteriodes*. Enzyme Microb Technol 40:1124–1129

Mullally MM, Meisel H, FitzGerald RJ (1996) Synthetic peptides corresponding to a lactalbumin and β-lactoglobulin sequences with angiotensin-I-converting enzyme inhibitory activity. J Biol Chem 377:259–260

Portes E, Gardrat C, Castellan A (2007) A comparative study on the antioxidant properties of tetrahydrocurcuminoids and curcuminoids. Tetrahedron 63:9092–9099

Potier P, Maccario V, Giudicelli MB, Queneau Y, Dangles O (1999) Gallic esters of sucrose as a new class of antioxidants. Tetrahedron Lett 40:3387–3390

Roche M, Dufour C, Mora N, Dangles O (2005) Antioxidant activity of olive phenols Mechanistic investigation and characterization of oxidation products by mass spectrometry. Org Biomol Chem 3:423–429

Vermeirssen V, Van-Camp J, Verstraete W (2002) Optimization and validation of an angiotensin-converting enzyme inhibition assay for the screening of bioactive peptides. J Biochem Biophys Methods 51:75–87

Villano D, Fernandez-Pachon MS, Moy ML, Troncoso AM, Garca-Parrilla MC (2007) Radical scavenging ability of polyphenolic compounds towards DPPH free radical. Talanta 71:230–235

Wu JY, Liu SW (2002) Influence of alcohol concentration on lipase-catalyzed enantioselective esterification of racemic naproxen in isooctane under controlled water activity. Enzyme Microb Technol 26:124–130

Index

S. Divakar, *Enzymatic Transformation*,
DOI 10.1007/978-81-322-0873-0, © Springer India 2013

M

Printed by Printforce, the Netherlands